G. Toraldo di Francia (Ed.)

Onde superficiali

Lectures given at the
Centro Internazionale Matematico Estivo (C.I.M.E.),
held in Varenna (Como), Italy,
September 4-13, 1961

Springer

FONDAZIONE
CIME
ROBERTO CONTI

C.I.M.E. Foundation
c/o Dipartimento di Matematica "U. Dini"
Viale Morgagni n. 67/a
50134 Firenze
Italy
cime@math.unifi.it

ISBN 978-3-642-10981-2 e-ISBN: 978-3-642-10983-6
DOI:10.1007/978-3-642-10983-6
Springer Heidelberg Dordrecht London New York

Reprint of the 1st ed. C.I.M.E., Ed. Cremonese, Roma, 1961
With kind permission of C.I.M.E.

Printed on acid-free paper

Springer.com

CENTRO INTERNATIONALE MATEMATICO ESTIVO
(C.I.M.E)

Reprint of the 1st ed.- Varenna, Italy, September 4-13, 1961

ONDE SUPERFICIALI

PREMESSA

Le onde superficiali elettromagnetiche, pur essendo note da lungo tempo, hanno acquistato negli ultimi anni un'importanza notevole in un gran numero di applicazioni.

La loro teoria presenta problemi matematici di alto interesse. Non di rado si presentano anche curiose e dibattutissime difficoltà riguardo all'interpretazione fisica dei risultati matematici.

L'abbondante fioritura di studi sulle onde superficiali che si è avuta recentemente, si trova purtroppo sparsa nei periodici più disparati e riflette punti di vista molto diversi. Era sentitissimo il bisogno di una introduzione e di una messa a punto d'insieme per coloro che si vogliono dedicare all'argomento.

A questo scopo ha voluto rispondere il corso organizzato a Varenna dal Centro Internazionale Matematico Estivo dal 3 al 12 settembre 1961. In questi appunti, compilati dagli autori, sono condensate le lezioni del corso, che ebbe grande successo e fu accompagnato da molte interessanti discussioni.

Come coordinatore del corso tengo a ringraziare tutti gl'insegnanti che hanno portato il loro contributo e si sono sobbarcati alla fatica di mettere per iscritto le loro lezioni. Voglio anche rivolgere a nome di tutti gli studiosi della materia un vivo ringraziamento al C.I.M.E. ed in particolare al Direttore Prof. E.Bompiani ed al Segretario Prof. R.Conti per aver resa possibile la realizzazione del corso in modo così felice e proficuo.

Sono sicuro che queste lezioni rappresenteranno un

contributo utilissimo alla letteratura internazionale su questo ramo della matematica applicata.

G. Toraldo di Francia

CENTRO INTERNAZIONALE MATEMATICO ESTIVO

(C.I.M.E.)

C. M. ANGULO

A DISCONTINUITY PROBLEM ON SURFACE WAVES : THE EXCITATION OF A GROUNDED DIELECTRIC SLAB BY A WAVEGUIDE .

ROMA - Istituto Matematico dell'Università

A DISCONTINUITY PROBLEM ON SURFACE WAVES :
The excitation of a grounded dielectric slab by a waveguide .

C. M. ANGULO

Institute for Defense Analyses, Washington D.C.+)

Introduction

The present discussion illustrates the solution of one discontinuity problem associated with the excitation of surface waves. The concepts developed in previous lectures by Zucker and Felsen are used repetedly throughout the discussion. One important point to emphasize is the usefullness of the modal analysis method which enables us to set up immediately the transform equation to apply the Wiener-Hopf technique.

The problem is illustrated in figure 1 . The input energy is contained in the dominant TM ($H_y = 0$) mode of the partially filled waveguide propagating from $y = +\infty$ to $y = 0$. The dimensions of the guide and the thickness of the slab are restricted to the range for which only one surface wave (the lowest)[1] exists along the slab and only one mode (the dominant TM) can propagate inside the partially filled waveguide. These conditions are [2] :

$$Kd < \pi (\varepsilon)^{-1/2} \tag{1a}$$

$$Kh < \arctan\left\{-(\varepsilon)^{-1/2}\tan\left[(\varepsilon)^{1/2}Kd\right]\right\} \tag{1b}$$

$$0 < \arctan\left\{-(\varepsilon)^{-1/2}\tan\left[(\varepsilon)^{1/2}Kd\right]\right\} < \pi \tag{1c}$$

+) On leave of absence from Brown University, Providence R.I.

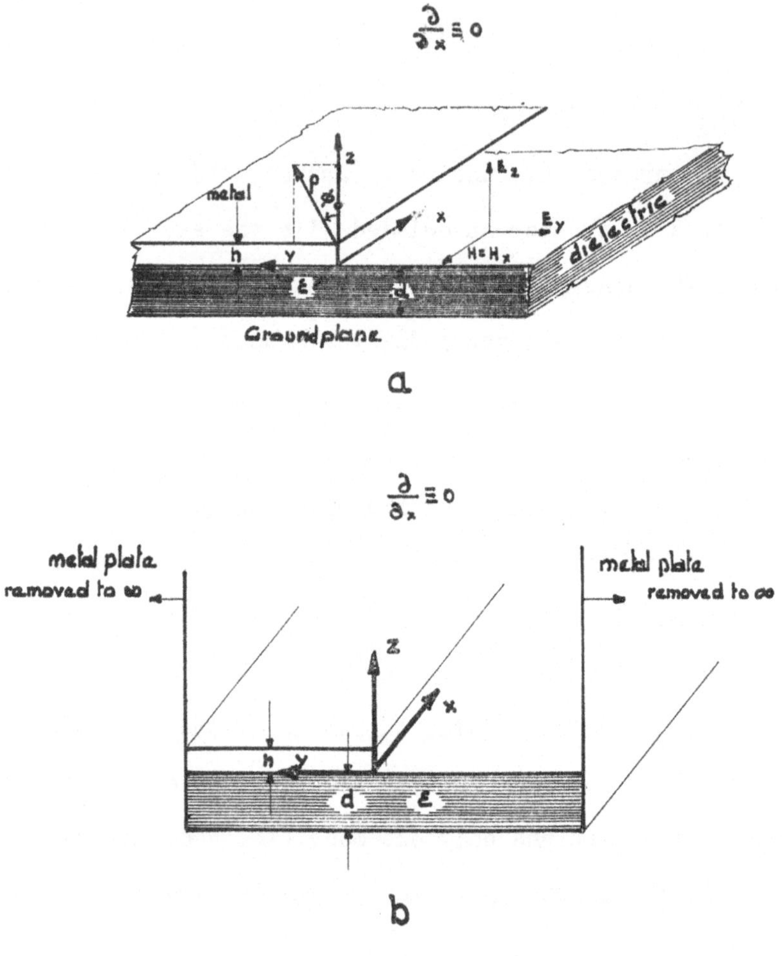
∂/∂x ≡ 0
metal
z
ρ
φ
x
E_z
E_y
dielectric
h
y
H=H_x
ε
d
Groundplane
a
∂/∂x ≡ 0
metal plate
removed to ∞
metal plate
removed to ∞
z
x
h
y
d
ε
b

Fig. 1

where $K = \omega(\mu_o \varepsilon_o)^{1/2}$, ε is the relative permittivity, d is the thickness of the dielectric, and d + h is the height of the parallel plate waveguide as indicated in Fig.1(a).

Because of the discontinuity at y = 0 where the upper plate is terminated, the energy incident upon the discontinuity will be partly reflected back into the waveguide, partly transmitted to the surface wave in the grounded dielectric slab and partly radiated. We are interested in finding the three power ratios for different values of Kd and Kh.

Since the structure shown in Fig.1(a) does not vary in the x direction and the incident wave is the lowest TM mode in the partially filled waveguide, all the fields excited will be independent of x and will have $E_x = H_y = H_z = 0$.

The structure shown in Fig.1(a) is regarded mathematically as a homogeneous parallel plate air waveguide (with walls at $y = \pm\infty$) and extending from z = 0 to $z = +\infty$ connected to a homogeneous parallel plate waveguide of length d, filled with dielectric of relative permittivity ε (also with walls at $y = \pm\infty$) and terminated by an electric wall at z = -d. Inside the first waveguide there is an obstacle, a semi-infinite perfectly conducting plane, placed at z = h from y = 0 to $y = +\infty$. Fig.1(b) illustrates the above description. By removing the y = const. walls to infinity, the structure of Fig.1(a) is obtained.

The modal analysis of a parallel plate waveguide with walls at infinity represents the transversal fields (E_y and H_x, in our case) in terms of their Fourier transforms in the cross-section (variable y, in our case). In the successive sections we shall proceed as follows :

1) The equation relating the Fourier transforms of the fields at the semi-infinite obstacle will be derived by the modal analysis method.

2) The Wiener-Hopf technique will be applied to the solution of the equation obtained, and the exact fields will be expressed as the results of integrations in the complex plane.[3-5]

3) These integrals will be evaluated only at points far away from the discontinuity. The evaluation will be carried out by analyzing the relationship between the singularities of the integrands and the analytical forms of the far fields, which are known. In fact, for $y \ll 0$ the principal contribution on the surface of the slab must be the principal surface wave propagating along a grounded dielectric slab, and for $y \gg 0$ and $z < h$ the fields must be those of the dominant mode in a parallel plate waveguide partially filled with dielectric.

THE EQUATION FOR THE FOURIER TRANSFORMS OF THE FIELDS AT PLANE z = h

We first separate H_x and E_y into the incident and scattered fields; in a second step, we find the expressions for the Fourier transforms of the scattered fields for $z < h$ and $z > h$; and finally we match the boundary conditions at $z = h$ in terms of the Fourier transforms.

Let us represent the fields everywhere as :

$$H_x = H_{ox} + \mathcal{H}_x \tag{2a}$$

$$E_y = E_{oy} + \mathcal{E}_y \tag{2b}$$

where E_{oy} and H_{ox} are the components of the dominant TM mode of the partially filled parallel waveguide and $\mathcal{H}_x$ and $\mathcal{E}_y$ are the scattered fields. For $-d < z < h$ we have

$$H_{ox} = \left\{ \frac{\cosh\left[K(z-h)s'\right]}{\cosh(Khs')} u(z) + \frac{\cos\left[K(z+d)r'\right]}{\cos(Kdr')} u(-z) \right\} \times \tag{3a}$$

$$\times \exp\left[jK(1+s'^2)^{1/2} y\right]$$

$$E_{oy} = -j \frac{Ks' \tanh(Khs')}{\omega \epsilon \epsilon_0} \left\{ \frac{\sinh\left[K(z-h)s'\right]}{\sinh(Khs')} u(z) - \right.$$

$$\left. - \frac{\sin\left[K(z+d)r'\right]}{\sin(Kdr')} u(-z) \right\} \exp\left[jK(1+s'^2)^{1/2} y\right] \tag{3b}$$

For $h < z$ we have

$$H_{ox} = 0 \,. \tag{3c}$$

$$E_{oy} = 0 \,. \tag{3d}$$

The time dependence is taken as $e^{i\omega t}$.

The function u(z) represents Heaviside's unit step function, zero for negative argument and one for positive arguments. The quantities r' and s' are the modulus of the wave numbers in the OZ direction in the dielectric and in the air respectively normalized with respect to K for the incident mode. They are the solutions of the following equations

$$r'^2 + s'^2 = \varepsilon - 1 \tag{4a}$$

$$\varepsilon \frac{s'}{r'} = \frac{\tan\,(Kdr')}{\tanh\,(Khs')} \tag{4b}$$

We will find below the quantity s similar to s'.s represents the modulus of the wavenumber in the OZ direction, in the air, normalized with respect to K for the lowest TM surface wave in a grounded dielectric slab.

The scattered fields can be represented by their Fourier transforms :

$$\mathcal{H}_x(y,z) = \frac{1}{(2\pi)^{1/2}} \int_{-\infty}^{+\infty} I(\eta,z) e^{-j\eta y} d\eta \tag{5a}$$

$$\mathcal{E}_y(y,z) = \frac{-1}{(2\pi)^{1/2}} \int_{-\infty}^{+\infty} V(\eta,z) e^{-j\eta y} d\eta \tag{5b}$$

$$I(\eta,z) = \frac{1}{(2\pi)^{1/2}} \int_{-\infty}^{+\infty} \mathcal{H}_x(y,z) e^{j\eta y} dy \tag{5c}$$

$$V(\eta,z) = \frac{-1}{(2\pi)^{1/2}} \int_{-\infty}^{+\infty} \mathcal{E}_y(y,z) e^{j\eta y} dy \tag{5d}$$

Maxwell's equations require that the transforms be solutions of the transmission line equations :

$$\frac{dV(\eta,z)}{dz} = -j\,\xi\, ZI(\eta,z) \tag{6a}$$

$$\frac{dI(\eta,z)}{dz} = -j\,\xi\, YV(\eta,z) \tag{6b}$$

for $z > h$ and for $h > z > -d$.

In the air,

$$\xi = \xi_a = (K^2 - \eta^2)^{1/2} \tag{6c}$$

$$Y_a = \frac{1}{Z_a} = \omega\varepsilon_0 (K^2 - \eta^2)^{-1/2} \tag{6d}$$

In the dielectric

$$\xi = \xi_d = (K^2\varepsilon - \eta^2)^{1/2} \tag{6e}$$

$$Y_d = \frac{1}{Z_d} = \omega\varepsilon_0\varepsilon (K^2\varepsilon - \eta^2)^{-1/2} . \tag{6f}$$

If we recall Fig.1, we see that the solutions for V and I can be written immediately from the theory of transmission lines for the two regions $z > h$ and $z < h$, as follows : For $z>h$,

$$V(\eta,z) = V(\eta, h_+)\exp\left\{-j\,\xi_a(z-h)\right\} \tag{7a}$$

$$I(\eta,z) = Y_a V(\eta,h_+)\exp\left\{-j\,\xi_a(z-h)\right\} . \tag{7b}$$

For $h > z > 0$,

$$V(\eta ,z) = V(\eta ,h_-) \cos \xi_a(z-h) -$$

$$-jZ_a I(\eta ,h_-)\sin \xi_a(z-h) \qquad (8a)$$

$$I(\eta ,z) = I(\eta ,h_-)\cos \xi_a(z-h) -$$

$$-jY_a V(\eta ,h_-)\sin \xi_a(z-h) \qquad (8b)$$

Finally, for $0 > z > -d$,

$$V(\eta ,z) = \left[V(\eta ,h_-)\cos \xi_a h + jZ_a I(\eta ,h_-)\sin \xi_a h\right] \cdot$$

$$\cdot \frac{\sin \xi_d(z + d)}{\sin \xi_d d} \qquad (8c)$$

$$I(\eta ,z) = \left[I(\eta ,h_-)\cos \xi_a h + jY_a V(\eta ,h_-)\sin \xi_a h\right] \cdot$$

$$\cdot \frac{\cos \xi_d(z + d)}{\cos \xi_d d} \qquad (8d)$$

where

$$I(\eta ,h_-) = -jY_a V(\eta ,h_-) \frac{\frac{\xi_d}{\varepsilon} \tan(\xi_a h)\tan(\xi_d d) - \xi_a}{\frac{\xi_d}{\varepsilon} \tan(\xi_d d) + \xi_a \tan(\xi_a h)} . \qquad (8e)$$

The relationship between the values of V and I at $z = h_-$ and at $z = h_+$ are obtained from the boundary conditions of $\mathcal{H}_x$ and $\mathcal{E}_y$ at $z = h$. Let us first define the following new quantities :

$$V^+(\eta ,h) = \frac{-1}{(2\pi)^{1/2}} \int_0^{\infty} \mathcal{E}_y(y, h)e^{j\eta y} dy \qquad (9a)$$

$$V^-(\eta ,h) = \frac{-1}{(2\pi)^{1/2}} \int_{-\infty}^{0} \mathcal{E}_y(y, h)e^{j\eta y} dy \qquad (9b)$$

$$\mathcal{J}^{+}(\eta,h) = \frac{1}{(2\pi)^{1/2}} \int_0^{\infty} \left[\mathcal{H}_x(y,h_+) - \mathcal{H}_x(y,h_-)\right] e^{j\eta y} dy \qquad (9c)$$

$$\mathcal{J}^{-}(\eta,h) = \frac{1}{(2\pi)^{1/2}} \int_{-\infty}^{0} \left[\mathcal{H}_x(y,h_+) - \mathcal{H}_x(y,h_-)\right] e^{j\eta y} dy \qquad (9d)$$

It is obvious that

$$V(\eta,h_+) = V(\eta,h_-) = V^{+}(\eta,h) + V^{-}(\eta,h) \qquad (10a)$$

and

$$I(\eta,h_+) - I(\eta,h_-) = \mathcal{J}^{+}(\eta,h) + \mathcal{J}^{-}(\eta,h). \qquad (10b)$$

Two constants will appear very often in the equations below, so for convenience we will represent them as follows :

$$a_1 = (1 + s^2)^{1/2} \qquad a_1 > 0$$

$$a_2 = (1 + s'^2)^{1/2} \qquad a_2 > 0 \ .$$

From the remaining boundary conditions at $z = h$, we obtain the following results :

$$\mathcal{E}_y = 0 \qquad \text{for} \qquad y > 0 \, , \qquad (11a)$$

therefore

$$V^{+}(\eta,h) = 0 \, ; \qquad (11b)$$

and

$$\mathcal{H}_x(y,h_+) - \mathcal{H}_x(y,h_-) = H_{ox}(y,h_-) \quad \text{for} \quad y < 0 \, , \qquad (12a)$$

therefore

$$\mathcal{J}^{-}(\eta,h) = -j \frac{\operatorname{sech}(Khs')}{(2\pi)^{1/2}\left[\eta + Ka_2\right]} \qquad (12b)$$

provided Imag η < Imag $(-Ka_2)$.

Therefore, all the boundary conditions at z = h are satisfied if

$$\mathcal{J}^{+}(\eta,h) = \frac{2\,\omega\varepsilon_0 (K^2a_1^2 - \eta^2)}{(K^2-\eta^2)^{1/2}\,(K^2a_2^2-\eta^2)}\, G(\eta)V^{-}(\eta,h) + j\,\frac{\operatorname{sech}(Khs')}{(2\pi)^{1/2}\,(\eta + Ka_2)} \tag{13a}$$

where

$$G(\eta) = \frac{-j + \tan\left[h(K^2-\eta^2)^{1/2}\right]}{2} \times \frac{K^2a_2^2 - \eta^2}{K^2a_1^2-\eta^2} \times \tag{13b}$$

$$\times \frac{1 + j\,\dfrac{(K^2\varepsilon - \eta^2)^{1/2}}{\varepsilon(K^2-\eta^2)^{1/2}}\tan\left[(K^2\varepsilon-\eta^2)^{1/2}d\right]}{\tan\left[(K^2-\eta^2)^{1/2}h\right] + \dfrac{(K^2\varepsilon-\eta^2)^{1/2}}{\varepsilon(K^2-\eta^2)^{1/2}}\tan\left[(K^2\varepsilon-\eta^2)^{1/2}d\right]}$$

The quantity s is the modulus of the wavenumber in the OZ direction in the air normalized with respect to K for the lowest TM surface wave in the grounded dielectric slab.

A study of the behavior of the functions in (13a) permits us to apply the Wiener-Hopf technique and solve for $\mathcal{J}^{+}$ and V^{-}.

C.M.Angulo

THE SOLUTION OF THE EQUATION FOR THE TRANSFORMS

The behavior of $\mathcal{J}^{+}(\eta,h)$ and $V^{-}(\eta,h)$ in the complex η plane is determined by the asymptotic behavior of $\mathcal{H}_x$ and $\mathcal{E}_y$ as well as by the singularities of the transformed Kernel

$$\frac{2\,\omega\,\varepsilon_0(K^2 a_1^2 - \eta^2)}{(K^2 - \eta^2)^{1/2}(K^2 a_2^2 - \eta^2)} G(\eta) \,. \qquad (14)$$

We will come back later to (14). Let us proceed now with a physical derivation of the dominant terms of the far fields.

In our problem we can obtain all of the excited fields from the x component of the magnetic field. The problem may be compared with the two-dimensional field excited along a grounded dielectric slab by a magnetic line source along the OX axis. If $-\pi/2 < \phi_0 < 0$ and ρ is very large, we will not be able to notice any difference between a magnetic line and a terminated parallel plate waveguide propagating the lowest E mode. Therefore, the nature of the solution for both problems is the same for that region of space. However, as ϕ_0 increases, the angular dependence will be different for the two problems.

In the case of the magnetic line, we would have only the surface wave for $\phi_0 \approx -\pi/2$, all other terms being of order $\rho^{-3/2}$ or lower. As ϕ_0 increases, the surface-wave contribution becomes negligible and the dominant term varies like $\rho^{-1/2}$ (always for large ρ). Finally, when ϕ_0 grows to $\pi/2$, the $\rho^{-1/2}$

terms are not present and we have only terms of order $\rho^{-3/2}$ or lower and the surface wave.

For a grounded dielectric slab excited by a parallel plate waveguide, we will have also the surface wave and terms of order $\rho^{-3/2}$ and lower if $\phi_o \approx -\pi/2$. As ϕ_o increases, the surface wave contribution becomes negligible and the dominant term varies like $\rho^{-1/2}$. Finally, when ϕ_o increases to $\pi/2$, the fields will go to zero as $\rho^{-3/2}$ at least and the surface wave will not reappear if we remain outside the waveguide.

It is therefore justified to write the form of the far fields for $K\rho \gg 1$ and $\pi/2 \geqslant \phi_o \geqslant -\pi/2$ as follows :

$$H_x = \mathcal{H}_x = g(\phi_o)\frac{e^{-jK\rho}}{\rho^{1/2}}\cos\phi_o + C\left[\frac{\cos\left[Kr(z+d)\right]}{\cos(Krd)}u(-z) + e^{-Ksz}u(z)\right]\Psi(\phi_o)\cdot\exp(jKya_1) \tag{15a}$$

$$E_y = \mathcal{E}_y = -\left(\frac{\mu_o}{\varepsilon_o}\right)^{1/2} g(\phi_o)\frac{e^{-jK\rho}}{\rho^{1/2}}\cos\phi_o + jsC\left(\frac{\mu_o}{\varepsilon_o}\right)^{1/2}\left[\frac{\sin\left[Kr(z+d)\right]}{\sin(Krd)},u(-z) + e^{-Ksz}u(z)\right]\Psi(\phi_o)\exp(jKya_1) \tag{15b}$$

$$\Psi(\phi_o) = 0 \quad \text{if} \quad \frac{\phi_o}{2} > -\arctan(a_1 - s) \tag{15c}$$

$$\Psi(\phi_o) = 1 \quad \text{if} \quad \frac{\phi_o}{2} < -\arctan(a_1 - s) \tag{15d}$$

The transmission coefficient to the surface wave is represented by C. The new quantities ϕ_o and ρ are the usual cylindrical coordinates illustrated in Fig.1(a) : r is the normalized wavenumber in the dielectric in the OZ direction for the lowest TM surface wave in the grounded dielectric slab. Finally $g(\phi_o)$ is a function of the observation angle ϕ_o.

Inside the partially filled parallel plate waveguide and far away from the discontinuity, we have only the incident and reflected wave associated with the only propagating mode.

Therefore,

$$\mathcal{H}_x = BH_{ox} \exp\left\{-j2Kya_2\right\} \qquad (16a)$$

for $Ky \gg 1$ and $-d < z < h$;

$$\mathcal{E}_y = -BE_{oy} \exp\left\{-j2Kya_2\right\} \qquad (16b)$$

for $Ky \gg 1$ and $-d < z < h$.

B is the reflection coefficient. The quantities K , Ka_1 and Ka_2 must have small negative imaginary parts for dissipative media.

From a detailed examination of the singularities and zeros of (14) and from the asymptotic expression of the fields given in (15) and (16), it follows that $V^-(\eta,h)$ is analytic in the lower half of the η plane for Imag $\eta <$ Imag $(-Ka_1)$ and its singularities are a branch point at $\eta = -K$ and a simple pole at $\eta = -Ka_1$.

It also follows that $\mathcal{J}^+(\eta,h)$ is analytic in the upper half of the η plane for Imag $\eta >$ Imag (Ka_2) and its singularities are a branch point at $\eta = K$, a simple pole at $\eta = Ka_2$, and a countable infinite number of poles on the negative imaginary axis $\eta = -j|\eta_i|$ (*).

In a slightly dissipative medium, γ^+, V^- and $G(\eta)$ are analytic in a common narrow strip of width $2w_d$ along the real axis, where

$$0 < w_d < |\text{Imag } K| \tag{17a}$$

$$0 < w_d < |\text{Imag } (Ka_2)| \tag{17b}$$

$$0 < w_d < |\text{Imag } (Ka_1)| . \tag{17c}$$

The regions where γ^+ and V^- are analytic, their singularities relevant to the integration, the overlapping strip and the branch cuts, are shown in Fig.2. We restrict ourselves to remain on the Riemann sheet where

$$\text{Imag } (K^2 - \eta^2)^{1/2} < 0$$

We now decompose the function G:

$$G(\eta) = \exp\left\{\gamma^-(\eta) - \gamma^+(\eta)\right\} \tag{18a}$$

where

$$\gamma^-(\eta) = \frac{-1}{2\pi j} \int_{-\infty + jw_d}^{\infty + jw_d} \frac{\ln G(\xi)}{\xi - \eta} \, d\xi \tag{18b}$$

(*) <u>Note:</u> $\pm j|\eta_i|$ are the roots of

$$\tan\left[h(K^2 - \eta^2)^{1/2}\right] + \frac{(K^2\varepsilon - \eta^2)^{1/2}}{\varepsilon(K^2 - \eta^2)^{1/2}} \tan\left[(K^2\varepsilon - \eta^2)^{1/2} d\right] = 0$$

encluding $\eta = \pm Ka_2$.

is analytic for

$$\text{Imag}\,\eta < w_d$$

and

$$\gamma^{+}(\eta) = \frac{-1}{2\pi j}\int_{-\infty-jw_d}^{\infty-jw_d} \frac{\ln G(\xi)}{\xi-\eta}\,d\xi$$

is analytic for Imag $\eta > -w_d$.

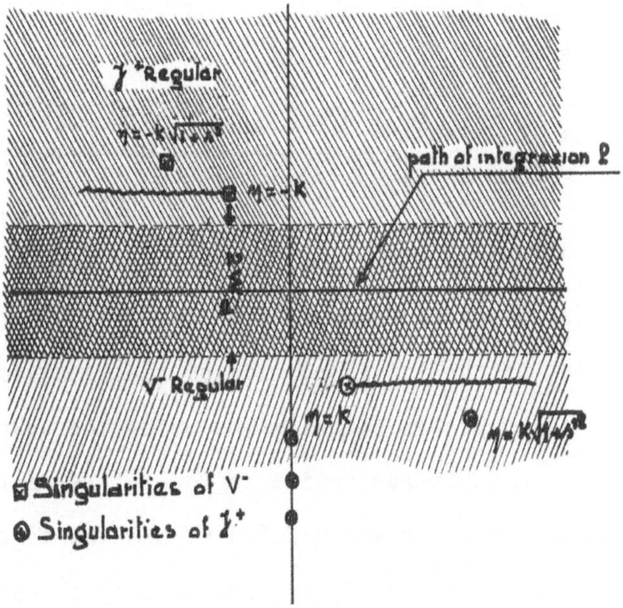

Fig.2 - The η-plane

Substituting (18a) into (13a) we can now group the terms with the same regions of regularity, as follows :

$$\mathcal{J}^{+}(\eta,h)\times\frac{(K-\eta)^{1/2}(\eta-Ka_2)\exp\{\gamma^{+}(\eta)\}}{\eta-Ka_1}$$

$$-j\exp\{\gamma^{+}(\eta)\}\frac{\operatorname{sech}(Khs')}{(2\pi)^{1/2}}\times\frac{(K-\eta)^{1/2}(\eta-Ka_2)}{(\eta+Ka_2)(\eta-Ka_1)}$$

$$+j\exp\{\gamma^{+}(-Ka_2)\}\frac{2\operatorname{sech}(Khs')}{(2\pi)^{1/2}}\times\frac{(K+Ka_2)^{1/2}}{\eta+Ka_2}\times\frac{a_2}{a_1+a_2}$$

$$=\frac{2\omega\varepsilon_0}{(K+\eta)^{1/2}}\times\frac{\eta+Ka_1}{\eta+Ka_2}\times V^{-}(\eta,h)\times\exp\{\gamma^{-}(\eta)\}$$

$$+j\exp\{\gamma^{+}(-Ka_2)\}\frac{2\operatorname{sech}(Khs')}{(2\pi)^{1/2}}\times\frac{(K+Ka_2)^{1/2}}{\eta+Ka_2}\times\frac{a_2}{a_1+a_2}$$

The last terms on each side of the previous equation are identical. They have been added in order to eliminate the pole at $\eta=-Ka_2$ of the left hand side of the equation.

The asymptotic behavior $\eta\to\infty$ of the unknown functions $\mathcal{J}^{+}$ and V^{-} in their respective regions of regularity is given by the behavior of the unknown field around the edge [6]. The physical requirement that the field scattered by the edge must have a finite amount of energy requires:

$$\frac{V^{-}(\eta,h)}{\sqrt{\eta}}\to 0 \quad \text{and} \quad \sqrt{\eta}\,\mathcal{J}^{+}(\eta,h)\to 0$$

as $\eta\to\infty$ in the region of regularity. These conditions are identical to those for the scattering of a plane wave by half a plane. [7]

We can now proceed with the customary reasoning of the Wiener-Hopf technique. The left hand side of the above equation is analytic in the upper half of the η plane including the narrow strip $|\,|\text{Imag}\ \eta| < w_d$ and the right hand side is analytic in the lower half including the narrow strip. Therefore, they must be analytic continuation of each other representing an entire function of η. Furthermore, both sides approach zero as η approaches infinity. Thus, the entire function is zero. Equating both sides to zero we obtain

$$\mathcal{J}^{+}(\eta,h) = \frac{j\ \text{sech}(Khs')}{(2\pi)^{1/2}(\eta + Ka_2)} \times \left\{1 - \frac{2a_2\left[K + Ka_2\right]^{1/2}(\eta - Ka_1)}{(K-\eta)^{1/2}(a_1 + a_2)(\eta - Ka_2)} \cdot \exp\left[\gamma^{+}(-Ka_2) - \gamma^{+}(\eta)\right]\right\} \quad (19a)$$

and

$$V^{-}(\eta,h) = \frac{-\,j\ \text{sech}\,(Khs')}{\omega\varepsilon_0(2\pi)^{1/2}(\eta + Ka_1)} \times \quad (19b)$$

$$\times \frac{(K+\eta)^{1/2}\, a_2(K + Ka_2)^{1/2}}{a_2 + a_1} \exp\left\{\gamma^{+}(-Ka_2) - \gamma^{-}(\eta)\right\}$$

CALCULATION OF THE FAR FIELDS

The "voltage" is now known for $z = h$, since

$$V(\eta, h) = V^{-}(\eta, h) . \tag{20a}$$

As for the "current" transform, we recall (7b) for $z = h$:

$$I(\eta, h_{+}) = Y_a V(\eta, h) \tag{20b}$$

Therefore $I(\eta, h_{+})$ is known. Moreover, (10b), (12b) and (19a) yield $I(\eta, h_{-})$, once $I(\eta, h_{+})$ has been found. Therefore, $I(\eta, h_{+})$ and $I(\eta, h_{-})$ are now both known.

The knowledge of V and I for $z = h$ gives us the expressions for V and I **anywhere**, as indicated in (7) and (8).

Finally, the inversion of V and I by (5) yields the exact expressions for the fields everywhere.

However, this solution everywhere is only formal, since the inversion of the transforms is practically impossible. Nevertheless, we can obtain all the information that we want by carrying out the inversion for the far fields at points of observation for which the method of steepest descents is easier to apply, and comparing the results with (15) and (16). In this way we can obtain the expressions for the coefficients B, C and $g(\phi_o)$ which give us the complete knowledge of the far fields.

The inverse transforms of special interest to us are :

$$E_y(y, h) = \mathcal{E}_y(y, h) = \frac{-1}{(2\pi)^{1/2}} \int_{-\infty}^{+\infty} V^{-}(\eta, h) e^{-j\eta y} d\eta \tag{21a}$$

for $z = h$, $y < 0$;

$$H_x(y, h_+) - H_x(y, h_-) + H_{ox}(y, h) = \mathcal{H}_x(y, h_+) - \mathcal{H}_x(y, h_-) =$$

$$= \frac{1}{(2\pi)^{1/2}} \int_{-\infty}^{+\infty} \left[\mathcal{J}^+(\eta, h) - \frac{j \operatorname{sech}(Khs')}{(2\pi)^{1/2}(\eta + Ka_2)} \right] e^{-j\eta y} d\eta \quad (21b)$$

for $z = h$, $y > 0$; and

$$E_y(y,z) = \mathcal{E}_y(y,z) = \frac{-1}{(2\pi)^{1/2}} \int_{-\infty}^{+\infty} V^-(\eta, h) e^{-j\eta y} \exp\left\{-j(K^2-\eta^2)^{1/2}(z-h)\right\} d\eta \quad (21c)$$

for $z > h$.

The path of integration is indicated in Fig.2.

The integrals are evaluated for the far fields by the method of steepest descents for the limiting case of zero dissipation. For convenience these integrations are not carried out in the η plane but in a new ν plane. The coordinates are also changed from cartesian to polar coordinates, as illustrated in Fig.1(a).

$$\eta = K \sin\nu \qquad (K^2 - \eta^2)^{1/2} = K \cos\nu \quad (22a)$$

$$y = \rho \sin\phi_o \qquad z - h = \rho \cos\phi_o . \quad (22b)$$

With this change, the integrals (21) become

$$E_y(y, h) = \frac{-K}{(2\pi)^{1/2}} \int_{\Lambda} V^-(K \sin\nu, h) . \exp\left\{-jK\rho \cos(\nu + \frac{\pi}{2})\right\} \cos\nu \, d\nu \quad (23a)$$

for $z = h$, $y < 0$;

$$H_x(y, h_+) - H_x(y,h_-) + H_{ox}(y,h_-) =$$

$$= \frac{K}{(2\pi)^{1/2}} \int_{\Lambda} \left[\gamma^+(K \sin \nu, h) - j \frac{\operatorname{sech}(Khs')}{K(2\pi)^{1/2}(\sin \nu + a_2)} \right] \times$$

$$\times \cos \nu \quad \exp\left\{ - K \rho\, j \cos\left(\nu - \frac{\pi}{2}\right)\right\} \, d\nu \qquad (23b)$$

for $z = h$, $y > 0$; and

$$E_y(y, z) = \frac{-K}{(2\pi)^{1/2}} \int_{\Lambda} V^-(K \sin \nu, h) \times$$

$$\times \exp\left\{ -jK \rho \cos(\nu - \phi_0)\right\} \cos \nu \, d\nu \qquad (23c)$$

for $z > h$.

The path of integration is shown in Fig.3.

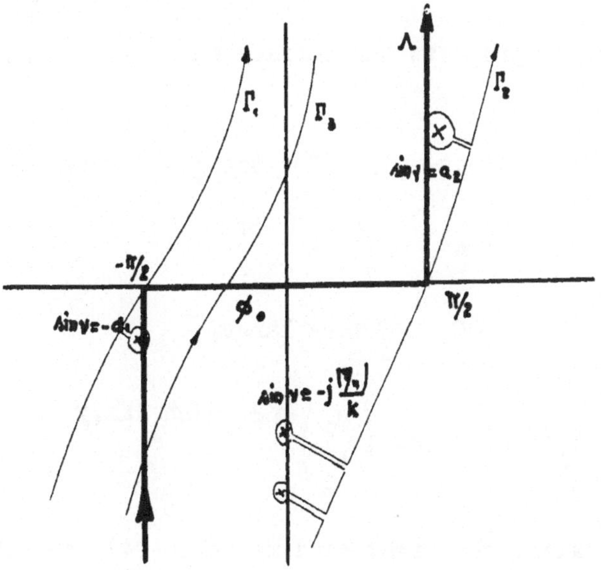

Fig.3 - The ν -plane

The saddle point for (23a) is obviously at $\nu = -\pi/2$, and the steepest descent path is the path Γ_1 in Fig.3. The integration indicated in (23a) is carried out along the path Γ_1 and not along the path Λ. The result is equal to $2\pi j$ times the residue of the integrand at the pole $\sin\nu = -a_1$ plus the asymptotic series obtained from the steepest descent integration. However, because of the factor $\cos\nu$, the $\rho^{-1/2}$ order contribution of the expansion is zero since $\cos(-\pi/2) = 0$. Therefore, if we neglect terms of order $\rho^{-3/2}$ or lower we can write for large ρ, (i.e., $y \ll 0$)

$$E_y(y,h) = -(2\pi)^{1/2} j e^{jKa_1 y} \lim_{\eta \to -Ka_1} \left\{ (\eta + Ka_1)\, V^-(\eta, h) \right\}. \tag{24a}$$

This represents the surface wave excited along the grounded slab.

If we evaluate (15b) for the given observation point

$$z = h \qquad \phi_0 = -\frac{\pi}{2}$$

$$y \ll 0 \qquad \rho \text{ very large,}$$

we obtain

$$E_y(y,h) = jsC \left(\frac{\mu_0}{\varepsilon_0}\right)^{1/2} e^{-Ksh}\, e^{jKa_1 y} \tag{24b}$$

Equating the right hand members (24), one obtains :

$$C = -\frac{(2\pi)^{1/2}}{s} \left(\frac{\varepsilon_0}{\mu_0}\right)^{1/2} \lim_{\eta \to -Ka_1} \left\{ (\eta + Ka_1)\, V^-(\eta, h) \right\} e^{Ksh}$$

The evaluation of the limit yields finally :

$$C = j \frac{a_2(1 - a_1)^{1/2}(1 + a_2)^{1/2}}{s(a_1 + a_2)} \times \operatorname{sech}(Khs') \times$$

$$\times \exp\left\{ Khs + \gamma^{+}(-Ka_2) - \gamma^{-}(-Ka_1) \right\} . \qquad (25a)$$

The saddle point for (23b) is at $\nu = \pi/2$ and the steepest descent path is Γ_2 in Fig.3. The residues at the poles on the imaginary axis of the ν plane decay exponentially with an increasing positive y and are negligible for $Ky \gg 1$. These poles correspond to the non-propagating ordinary modes in the parallel plate waveguide with the dielectric slab, and we expected a negligible contribution for $Ky \gg 1$. The asymptotic expansion does not contain a term of order $\rho^{-1/2}$ for identical reasons that the integral discussed in the previous paragraph did not. The dominant contribution to the integral (23b) is therefore :

$$\mathcal{H}_x(y, h_+) - \mathcal{H}_x(y, h_-) = -(2\pi)^{1/2} je^{-jKa_2 y} \times$$

$$\times \lim_{\eta \to Ka_2} \left\{ (\eta - Ka_2) \left[\mathcal{J}^{+}(\eta, h) - \frac{j \operatorname{sech}(Khs')}{(2\pi)^{1/2}(\eta + Ka_2)} \right] \right\} (26a)$$

where only terms of order $\rho^{-3/2}$ or lower have been neglected and $Ky \gg 1$. The expression (26a) clearly represents the reflected surface wave in the parallel plate waveguide evaluated at the upper plate.

If we evaluate (15a) and (16a) at the point of observation

$$z = h \qquad \phi_0 = \frac{\pi}{2}$$

$$Ky \gg 1 \qquad \rho \text{ very large},$$

and subtract, we obtain

$$\mathcal{H}_x(y,h_+) - \mathcal{H}_x(y,h_-) = g\left(\frac{\pi}{2}\right)\frac{e^{-jk\rho}}{\rho^{1/2}} - B\,\frac{e^{-jKa_2y}}{\cosh(Khs')} \tag{26b}$$

The identity of (26a) and (26b) requires

$$g\left(\frac{\pi}{2}\right) = 0$$

and

$$B = (2\pi)^{1/2} j \cosh(Khs') \lim_{\eta \to Ka_2} \left\{ (\eta - Ka_2)\left[\mathcal{G}^+(\eta,h) - \frac{j\,\mathrm{sech}(Khs')}{(2\pi)^{1/2}(\eta + Ka_2)} \right] \right\} \tag{26c}$$

Finally

$$B = \frac{(1+a_2)^{1/2}(a_2-a_1)}{(1-a_2)^{1/2}(a_2+a_1)} \exp\left\{ \gamma^+(-Ka_2) - \gamma^+(Ka_2) \right\} \tag{27}$$

The evaluation of the integral (23c) is carried out in the same way as with the two discussed above. The saddle point is at $\nu = \phi_0$ and the path of integration is deformed now from Λ to Γ_3 in Fig.3.

The observation angle may vary $-\pi/2 < \phi_0 < \pi/2$; and depending on the value of ϕ_0, we will have or will not have a residue contribution from the pole at $\sin \nu = -a_1$. This contribution is the surface wave excited on the grounded dielectric slab. The asymptotic series will contribute this time with a term of the order $\rho^{-1/2}$ provided $|\phi_0| < \pi/2$. The dominant term of the expansion plus the pole contribution yield

$$E_y(y,z) = -(1+j)(K/2\rho)^{1/2}\, e^{-jK\rho} \cos\phi_o V^-(K\sin\phi_o, h) -$$

$$-j(2\pi)^{1/2} \exp\left\{jKa_1 y - sK(z-h)\right\} u\left[-\frac{\phi_o}{2} - \tan^{-1}(a_1 - s)\right] \times$$

$$\times \lim_{\eta \to -Ka_1} \left\{(\eta + Ka_1)\, V^-(\eta, h)\right\} \qquad (28a)$$

where $z > h$, $K\rho \gg 1$, $-\pi/2 < \phi_o < \pi/2$, and $u(x)$ is the Heaviside unit step function. If the observation angle is $\phi_o > -2\tan^{-1}(a_1 - s)$, i.e. if we are in the region with no surface wave :

$$E_y(y,z) = -(1+j)(K/2\rho)^{1/2} e^{-jK\rho} \cos\phi_o V^-(K\sin\phi_o, h). \qquad (28b)$$

We can evaluate now (15b) for

$$z > 0 \qquad \phi_o > -2\tan^{-1}(a_1 - s)$$

$$\rho \text{ very large}$$

and obtain :

$$E_y(y,z) = -(\mu_o/\varepsilon_o)^{1/2} g(\phi_o) \cos\phi_o \frac{e^{-jK\rho}}{\rho^{1/2}} \qquad (28c)$$

The identity of (28b) and (28c) requires

$$g(\phi_o) = (1+j)\, \frac{\omega\,\varepsilon_o}{(2K)^{1/2}}\, V^-(K\sin\phi_o, h).$$

Replacing V^- by its value, one obtains finally

$$g(\phi_o) = \frac{(1-j)(1+\sin\phi_o)^{1/2} a_2(1+a_2)^{1/2}\operatorname{sech}(Khs')}{2(\pi K)^{1/2}(\sin\phi_o+a_1)(a_2+a_1)} \times$$

$$\times \exp\left\{ \gamma^+(-Ka_2) - \gamma^-(K\sin\phi_o)\right\} \tag{29}$$

With the knowledge of B,C and $g(\phi_o)$, we can complete (15) and (16). Therefore, we have the far fields everywhere. The computations of γ^+ and γ^- involves integration in a complex plane and carefull definitions of the paths of integration, but there are no serious difficulties and the integrations have no special importance in the physics of the problem. The component E_z is not evaluated explicitly. It is related simply to H_x through the relation

$$E_z = \frac{j}{\omega\varepsilon_o\,\varepsilon(z)}\,\frac{\partial H_x}{\partial y} \tag{30}$$

Because of the orthogonality of the modes of either the grounded dielectric slab or the partially filled parallel plate waveguide, the powers carried by the launched surface wave and the radiated power are not coupled to each other. The reflected power, the radiated power and the power in the surface wave per unit incident power can be found individually from Poynting's theorem.

The percentage of power reflected is :

$$P_{ref} = \frac{(1+a_2)^2(a_2-a_1)^2}{(a_2-1)^2(a_2+a_1)^2}\exp\left\{\frac{-4a_2}{\pi}\int_0^1 \frac{\Omega(x)}{1+s'^2-x^2}\,dx\right\} \tag{31a}$$

The percentage of power transmitted is :

$$P_{trans} = \frac{4(a_1-1)\, a_2(a_2+1)\, a_1 \operatorname{sech}^2(Khs)}{s(a_2-1)(a_2+a_1)^2 \left[1+\tanh(Khs)\right]^2} \times \tag{31b}$$

$$\times \exp\left\{2Khs - \frac{1}{2} \ln\frac{a_1+1}{a_1-1} - \frac{2}{\pi}\int_0^1 \left[\frac{a_2}{a_2^2 - x^2} - \frac{a_1}{a_1^2 - x^2}\right]\Omega(x)dx\right\}$$

$$\Omega(x) = Kh(1-x^2)^{1/2} + \arctan\left\{\frac{(\varepsilon - x^2)^{1/2}}{\varepsilon(1-x^2)^{1/2}} \tan\left[Kd(\varepsilon - x^2)^{1/2}\right]\right\} \tag{31c}$$

The percentage of power radiated per unit angle in the direction ϕ_o (see Fig.1) is :

$$P_{rad} = \frac{2a_2(a_2+1)(s'^2-s^2)(1+\sin\phi_o)(s^2+\cos^2\phi_o)}{\pi\,(a_1+\sin\phi_o)^2(a_2+a_1)^2(a_2-1)} \times \cos^2\phi_o$$

$$\times \frac{\left|\tan(Khcos\,\phi_o) + \frac{(\varepsilon - \sin^2\phi_o)^{1/2}}{\varepsilon\cos\phi_o} \tan\left[Kd\,(\varepsilon - \sin^2\phi_o)^{1/2}\right]\right|}{\left|\sec(Khcos\,\phi_o)\right|(s'^2+\cos^2\phi_o)\left\{1+ \frac{\varepsilon - \sin^2\phi_o}{\varepsilon^2\cos^2\phi_o}\tan^2\left[Kd(\varepsilon - \sin^2\phi_o)^{1/2}\right]\right\}^{1/2}}$$

$$\times \exp\left[\frac{-2a_2}{\pi}\int_0^1 \frac{\Omega(x)\, dx}{a_2^2 - x^2} + \frac{2\sin\phi_o}{\pi}\, \mathcal{P}\int_0^1 \frac{\frac{\pi}{2} - \Omega(x)}{\sin^2\phi_o - x^2}\, dx\right] \tag{31d}$$

Symbol $\mathcal{P}$ in (31d) stands for Cauchy's Principal Value.

The values of equations (31) can be calculated without difficulties.

C.M.Angulo

B I B L I O G R A P H Y

1) C.M.Angulo and W.S.C.Chang, "On the Excitation of Surface Waves in Dielectric Slabs, Part I : Excitation by a Magnetic Line", Div. of Engrg., Brown Univ., Providence, R.I., Rept. No. AF 1391/2; June, 1956. AFCRC-TN 56-950, ASTIA Doc. No. AD 110141.

2) C.M.Angulo and W.S.C.Chang, "The Launching of Surface Waves by a Parallel Waveguide", Div. of Engrg., Brown Univ., Providence, R.I., Rept.No. 1391/5; April, 1957. AFCRC-TN-57-365, ASTIA Doc. No. AD 117059, pp.38-46.

3) R.Paley and N.Wiener, "Fourier Transform in Complex Domain", Amer.Mathematical Soc. Colloquium Publications, vol. XIX, pp. 49-58; 1943.

4) P.M.Morse and H.Feshbach, "Methods of Theoretical Physics", McGraw-Hill Book Co., New York, N.Y., vol.1, pp.960-992; 1953.

5) For results of a similar analysis for the same configuration without the dielectric slab, see N.Marcuvitz, "Waveguide Handbook", McGraw-Hill Book Co. Inc., New York, N.Y., Radiation Laboratory Series, vol.10, pp.179-186; 1951.

6) Morse and Feshbach, op.cit., p.462.

7) E.T.Copson, Quarterly J. of Math. (Oxford), vol.17, no.65, pp.19-34; March, 1946.

CENTRO INTERNAZIONALE MATEMATICO ESTIVO

(C.I.M.E.)

C. J. B O U W K A M P

Philips Research Laboratories

N.V. Philips' Gloeilampenfabrieken

Eindhoven-Netherlands

NOTES ON THE CONFERENCE

ROMA - Istituto Matematico dell'Università

NOTES ON THE CONFERENCE

by C.J.BOUWKAMP

Philips Research Laboratories

N.V. Philips' Gloeilampenfabrieken

Eindhoven-Netherlands

1. CHOICE OF TIME FACTOR.

Right at the beginning of the conference, it was pointed out that some authors use the time factor exp(iωt), while others, as Prof. Bremmer, use exp(-iωt). In addition, Prof. Graffi preferred the j to the i for indicating the "imaginary unit". Why then do we scientists (as we all claim to be) lack uniformity in our notations and conventions?

First of all, it is safe to state that mathematicians generally use i and that most electrical engineers use j for the imaginary unit. In the latter case no confusion arises when the engineer denotes electric current by i. The question i versus j therefore is a trivial matter, which need not concern us any longer. However, I do object to the engineers' customary "definition" of j, if they introduce it as the square root of minus one, because j cannot be defined in terms of reals and the elementary arithmetic operations. Therefore, they should avoid "where $j = \sqrt{-1}$ ", but write instead "where j is the imaginary unit".

Secondly, what about the choice of sign, exp(iωt) versus exp(-iωt)? At first sight, this seems to be also a trivial difference, because it would come down to a transformation to complex-conjugated quantities. That is true, indeed, if we have got a result in the form of some complex number $a + ib$, where a and b are real numbers. However, before arriving at this result we may have had to handle complex functions that do not contain the imaginary unit explicity (cf. the proper branch of a multiply valued function as encountered in the preceding lectures).

Consequently, we should always indicate explicitly which of the two conventions is going to be used. In this respect I may warn you not to follow the example of Born's Lehrbuch der Optik. In this famous and marvellous book on optics, the author applies different time factors in different chapters. In some chapters the time factor is not stated but, worst of all, sometimes results of one chapter are taken over in other chapters without realizing that a different time factor was used there. As a consequence, the book of Born is marred by errors; if, in addition, left- and right-handed coordinate systems are used simultaneously (I know of at least one such book) all results in electromagnetic theory should be suspected.

Electrical engineers stick to their $\exp(j\omega t)$, and rightly so. If they would turn to $\exp(-j\omega t)$, all text-books on electrical engineering would become a nuisance to them. So let these engineers use $\exp(j\omega t)$ for ever!

Theoreticians, like Bremmer and myself, for example, heartily agree with the $\exp(j\omega t)$ of electrical engineers. However, we disagree with them if they are using $\exp(i\omega t)$.

I shall now indicate why, in my opinion, preference should be given to $\exp(-i\omega t)$ for theoretical work in electromagnetics and the like, if there is a choice to be made out of four possibilities:

$$\exp(i\omega t)\ ,\quad \exp(-i\omega t)\ ,\quad \exp(j\omega t)\quad \exp(-j\omega t)$$

First of all, we vote for the i instead of the j , in honour of the mathematician's use of i (we are theorists anyway) and we would not steal the j from the engineering practice because there j serves a most useful purpose. So that leaves choosing between $\exp(i\omega t)$ and $\exp(-i\omega t)$. Why then do we vote for the latter?

Sommerfeld, that illustrious master of classical mathematical physics, always used the time factor $\exp(-i\omega t)$ in his many important papers. In one of his books on theoretical physics, he emphasizes the reason why he does so. It is simply this: A plane wave (wave number k) propagating in the positive x -direction is represented by $\exp(+ikx)$, or shorter, $\exp(ikx)$, if the time factor (suppressed) is understood to be $\exp(-i\omega t)$. Under the same circumstances, a plane wave propagating in the negative x -direction is represented by $\exp(-ikx)$. Therefore, positive (negative) x -direction is naturally reflected by a plus (minus) sign in the space exponential. I could elaborate on this point by showing that we also avoid many minus signs in our common formulae, therefore help the printer (note: if the sign of a quantity is plus, we usually omit it) and make the formulae looking better. I will not do so, however, because in my opinion there is an even better reason to use $\exp(-i\omega t)$. (Sommerfeld must have been aware of this)*.

---

*In the discussion, Prof.Toraldo rightly pointed out that also in quantum mechanics $\exp(-i\omega t)$ is the usual time factor.

As you have seen during these lectures, the complex number k often occurs in our analysis. This k lies in the (first half of the) first quadrant of the complex k-plane if $\exp(-i\omega t)$ is used. On the other hand, it lies in the fourth quadrant in the case of $\exp(i\omega t)$. Without doubt, it is easier to work in the first than in the fourth quadrant, also from the educational point of view. Also, and I emphasize, the first quadrant is preferred to the fourth if one is going to define the proper branch of multiply valued functions as occurring in our radiation problems.

In conclusion, I suggest that electrical engineers should use $\exp(j\omega t)$ and theoreticians should use $\exp(-i\omega t)$. If all workers in the field should stick to these conventions, a source of many possibile errors would have been eliminated.

2. RADIATING DIPOLE IN THE INTERFACE OF TWO MEDIA.

In connection with a question of Prof. Boella, I should like to comment on the Sommerfeld problem: the radiation of a vertical electric dipole over plane earth if the dipole is situated in the interface between air and earth. If not stated otherwise, I use Bremmer's notation. The time factor is $\exp(-i\omega t)$; r, ϕ, z denote right-handed cylindrical coordinates and x, y, z right-handed rectangular coordinates.

In his first paper on the subject, which appeared in Ann.Physik 28 (1909) pp.665-736, Sommerfeld introduced a scalar Π different from that which is common at present.

I shall denote Sommerfeld's Π by S and reserve the symbol Π for a scalar wave function obtained from Bremmer's Π by dividing the latter by $-i\omega$; $\Pi_{Bremmer} = -i\omega \Pi$. In terms of S , the electromagnetic field of the dipole is given by

$$H_\phi = -\frac{\partial S}{\partial r} ,$$

$$E_r = \frac{i\omega\mu}{k^2} \frac{\partial^2 S}{\partial r\, \partial z} ,$$

$$E_z = \frac{i\omega\mu}{k^2} \left[k^2 S + \frac{\partial^2 S}{\partial z^2} \right] ,$$

with $k^2 = \omega^2 \varepsilon\mu + i\omega\mu\sigma.$

The boundary conditions at the interface $z = 0$ are expressed as follows:

$$S \quad \text{and} \quad \frac{\mu}{k^2} \frac{\partial S}{\partial z} \quad \text{continuous at } z = 0 .$$

Therefore, the connection between Sommerfeld's S and our Π is :

$$S = \frac{k^2}{i\omega\mu} \Pi .$$

Let R measure the distance from the point of observation $(x, y, \dot{z})$ to the origin:

$$R = \sqrt{x^2 + y^2 + z^2} ,$$

and let S_0 be defined by

$$S_0 = \frac{e^{ikR}}{R} ;$$

S_0 then represent a " unit dipole " radiating in a homogeneous medium with wave number k . It is important to note that k should be that square root of k^2 which lies in the first quadrant. To be more precise,

$$0 \leqslant \arg k \leqslant \frac{\pi}{4} .$$

Sommerfeld's formulation of the boundary-value problem is as follows:

(1) $\Delta S_1 + k_1^2 S_1 = 0 \quad (z > 0), \quad \Delta S_2 + k_2^2 S_2 = 0 \quad (z < 0) ;$

(2) $S_1 = S_2$ and $\frac{\mu_1}{k_1^2} \frac{\partial S_1}{\partial z} = \frac{\mu_2}{k_2^2} \frac{\partial S_2}{\partial z}$ at $z = 0$;

(3) $S_1 \to 0$ if $R_1 \to \infty \; (z \geqslant 0), \quad S_2 \to 0$ if $R_2 \to \infty \; (z \leqslant 0);$

(4) $S_1 - \frac{e^{ik_1R_1}}{R_1} \; (z \geqslant 0), \quad S_2 - \frac{e^{ik_2R_2}}{R_2} \; (z \leqslant 0)$ and

their first-order derivatives finite and continuous including $R_{1,2} = 0$.

As to (3) I may remark that it is my interpretation of Sommerfeld's badly phrased condition at infinity. Condition (3) is a version of the radiation condition which only applies if both k_1 and k_2 are complex. In the 1909-paper Sommerfeld proved existence and uniqueness of the solution under these conditions (his proof is not complete, but can be made so). It is important to verify that conditions (2) and (4) are compatible with each other. Since $\exp(ikR) \to 1$ as $R \to 0$, the singularity in both S_1 and S_2 is $1/R$. Conditions (2) can and must hold for the singular parts of S_1 and S_2 as well as for the remaining regular parts separately.

Let us now turn to the question of how conditions (1) to (4) should be modified if we start with the function Π. The first three are easy:

$$(1') \quad \Delta\Pi_1 + k_1^2\Pi_1 = 0 \quad (z > 0), \quad \Delta\Pi_2 + k_2^2\Pi_2 = 0 \quad (z < 0) \; ;$$

$$(2') \quad \frac{k_1^2}{\mu_1}\Pi_1 = \frac{k_2^2}{\mu_2}\Pi_2 \quad \text{and} \quad \frac{\partial\Pi_1}{\partial z} = \frac{\partial\Pi_2}{\partial z} \quad \text{at } z = 0 \; ;$$

$$(3') \quad \Pi_1 \to 0 \text{ if } R_1 \to \infty \ (z \geqslant 0), \quad \Pi_2 \to 0 \text{ if } R_2 \to \infty \ (z \leqslant 0).$$

However, in reformulating (4) one should be careful. It is wrong to require finiteness of

$$(4'') \qquad \Pi_1 - \frac{e^{ik_1R_1}}{R_1} \quad (z \geqslant 0) \quad \text{and} \quad \Pi_2 - \frac{e^{ik_2R_2}}{R_2} \quad (z \leqslant 0),$$

etc.,

because such a requirement would not be compatible with (2'). Instead, we should observe that the singularities in the two half-space must have different coefficients C_1 and C_2 with $C_1 + C_2 = 2$ such that their ratio is equal to the ratio of the corresponding values of μ / k^2. That is to say,

$$C_1 = \frac{2\,\mu_1\,k_2^2}{\mu_1 k_2^2 + \mu_2 k_1^2}, \qquad C_2 = \frac{2\,\mu_2\,k_1^2}{\mu_1\,k_2^2 + \mu_2\,k_1^2}.$$

Therefore, condition (4) is to be replaced not by (4") but by

$$(4') \qquad \Pi_1 - \frac{C_1}{R_1}\,e^{ik_1R_1} \quad (z \geqslant 0), \qquad \Pi_2 - \frac{C_2}{R_2}\,e^{ik_2R_2} \quad (z \leqslant 0)$$

and their first-order derivatives finite and continuous including $R_{1,2} = 0$.

If you will look up the pertinent chapter in Stratton's well-known book on electromagnetic theory, you will observe that Stratton, using Π, was not aware of condition (4'). Instead of requiring (4') he implies condition (4"). Notwithstanding this, Stratton does get the correct answer. The fact that Stratton did not observe the inconsistency in his treatment is due to another failure: he did not verify that the solution finally obtained does indeed satisfy all conditions required in the Ansatz.

For a further discussion, it is necessary to quote Sommerfeld's result of 1909. If we assume $\mu = \mu_1 = \mu_2 = 1$, it is

$$S_1 = \int_0^\infty \frac{k_1^2 + k_2^2}{N} J_0(\lambda r)\, e^{-z\sqrt{\lambda^2 - k_1^2}}\, \lambda\, d\lambda \quad ,(\text{singularity } \tfrac{1}{R}),$$

$$S_2 = \int_0^\infty \frac{k_1^2 + k_2^2}{N} J_0(\lambda r)\, e^{z\sqrt{\lambda^2 - k_2^2}}\, \lambda\, d\lambda \quad ,(\text{singularity } \tfrac{1}{R}),$$

in which the square roots are such that their real parts are positive and where N is defined by

$$N = k_1^2 \sqrt{\lambda^2 - k_2^2} + k_2^2 \sqrt{\lambda^2 - k_1^2} \ .$$

In addition, let me also state the corresponding result in the Π-formulation:

$$\Pi_1 = \int_0^\infty \frac{2\,k_2^2}{N} J_0(\lambda r)\, e^{-z\sqrt{\lambda^2 - k_1^2}} \lambda d\lambda\ , \quad (\text{sing. } \frac{2k_2^2}{k_2^2+k_1^2} \frac{1}{R}),$$

$$\Pi_2 = \int_0^\infty \frac{2\,k_1^2}{N} J_0(\lambda r)\, e^{z\sqrt{\lambda^2 - k_2^2}} \lambda d\lambda \quad (\text{sing. } \frac{2k_1^2}{k_2^2+k_1^2}.\frac{1}{R}) .$$

Now, as you will remember, Bremmer assumed the dipole <u>above</u> the interface between air and earth. We shall shortly indicate what happens if the height h of the dipole tends to zero. In the limit the dipole may be said to be located in the interface, and it is quite natural to expect that the limiting form of the wave function is identical with that obtained if the source is taken in the interface right from the start. <u>This is true, but only so for the Π-formulation</u>. However, if we start in the S-formulation with a dipole above the interface, and let h tend to zero, we get

$$S_1' = \int_0^\infty \frac{2k_2^2}{N} J_0(\lambda r)\, e^{-z\sqrt{\lambda^2-k_1^2}} \lambda\, d\lambda\ , \quad (\text{sing. } \frac{2k_2^2}{k_2^2+k_1^2}.\frac{1}{R}),$$

$$S_2' = \int_0^\infty \frac{2\,k_2^2}{N} J_0(\lambda r)\, e^{z\sqrt{\lambda^2-k_2^2}} \lambda\, d\lambda\ , \quad (\text{sing. } \frac{2k_2^2}{k_2^2+k_1^2}.\frac{1}{R} .$$

These formulae <u>are not identical</u> with Sommerfeld's 1909 results. They differ in that Sommerfeld's result is obtained from the primed quantities through multiplication by

$$\frac{k_1^2 + k_2^2}{2\,k_2^2} .$$

We also note that Π_1 equals S_1', but Π_2 is not equal to S_2'.

In view of this peculiar effect, it is obvious that the Π-formulation is preferable to the S-formulation. In one of his later papers (1926) Sommerfeld does indeed use the Π-formulation. But in that paper, and also in the first edition of the book of Frank and Von Mises, the singularities at the origin are not correctly ascertained. It is only in the new edition (see page 920) that Sommerfeld got the final correct formulation. (Sommerfeld gave credit to Fock for the correction.)

In concluding this section, I should also remark that Weyl used the S-formulation but (sofar as I remember correctly without having the original paper at hand) he started with the dipole above the interface. Hence, to make his result identical with Sommerfeld's, he had to multiply it by $(k_1^2 + k_2^2)/2k_2^2$, which, I believe, clarifies the difficulty observed by Prof. Boella.

3. THE ZEROS OF N (λ)

During this conference we have been frequently confronted with functions that need special care because they are not single-valued in the complex plane. A simple example is

$$f(\lambda) = \sqrt{\lambda^2 - k^2} ,$$

where k is a fixed complex constant ($0 \leq \arg k \leq \pi/4$, say) and λ varies in the complex λ-plane. To handle such functions, it is appropriate to introduce cuts in the λ-plane such that in the cut plane the function becomes single-valued. In our example, $f(\lambda)$ has two singularities, so-called branch points, which lie at $\lambda = \pm k$. If we cut the λ-plane along a curve connecting these two branch points we can define $f(\lambda)$ as a single-valued function in the whole cut λ-plane.

If, as in Sommerfeld's problem, we require that $\mathrm{Re}[f(\lambda)]$ shall be positive in the cut plane, the above curve or cut has to be chosen adequately, that is, the real part of $f(\lambda)$ should be equal to zero along the cut. In the case of $f(\lambda)$, the cut turns out to be part of an orthogonal hyperbola lying in the first and third quadrants. One branch cut connects $\lambda = k$ with $\lambda = i\infty$, the other $\lambda = -k$ with $\lambda = -i\infty$. If we walk in positive direction along the cut around the branch point, $f(\lambda)$ varies from $-i\infty$ to $i\infty$ (has real part zero). In the direction of the positive real axis, $f(\lambda)$ tends to λ as $\lambda \to \infty$; in the direction of the negative real axis, $f(\lambda)$ tends to $-\lambda$ as $\lambda \to -\infty$. Also, $f(0) = -ik$, and $f(\lambda) = f(-\lambda)$. If $k > 0$, the branch cuts reduce to straight segments (imaginary axis plus segment from $-k$ to k).

As you will remember, the following combination of two different square roots plays an important role in Sommerfeld's theory:

$$N(\lambda) = k_1^2 \sqrt{\lambda^2 - k_2^2} + k_2^2 \sqrt{\lambda^2 - k_1^2} .$$

We assume that k_1 and k_2 again are situated in the first half of the first quadrant (they may be real), and that both square roots have their real parts positive. This is possible by introducing four cuts, two for each square root as in the case of $f(\lambda)$. Of course, $N(\lambda)$ is more complicated than $f(\lambda)$ but it is not difficult to evaluate N at an arbitrary point λ.

It is known that, in general, $N(\lambda)$ possesses two simple zeros in the cut λ-plane, one of them giving rise to the Zenneck surface wave in Sommerfeld's analysis. For later purposes I want precise information on the location of this zero, and I want to ascertain what happens to this zero if k_1 and k_2 both are real (question of Prof.Eckart).

From $N(\lambda) = 0$ follows $\left[k_1^2 \sqrt{\lambda^2 - k_2^2}\right]^2 = \left[-k_2^2 \sqrt{\lambda^2 - k_1^2}\right]^2$, that is $k_1^4(\lambda^2 - k_2^2) = k_2^4(\lambda^2 - k_1^2)$, so that

$$\lambda^2 = \frac{k_1^2 k_2^2}{k_1^2 + k_2^2} .$$

What we have proved is: if $N(\lambda)$ has a zero, this must be of the form

$$\pm \frac{k_1 k_2}{\sqrt{k_1^2 + k_2^2}} .$$

Now, let us define $\sqrt{k_1^2 + k_2^2}$ as a number that lies in the first quadrant, and then define

$$h = \frac{k_1 k_2}{\sqrt{k_1^2 + k_2^2}} .$$

It can be shown that h lies inside the closed triangle Ok_1k_2 at a distance from the origin not greater than $\min(|k_1|, |k_2|)$ (Niessen-Van der Pol, Kahan-Eckart).

I shall prove that h is a zero of $N(\lambda)$, in the case that k_1 is real and k_2 finite but not real. We have

$$h^2 = \frac{k_1^2 k_2^2}{k_1^2 + k_2^2} , \quad h^2 - k_1^2 = \frac{-k_1^4}{k_1^2 + k_2^2} , \quad h^2 - k_2^2 = \frac{-k_2^4}{k_1^2 + k_2^2} .$$

Therefore

$$\sqrt{h^2-k_1^2} = \pm \frac{i\,k_1^2}{\sqrt{k_1^2+k_2^2}}\ , \qquad \sqrt{h^2-k_2^2} = \pm \frac{i\,k_2^2}{\sqrt{k_1^2+k_2^2}}\ .$$

We must choose the correct sign from the fact that both $\sqrt{h^2-k_1^2}$ and $\sqrt{h^2-k_2^2}$ should have positive real part. Let $\theta = \arg(k_1^2+k_2^2)$. Then $0 < \theta < \frac{\pi}{2}$, and

$$\arg \frac{i\,k_1^2}{\sqrt{k_1^2+k_2^2}} = \frac{\pi}{2} + 0 - \frac{1}{2}\theta\ ,$$

so that

$$\frac{\pi}{4} < \arg \frac{i\,k_1^2}{\sqrt{k_1^2+k_2^2}} < \frac{\pi}{2}$$

Therefore $\frac{i\,k_1^2}{\sqrt{k_1^2+k_2^2}}$ has a <u>positive</u> real part. Consequently

$$\sqrt{h^2-k_1^2} = \frac{+i\,k_1^2}{\sqrt{k_1^2+k_2^2}}\ .$$

In the same way,

$$\arg \frac{i\,k_2^2}{\sqrt{k_1^2+k_2^2}} = \frac{\pi}{2} + \arg(k_2^2) - \frac{1}{2}\theta \ .$$

But $\arg(k_2^2) - \frac{1}{2}\theta > 0$ and $< \frac{\pi}{2}$, so that

$$\frac{\pi}{2} < \arg \frac{i\,k_2^2}{\sqrt{k_1^2+k_2^2}} < \pi \ .$$

Hence $\dfrac{i\,k_2^2}{\sqrt{k_1^2+k_2^2}}$ has a <u>negative</u> real part, and consequently

$$\sqrt{h^2 - k_2^2} = \frac{-ik_2^2}{\sqrt{k_1^2+k_2^2}} \ .$$

Under these circumstances

$$k_1^2 \sqrt{h^2 - k_2^2} + k_2^2 \sqrt{h^2 - k_1^2} = 0 \ , \quad \text{indeed.}$$

To repeat: if $k_1 > 0$, $\mathrm{Re}(k_2^2) \geqslant 0$, $\mathrm{Im}(k_2^2) > 0$, then $N(\lambda)$ has (simple) zeros at $\lambda = \pm h$.

<u>Remarks</u> : If k_1 and k_2 are both real, h lies on the cut, and h ceases to be a zero of $N(\lambda)$. If $|k_2| \to \infty$, then $h \to k_1$; in practice h lies near k_1 . In the neighbourhood of h we have

$$\frac{1}{N(\lambda)} = \frac{ik_1 k_2}{k_2^4 - k_1^4} \cdot \frac{1}{\lambda - h} + \text{regular terms.}$$

4. CONDITION AT INFINITY

Sommerfeld's Ansatz for solving his problem is the integral representation of the point singularity

$$S_0 = \frac{e^{ikR}}{R} = \int_0^\infty \frac{\lambda}{\sqrt{\lambda^2 - k^2}} J_0 (\lambda r)\ e^{-|z| \sqrt{\lambda^2 - k^2}}\ d\lambda.$$

Here the integration is understood along the positive real axis ($0 \leq \arg k \leq \pi/4$) . If $k > 0$, then the path of integration should be indented below k. The integral representation remains valid if the path is deformed such that the branch cut is not crossed.

Consider again the problem treated in sec. 2, if $k_1 > 0$ and k_2 is finite but not real. Consider a path of integration, W , from 0 to ∞ such that it passes below k_1 and k_2 but above the point h , the one zero of $N(\lambda)$ in the first quadrant.

If we take the same integrands as in the case of S_1 and S_2, we get (integrating along W) two potentials Σ_1 and Σ_2, say. They satisfy conditions (1), (2) and (4). Therefore, if we had omitted a condition like (3), the new solution Σ and the old solution S would have equal rights in that either could be used as the solution of Sommerfeld's problem. Of course, S and Σ are different. This difference is just the value of the residue at h, which gives a standing type of Zenneck surface wave. As a matter of fact, for example in the air,

$$S_1 - \Sigma_1 = f(k_1, k_2)\, J_0(hr)\, e^{-z\sqrt{h^2-k_1^2}}.$$

So you see that both S and Σ cannot satisfy the radiation condition (3) at infinity. Only one therefore is correct, which of course is S_1. But, if we should take k_1 and k_2 both real from the start, we would have no means to find out which one, S_1 or Σ_1, would be the correct one. That is, not by trivial means. We should then use the deeper result that only for the integration along the positive real axis do the eigenfunctions form a complete system, while satisfying well-posed boundary conditions at infinity (proof, what conditions?).

In a different approach one might wish to formulate the appropriate radiation condition at infinity for both k_1 and k_2 real. Such condition could be obtained by asymptotical analysis of S as $R \to \infty$ (if possible, asymptotic approximations uniform in the full closed range of the polar angle θ with $0 \leq \theta \leq \pi$ should be searched for). If we then could give a uniqueness proof, everything would be nice. However, so far as I know, this programme has never been carried out in detail. My own unpublished results are not as I should like.

In conclusion I may refer to a paper of Haug, Z. Naturforsch. 7a(1952) pp.501-505, where the same problem is dealt with.

CENTRO INTERNAZIONALE MATEMATICO ESTIVO

(C.I.M.E.)

H, B R E M M E R

ELECTROMAGNETIC WAVE PROPAGATION AROUND THE EARTH

ROMA - Istituto Matematico dell'Università

ELECTROMAGNETIC WAVE PROPAGATION AROUND THE EARTH

by

H. BREMMER

I. The field of a vertical dipole in a homogeneous atmosphere above a plane homogeneous earth (Sommerfeld Problem).

Ia. Derivation of the rigorous solution.

Historically this is the first problem dealing with the effect of the earth on radio-wave propagation. The atmosphere and the earth are both assumed to be homogeneous. In each of them Maxwell's equations read in M.K.S. units:

$$\begin{aligned} &\operatorname{curl} \vec{E} + \mu \frac{\partial \vec{H}}{\partial t} = \vec{0}\,, \\ &\operatorname{curl} \vec{H} - \varepsilon \frac{\partial \vec{E}}{\partial t} - \sigma \vec{E} = \vec{J}\,, \end{aligned} \tag{1}$$

where $\vec{J}$ is the current density due to the sources, and $\sigma\vec{E}$ that due to currents induced according to Ohm's law in a medium with conductivity σ. For harmonic time dependence, with all field components proportional to $e^{-i\omega t}$, (1) reduces to :

$$\begin{aligned} &\operatorname{curl} \vec{E} - i\omega\mu \vec{H} = \vec{0}\,, \\ &\operatorname{curl} \vec{H} + i\omega\varepsilon_{eff} \vec{E} = \vec{J}\,, \end{aligned} \tag{2}$$

where

$$\varepsilon_{eff} = \varepsilon + \frac{i\sigma}{\omega}\,.$$

In each of the homogeneous half spaces (2) can be solved by substituting :

$$\vec{E} = i\omega\, \varepsilon_{eff}\, \mu \vec{\Pi} + \frac{i}{\omega} \text{grad div } \vec{\Pi} ,$$
$$\vec{H} = \varepsilon_{eff} \text{ curl } \vec{\Pi} . \tag{3}$$

The first equation (2) is then satisfied automatically, the second if, and only if, the Hertzian vector $\vec{\Pi}$ satisfies the scalar wave equation

$$(\nabla^2 + k^2)\, \vec{\Pi} = - \frac{\vec{J}}{\varepsilon_{eff}} ,$$

in which the wave number k is defined as the square root in the first quadrant of :

$$k^2 = \omega^2 \mu\, \varepsilon_{eff} = \omega^2 \mu\varepsilon + i\, \omega\, \mu\sigma . \tag{4}$$

In view of the scalar wave equation we may replace (3) by

$$\vec{E} = \frac{i}{\omega}(\text{curl curl } \vec{\Pi} - \frac{\vec{J}}{\varepsilon_{eff}}) ,$$
$$\vec{H} = \varepsilon_{eff} \text{ curl } \vec{\Pi} . \tag{3a}$$

Let the plane earth's surface be the xy plane $z = 0$. The Sommerfeld problem most considered concerns a short vertical dipole at a height h above the earth. This source may be located at $x = y = 0$, $z = h$, while its current-density distribution is represented by a vector with a z component only, which is given by

$$J_z = M \; \delta(x)\, \delta(y)\, \delta(z-h) \; . \qquad (5)$$

M is the momentum of the transmitter. The atmosphere $(z > 0)$ and the earth $(z < 0)$ can be characterized by constants k_1 and k_2. We solve the problem with the aid of a vertically directed Hertzian vector $\vec{\Pi}$. Its amplitude, Π say, is a scalar from which all field components are to be derived. In order to account for the boundary conditions at the earth's surface, viz. the continuity there of the horizontal components of $\vec{E}$ and $\vec{H}$, we need the expressions :

$$E_x = \frac{i}{\omega} \frac{\partial^2 \Pi}{\partial x \partial z} \quad , \qquad E_y = \frac{i}{\omega} \frac{\partial^2 \Pi}{\partial y \partial z} \; ,$$

$$H_x = \varepsilon_{eff} \frac{\partial \Pi}{\partial y} = \frac{k^2}{\omega^2 \mu} \frac{\partial \Pi}{\partial y} \quad ,$$

$$H_y = - \; \varepsilon_{eff} \frac{\partial \Pi}{\partial x} = - \frac{k^2}{\omega^2 \mu} \frac{\partial \Pi}{\partial x} \; .$$

The boundary conditions at the earth's surface $z = 0$ are therefore guaranteed for all values of x and y if $\partial\Pi/\partial z$ and $k^2\Pi$ are continuous there (assuming one and the same permeability μ_0 for both media). The other boundary conditions at infinity amount to the so-called radiation condition (propagation towards infinity). The problem is fixed in its scalar form by these boundary conditions, together with the two differential equations :

$$(\nabla^2 + k_1^2)\,\Pi = -\frac{M}{\varepsilon_{eff,1}}\,\delta(x)\,\delta(y)\,\delta(z-h)\,, \quad z>0 \qquad (6a)$$

$$(\nabla^2 + k_2^2)\,\Pi = 0\,. \qquad z<0 \qquad (6b)$$

In the absence of the earth (6a) holds throughout space. The primary field then existing, depends on the corresponding solution

$$\Pi_{pr} = \frac{M}{4\pi\,\varepsilon_{eff,1}}\,\frac{e^{ik_1R}}{R} \qquad (7a)$$

in which $R = \left\{x^2 + y^2 + (z-h)^2\right\}^{1/2}$ is the distance from the transmitter to the point of observation. In order to solve the problem in the presence of the earth, this primary field is expressed, following <u>Sommerfeld</u>[1), in terms of solutions of the homogeneous wave equation (6a) that are separated with respect to the cylindrical coordinates $\rho = (x^2 + y^2)^{1/2}$ and z :

$$\Pi_{pr} = \frac{M}{4\pi\,\varepsilon_{eff,1}}\int_0^\infty \frac{d\lambda\,\lambda}{\sqrt{\lambda^2-k_1^2}}\,J_0(\lambda\rho)\,e^{-\sqrt{\lambda^2-k_1^2}\,|z-h|} \qquad (7b)$$

The square root is assumed to have a positive real part.

The additional field produced in the half space $z>0$ by the presence of the earth may be derived from a corresponding contribution Π_{sec} to the scalar Π . This latter contribution is source-free in $z>0$, but it is generated at the earth's surface, and thus propagates towards

$z = +\infty$ throughout the half space $z > 0$. For an absorbing atmosphere ($\mathfrak{Im}\, k_1 > 0$), how ever small the absorption may be, this propagation property is guaranteed by a solution of the homogeneous wave equation $(\nabla^2 + k_1^2)\Pi = 0$ of the following form :

$$\Pi_{sec} = \frac{M}{4\pi\, \varepsilon_{eff,1}} \int_0^\infty \frac{d\lambda\, \lambda}{\sqrt{\lambda^2 - k_1^2}}\, \mathfrak{R}(\arcsin \frac{\lambda}{k_1}).J_0(\lambda\rho)e^{-\sqrt{\lambda^2-k_1^2}\,(h+z)} \qquad (z>0) \qquad (8)$$

The weighting function $\mathfrak{R}$ of λ will be determined presently.

On the other hand, the waves penetrating into the earth produce a source-free field there which propagates towards $z = -\infty$; its scalar Π_2 has to satisfy the homogeneous wave equation (6b). This is brought about by a solution of the form :

$$\Pi_2 = \frac{M}{4\pi\, \varepsilon_{eff,1}} \int_0^\infty \frac{d\lambda\, \lambda}{\sqrt{\lambda^2 - k_1^2}}\, .\mathfrak{D}(\arcsin \frac{\lambda}{k_1}).J_0(\lambda\rho)e^{-\sqrt{\lambda^2-k_1^2}\,h + \sqrt{\lambda^2-k_2^2}\,z} \qquad (z<0) \qquad (9)$$

in which $\sqrt{\lambda^2-k_2^2}$ is also assumed with a positive real part.

The weighting functions $\mathfrak{R}$ and $\mathfrak{D}$ follow from the continuity at $z=0$ of both $\partial\Pi/\partial z$ and $k^2\Pi$, or explicitly,

$$\left(\frac{\partial \Pi_{pr}}{\partial z}\right)_{z=+0} + \left(\frac{\partial \Pi_{sec}}{\partial z}\right)_{z=0} = \left(\frac{\partial \Pi_2}{\partial z}\right)_{z=-0}$$

$$k_1^2\,(\Pi_{pr} + \Pi_{sec})_{z=+0} = k_2^2\,(\Pi_2)_{z=-0} \;.$$

The evaluation of these expressions leads to relations which can be satisfied by dropping the integral signs and common factors of the integrands. We find

$$\sqrt{\lambda^2-k_1^2}\left\{1-\mathcal{R}(\arcsin\frac{\lambda}{k_1})\right\}=\sqrt{\lambda^2-k_2^2}\,.\mathcal{D}(\arcsin\frac{\lambda}{k_1})\ ,$$

$$k_1^2\left\{1+\mathcal{R}(\arcsin\frac{\lambda}{k_1})\right\}=k_2^2\,.\mathcal{D}(\arcsin\frac{\lambda}{k_1})\ ,$$

and thus

$$\mathcal{R}(\arcsin\frac{\lambda}{k_1})=\frac{k_2^2\sqrt{\lambda^2-k_1^2}-k_1^2\sqrt{\lambda^2-k_2^2}}{k_2^2\sqrt{\lambda^2-k_1^2}+k_1^2\sqrt{\lambda^2-k_2^2}}\ ,$$

$$\mathcal{D}(\arcsin\frac{\lambda}{k_1})=\frac{k_1^2}{k_2^2}(1+\mathcal{R})=\frac{2\,k_1^2\sqrt{\lambda^2-k_1^2}}{k_2^2\sqrt{\lambda^2-k_1^2}+k_1^2\sqrt{\lambda^2-k_2^2}} \qquad (10)$$

Substitution of these expression into (8) and (9), and a further application of (3), yields the final solution for all field components. A similar solution can be obtained for a vertical magnetic dipole; it is arrived at by assuming a right-hand side of the type (5) in the upper equation (2), instead of the lower one.

Ib. Alternative representations of the solution of the Sommerfeld Problem.

A representation derived by Weyl [2)] is obtained from the above results by substituting

$$J_0(\lambda\rho) = \frac{1}{2\pi}\int_0^{2\pi} d\varphi' e^{i\lambda(x\cos\varphi' + y\sin\varphi')} ,$$

and by passing to a new integration variable ϑ' (instead of λ) given by $\lambda = k_1 \sin\vartheta'$ and $\sqrt{\lambda^2 - k_1^2} = -i\,k_1\cos\vartheta'$. In the special case $z > h$ this procedure yields the following expression for the field in the atmosphere if we add together the contributions resulting from (7b) and (8):

$$\Pi = \frac{i\,k_1\,M}{8\pi^2\,\varepsilon_{eff,1}} \int_0^{2\pi} d\varphi' \int_0^{\frac{\pi}{2}-i\infty} d\vartheta'\sin\vartheta'\, e^{-ik_1 h\cos\vartheta'}\left\{1 + R(\vartheta')\, e^{2ik_1 h\cos\vartheta'}\right\} \times$$

$$\times\, e^{ik_1(x\sin\vartheta'\cos\varphi' + y\sin\vartheta'\sin\varphi' + z\cos\vartheta')} .$$

This double integral interprets the field as a sum of plane waves (instead of the cylindrical waves in the previous section) propagating in directions which form an angle ϑ' with the z - axis, and which are further determined by the azimuthal angle φ' around this axis. However, these propagation directions are necessarily partly complex which proves to be essential for the description of a field, that is not source-free. The first term between the brackets represents the primary field, the second term the contributions due to the reflection of the plane waves composing the primary field at the earth's surface. This reflection property is stressed by the significance of the parameter $\mathcal{R}$ as a Fresnel reflection coefficient. In fact, the latter can be put in the form :

$$\mathcal{R}(\vartheta') = \frac{k_2^2 \cos\vartheta' - k_1 \sqrt{k_2^2 - k_1^2 \sin^2\vartheta'}}{k_2^2 \cos\vartheta' + k_1 \sqrt{k_2^2 - k_1^2 \sin^2\vartheta'}} , \qquad (11a)$$

or also

$$\mathcal{R}(\vartheta') = \frac{k_2 \cos\vartheta' - k_1 \cos\vartheta''}{k_2 \cos\vartheta' + k_1 \cos\vartheta''} = \frac{\tan(\vartheta' - \vartheta'')}{\tan(\vartheta' + \vartheta'')} , \qquad (11b)$$

if ϑ' and ϑ'' are connected according to Snell's law $k_1 \sin\vartheta' = k_2 \sin\vartheta''$. Formula (11) constitutes the conventional form of Fresnel's reflection coefficient for a wave that propagates in directions making the angles ϑ' and ϑ'' with the normal to the interface in the above and the lower half space respectively .

Another modification of the original λ integrals derived by Sommerfeld results after substituting

$$J_0(\lambda\rho) = \frac{1}{2} H_0^{(1)}(\lambda\rho) - \frac{1}{2} H_0^{(1)}(\lambda\rho\, e^{\pi i}) ,$$

while passing to the variable $\lambda' = \lambda e^{\pi i}$ in the contribution depending on the second Hankel function. We then get, e.g. , the following integral for the secondary field (8), taking into account that $\mathcal{R}$ is an even function of λ :

$$\Pi_{sec} = \frac{M}{8\pi\ \varepsilon_{eff,1}} \int_{-\infty+io}^{\infty} \frac{d\lambda\ \lambda}{\sqrt{\lambda^2-k_1^2}} \mathcal{R}(\arc\sin\frac{\lambda}{k_1}).H_0^{(1)}(\lambda\rho)e^{-\sqrt{\lambda^2-k_1^2}(h+z)}$$

$(z>0)$ (12)

In the history of the Sommerfeld problem the latter representation has been the starting point for the numerous discussions on the question whether a special part of the solution may be considered as a surface wave. The integration path of (12) is then completed to a contour along the upper half of the λ -plane. The integral is next reduced to contributions arising from loops around cross-cuts of the functions $\sqrt{\lambda^2 - k_1^2}$ and $\sqrt{\lambda^2 - k_2^2}$, and a term due to the residue at the only pole of $\mathcal{R}$. The latter term has been interpreted as a surface wave but its numerical insignificance makes the discussion of its reality a purely academic question.

I.C.Numerical approximations.

The rigorous expressions considered so far are unsuitable for numerical computations. A practical formula is obtained as follows if transmitter and receiver are both on the ground ($h=0$, $z=+0$). The sum of (7b) and (8) then reduces to :

$$\Pi = \frac{k_2^2 M}{2\pi\varepsilon_{eff,1}} \int_0^{\infty} d\lambda\ \lambda \frac{J_0(\lambda\rho)}{k_2^2\sqrt{\lambda^2-k_1^2} + k_1^2\sqrt{\lambda^2-k_2^2}} \quad (13)$$

$(h=0,\ z=+0)$

The main contribution to the integral proves to result from the vicinity of $\lambda = k_1$. This is understood physically from the fact that $\lambda = k_1$ corresponds to $\vartheta' = \pi/2$ in Weyl's representation, that is to those waves (travelling horizontally) which can reach the receiver (under the conditions assumed here) directly over the earth's surface. This suggests that a good approximation might be obtained by replacing the second term in the denominator of (13) by

$$k_1^2 \sqrt{k_1^2 - k_2^2} = -i\, k_1^2 \sqrt{k_2^2 - k_1^2} = -i\, b \quad , \quad \text{say.}$$

We then transform (13) into the double integral :

$$\Pi \sim \frac{k_2^2 M}{2\pi\, \varepsilon_{eff,1}} \int_0^\infty d\lambda\, \lambda\, J_o(\lambda\rho) \int_0^\infty ds\, e^{-k_2^2 \sqrt{\lambda^2 - k_1^2}\, s + ibs} =$$

$$= -\frac{M}{2\pi\, \varepsilon_{eff.1}} \int_{s=0}^{s=\infty} e^{ibs} . d \left\{ \int_0^\infty d\lambda\, \frac{\lambda \cdot J_o(\lambda\rho)}{\sqrt{\lambda^2 - k_1^2}}\, e^{-k_2^2 \sqrt{\lambda^2 - k_1^2}\, s} \right\} .$$

The λ integral entering here is of the type (7b) which equals the elementary expression (7a). Its evaluation yields :

$$\Pi \sim -\frac{M}{2\pi\,\varepsilon_{eff,1}} \int_{s=0}^{s=\infty} e^{ibs} \,.\, d\, \frac{e^{i k_1 \sqrt{\rho^2 + k_2^4 s^2}}}{\sqrt{\rho^2 + k_2^4 s^2}} =$$

$$= \frac{M}{2\pi\,\varepsilon_{eff,1}} \left\{ \frac{e^{ik_1\rho}}{\rho} + ib \int_0^\infty \frac{e^{ik_1\sqrt{\rho^2+k_2^4 s^2}+ibs}}{\sqrt{\rho^2+k_2^4 s^2}}\, ds \right\}.$$

The main contribution now arises from the vicinity of $s=0$ so that the exponent may be cut off after the second term of the power expansion with respect to s, whereas we may take $s=0$ in the denominator. We find :

$$\Pi \sim \frac{M}{2\pi\,\varepsilon_{eff,1}} \frac{e^{ik_1\rho}}{\rho} \left\{ 1 + ib \int_0^\infty e^{i\frac{k_1 k_2^4}{2\rho} s^2 + ibs}\, ds \right\} =$$

$$= \frac{M}{2\pi\,\varepsilon_{eff,1}} \frac{e^{ik_1\rho}}{\rho} \left\{ 1 + 2\sqrt{\rho'}\; e^{-\rho'} \int_{\sqrt{\rho'}}^{i\infty} e^{u^2}\, du \right\},$$

an expression depending mainly on the so-called numerical distance

$$\rho' = \frac{i\, b^2 \rho}{2k_1 k_2^4} = i\, \frac{k_1^3 (k_2^2 - k_1^2)}{2\, k_2^4}\, \rho \quad .$$

This approximation, as well as its extension for non-vanishing heights h and z of the transmitter and receiver above the earth, can also be derived from Weyl's expression by applying a saddlepoint method.

II. The pulse solution corresponding to the Sommerfeld problem for a dielectric earth.

IIa Introduction.

By the pulse solution we understand the field that results from a transmitter the momentum of which suddenly jumps at $t=0$ from zero to a constant value M. This momentum is thus given the Heaviside unit function $M\,\mathcal{U}(t)$ $\left\{\mathcal{U}(x)\right.$ being unity for $x>0$ and zero for $\left.x<0\right\}$, instead of the former function $M\,e^{-i\omega t}$. We shall denote the Hertzian scalar for the pulse solution by $\Pi_{\int}(t)$, and that for the time harmonic function by Π_{ω}. The elementary relation (the integral of which should be interpreted as a Cesàro limit)

$$M\,e^{-i\omega t} = -i\omega \int_{-\infty}^{\infty} d\tau \; M\,\mathcal{U}(t-\tau)\, e^{-i\omega\tau} \quad ,$$

and the linearity of the problem involve the following connection between both solutions:

$$\Pi_{\omega} = -i\omega \int_{-\infty}^{\infty} d\tau \; \Pi_{\int}(t-\tau)\, e^{-i\omega\tau} \quad .$$

The time-harmonic solution being known, the pulse solution can therefore be derived in principle with the aid of an inverse Laplace transform.

This connection has been the starting point for Pekeris and Alterman [3], van der Pol and Levelt [4], and de Hoop and Frankena [5] who were able to get integrals of elliptic type for the pulse solution, but only under restricted conditions. The latter presume a negligible conductivity of both the atmosphere and the earth, and further a transmitter or receiver situated on the earth's surface when dealing with the field inside the earth. Finite conductivities involve complications which have hardly been discussed so far. The interest in the problem had been stimulated by Van der Pol's [6] discovery that for a transmitter and receiver both on the ground (still assuming negligible conductivities) the pulse solution degenerates to a most elementary function given by a square root. The analysis of the above mentioned authors is complicated by the multi-valuedness of the two square roots entering in Fresnel's reflection coefficient (I.10). The problem becomes simpler when starting right away from the time-dependent original Maxwell equations (I.1) instead of the time-harmonic equations (I.2). This will be worked out briefly in the next analysis which is given in detail elsewhere [7].

IIb. A derivation of the pulse solution from the time-dependent wave equation.

Our problem depends on two dielectrics ($\varepsilon = \varepsilon_1$ for $z > 0$, $\varepsilon = \varepsilon_2$ for $z < 0$). The pulse transmitter being at $x = y = 0$, $z = h$, the situation is described by the Maxwell equations :

$$(1)\quad \begin{cases} \operatorname{curl} \vec{E} + \mu_0 \dfrac{\partial \vec{H}}{\partial t} = \vec{0}\,, & \\ \operatorname{curl} \vec{H} - \varepsilon_1 \dfrac{\partial \vec{E}}{\partial t} = M\, \delta(x)\, \delta(y)\, \delta(z-h)\, \mathcal{U}(t)\, \vec{u}_z & (z > 0) \\ \operatorname{curl} \vec{H} - \varepsilon_2 \dfrac{\partial \vec{E}}{\partial t} = \vec{0}\,, & (z < 0) \end{cases}$$

$\vec{u}_z$ being a unit vector in the z direction.

Once again, the field components can be derived from a vertically directed Hertzian vector with an amplitude $\Pi(x, y, z, t)$. It leads to the following representation of the field in either medium ($\varepsilon = \varepsilon_1$ or ε_2):

$$\left.\begin{aligned} \vec{E} &= -\varepsilon\, \mu_0 \frac{\partial \Pi}{\partial t} \vec{u}_z + \int_{-\infty}^{t} d\tau \,.\, \operatorname{grad} \frac{\partial \Pi}{\partial z}\,, \\ \vec{H} &= \varepsilon \operatorname{curl} (\Pi \vec{u}_z)\,. \end{aligned}\right\} \quad (2)$$

In fact, this field is easily verified as a solution of (1) provided that Π satisfies the time dependent wave equations

$$\left(\nabla^2 - \mu_0\, \varepsilon_1 \frac{\partial^2}{\partial t^2} \right) \Pi = -\frac{M}{\varepsilon_1}\, \delta(x)\, \delta(y)\, \delta(z-h)\, \mathcal{U}(t) \qquad (z > 0) \quad (3a)$$

$$\left(\nabla^2 - \mu_0\, \varepsilon_2 \frac{\partial^2}{\partial t^2} \right) \Pi = 0\,. \qquad (z < 0) \quad (3b)$$

The primary field is obtained for $\varepsilon_2 = \varepsilon_1$, in which case the upper equation holds throughout space. Its solution reads :

$$\Pi_{pr} = \frac{M}{4\pi\varepsilon_1} \frac{\mathcal{U}(t - \sqrt{\mu_0 \varepsilon_1}\, R)}{R} \quad , \tag{4}$$

which can also be represented in the alternative form :

$$\Pi_{pr} = \frac{M\sqrt{\mu_0/\varepsilon_1}}{8\pi^2 i} \int_L \frac{d \cos\vartheta'}{\sqrt{(t - \sqrt{\mu_0\varepsilon_1}\,|z-h| \cos\vartheta')^2 - \mu_0\varepsilon_1\rho^2 \sin^2\vartheta'}} \tag{5}$$

The path of integration L is a contour consisting of the inner boundary of the complete right-hand half of the complex $\cos\vartheta'$-plane, completed by a loop surrounding the part $0 < \cos\vartheta' < 1$ of the real axis at an infinitesimal distance from it (see the fig.1).

The equivalence of (4) and (5) is demonstrated as follows. The only singularities enclosed by L are the branch points formed by the zeros of the denominatos. For $t^2 < \mu_0 \varepsilon_1 R^2$ the latter prove to be situated on the real $\cos \vartheta'$ - axis, between 0 and 1, and thus outside L. The integral then vanishes, as it should. However, when $t^2 > \mu_0 \varepsilon_1 R^2$, we find zeros situated at

$$\cos \vartheta' = u_{pr} \pm i\, v_{pr} , \tag{6}$$

where

$$u_{pr} = \frac{t\,|z-h|}{\sqrt{\mu_0 \varepsilon_1}\; R^2} , \qquad v_{pr} = \frac{\rho \sqrt{t^2 - \mu_0 \varepsilon_1 R^2}}{\sqrt{\mu_0 \varepsilon_1}\; R^2} . \tag{7}$$

For $t < 0$ both zeros (6) are to the left of the imaginary axis of the $\cos \vartheta'$ -plane, again outside L, but for $t > 0$ they are two conjugated complex quantities situated inside L. In the first case the integral vanishes once again (as it should for all negative t values), in the latter case it can be evaluated elementarily and it turns out to be identical with (3).

Just as the λ -integral (I.7b) for the time-harmonic solution is composed of elementary solutions of the time-harmonic wave equation, so the integral (5) constitutes a sum of corresponding solutions of the general homogeneous wave equation $(\nabla^2 - \mu_0 \varepsilon_1 \, \partial^2/\partial t^2)\Pi = 0$. For $z > h$ and $z < h$ these solutions are of the form

$$\frac{1}{\sqrt{\left\{t-t_0 \mp \sqrt{\mu_0 \varepsilon_1}\cos\vartheta'(z-z_0)\right\}^2 - \mu_0\varepsilon_1 \sin^2\vartheta'\, \rho^2}} \tag{8}$$

respectively. The upper (lower) sign refers to a contribution moving upwards (down), away from the horizontal level $z = h$ through the transmitting dipole. These propagation directions of (8) are made plausible by considering (for a given moment t) the locus of its infinities, that is of the zeros of the denominatos. For the complex $\cos\vartheta'$ - values along L these loci are found to be the circular cross-section of the cone

$$\frac{\rho}{z-z_0} = \pm \frac{|\mathcal{R}e \sin\vartheta'|}{\mathcal{R}e \cos\vartheta'}$$

with the moving horizontal plane :

$$z - z_0 = \pm \frac{(t-t_0)}{\sqrt{\mu_0 \varepsilon_1}} \frac{\mathcal{R}e \cos\vartheta'}{(\mathcal{R}e \cos\vartheta')^2 + (\mathcal{R}e \sin\vartheta')^2} .$$

Therefore, the locus of the infinities moves upwards (down), as it should, for the relevant values of $\cos\gamma$ which have a positive (possibly infinitesimal) real part.

These considerations suggest a representation for the secondary field generated in the upper half space $z > 0$ by the presence of the earth. The upward propagation of this field is guaranteed by the following integration over solutions of the general homogeneous wave equation for $z > 0$:

$$\Pi_{sec} = \frac{M}{8\pi^2 i}\sqrt{\frac{\mu_0}{\varepsilon_1}}\int_L \frac{\mathcal{R}(\vartheta')\; d\cos\vartheta'}{\sqrt{\left\{t-\sqrt{\mu_0\varepsilon_1}(h+z)\cos\vartheta'\right\}^2-\mu_0\varepsilon_1\rho^2\sin^2\vartheta'}}$$

$(z > 0)$ (9)

Similar to the situation in (I.8) the weighting function $\mathcal{R}(\vartheta')$ has to be derived from the boundary conditions. We therefore also need the expression for the field in the earth. The latter consists of contributions moving down which satisfy the homogeneous wave equation (3b). It can be represented by

$$\Pi = \frac{M}{8\pi^2 i}\sqrt{\frac{\mu_0}{\varepsilon_1}}\int_L \frac{\mathcal{D}(\vartheta')\; d\cos\vartheta'}{\sqrt{\left\{t-\sqrt{\mu_0\varepsilon_1}\,h\cos\vartheta'+\sqrt{\mu_0\varepsilon_2}\,z\cos\vartheta''\right\}^2-\mu_0\varepsilon_2\rho^2\sin^2\vartheta''}}$$

$(z < 0)$ (10)

We have introduced here a new variable $\cos\vartheta''$ connected with $\cos\vartheta'$ by the relation

$$\sqrt{\varepsilon_1}\;\sin\vartheta' = \sqrt{\varepsilon_2}\;\sin\vartheta'' . \qquad (11)$$

The sign of $\cos\vartheta''$ is defined by the introduction of a cross-cut (if $\varepsilon_2 > \varepsilon_1$) along the part of the imaginary $\cos\vartheta'$ - axis between

$$\cos\vartheta' = -i\sqrt{\frac{\varepsilon_2}{\varepsilon_1} - 1} \quad \text{and} \quad \cos\vartheta' = i\sqrt{\frac{\varepsilon_2}{\varepsilon_1} - 1},$$

and by the property that $\cos\vartheta' - \cos\vartheta'' \longrightarrow 0$ should hold at infinity. For the rest, the integration path for $\cos\vartheta''$ here proves to be identical with that of $\cos\vartheta'$; either variable may be used according to its suitability.

The required continuity of the horizontal field components along the interface $z = 0$ involves, in view of (2), the continuity there of $\varepsilon\Pi$ and $\partial\Pi/\partial z$. When equating the integrands of the expressions derived for these quantities from the sum of (5) and (9), and from (10), also applying (11), we get two equations for the unknown weighting functions $\mathcal{R}(\vartheta')$ and $\mathcal{D}(\vartheta')$. The solution reads :

$$\mathcal{R}(\vartheta') = \frac{\sqrt{\varepsilon_2}\cos\vartheta' - \sqrt{\varepsilon_1}\cos\vartheta''}{\sqrt{\varepsilon_2}\cos\vartheta' + \sqrt{\varepsilon_1}\cos\vartheta''},$$

$$\mathcal{D}(\vartheta') = \frac{\varepsilon_1}{\varepsilon_2}\left\{1 + \mathcal{R}(\vartheta')\right\} = \frac{2\,\varepsilon_1\cos\vartheta'}{\sqrt{\varepsilon_2}\left(\sqrt{\varepsilon_2}\cos\vartheta' + \sqrt{\varepsilon_1}\cos\vartheta''\right)} \qquad (12)$$

These expressions are identical with those obtained in (I.10) and (I.11b) if we take into account that the neglect of the conductivities and the assumption of a single permeability

μ_0 involve the wave numbers $k_1 = \omega\sqrt{\varepsilon_1\mu_0}$ and $k_2 = \omega\sqrt{\varepsilon_2\mu_0}$ in (I.4). We find that the Fresnel reflection coefficient maintains its significance of fixing the amplitude of a reflected wave even for the particular solutions (8) of the general wave equation.

IIc. Reduction of the pulse solution to elliptic integrals.

The expressions (9) and (10) can be considered as final representations of the pulse solution after substitution of (12). However, the infinite contour L can be contracted in various ways to a finite path of integration if we know the singularities of the integrand. Let us consider the secondary field (9) first. An examination of its integrand reveals the existence of the following (and no other) singularities of its integrand :

(1) branch points connected with the definition of $\cos\vartheta''$. These are situated at

$$\cos\vartheta' = \pm i\sqrt{\frac{\varepsilon_2}{\varepsilon_1} - 1} \tag{13}$$

and, therefore, are just outside L .

(2) Two real zeros of the square root in the denominator of (9), only existing if $t^2 < \mu_0\varepsilon_1 R^2_{sec}$. The quantity R_{sec} denotes the distance from the image (at $x = y = 0$, $z = -h$) of the transmitter (in the earth) to the point of observation. These zeros, situated at

$$\cos \vartheta' = \frac{t(z+h) \pm \rho \sqrt{\mu_0 \varepsilon_1 R^2_{sec} - t^2}}{\sqrt{\mu_0 \varepsilon_1}\ R^2_{sec}} ,$$

are on the real axis, but outside L .

(3) Two complex zeros of the denominator of (9), only for $t^2 > \mu_0 \varepsilon_1 R^2_{sec}$. Their positions can be indicated by $\cos \vartheta' = u_{sec} \pm i\, v_{sec}$ if we introduce the new parameters [compare (7)] :

$$u_{sec} = \frac{t(z+h)}{\sqrt{\mu_0 \varepsilon_1}\ R^2_{sec}} , \quad v_{sec} = \frac{\rho \sqrt{t^2 - \mu_0 \varepsilon_1 R^2_{sec}}}{\sqrt{\mu_0 \varepsilon_1}\ R^2_{sec}} .$$

For $t < 0$ these singularities are outside L , for $t > 0$ they are within this contour.

We infer that singularities inside L only occur if $t > \sqrt{\mu_0 \varepsilon_1}\ R_{sec}$. Therefore the field vanishes if $t < \sqrt{\mu_0 \varepsilon_1}\ R_{sec}$, that is before the arrival of the first disturbance (leaving the transmitter at $t = 0$) at the receiver via a reflection against the earth's surface . After this arrival the contour integral can be contracted to a loop just surrounding the straight section in the $\cos \vartheta'$-plane that connects the two singularities at

$u_{sec} - i\,v_{sec}$ and $u_{sec} + i\,v_{sec}$. In its turn this new contour integral can be reduced to a line integral along this section. This procedure results in :

$$\Pi_{sec} = \frac{M}{4\pi^2 i\varepsilon_1} \frac{\mathcal{U}(t - \sqrt{\mu_0\varepsilon_1}\,R_{sec})}{R_{sec}} \int_{u_{sec}-iv_{sec}}^{u_{sec}+iv_{sec}} \frac{\mathcal{R}(\vartheta')\,d\cos\vartheta'}{\sqrt{(\cos\vartheta' - u_{sec})^2 + v_{sec}^2}} ,$$

in which the sign of the square root is defined by the value along the right-hand side of the path of integration.

A further substitution $\cos\vartheta' = u_{sec} - i\,v_{sec}\cos\psi$ leads to the alternative form :

$$\Pi_{sec} = \frac{M}{2\pi^2\varepsilon_1} \frac{\mathcal{U}(t - \sqrt{\mu_0\varepsilon_1}\,R_{sec})}{R_{sec}} \,\mathcal{R}e \int_0^{\pi/2} \mathcal{R}\{\cos^{-1}(u_{sec} - iv_{sec}\cos\psi)\}\,d\psi$$

This expression is equivalent to a general formula derived by <u>De Hoop</u> and <u>Frankena</u> if it is applied to the pulse solution.

Another elliptic integral for the secondary field of the pulse solution can be derived by considering L as the difference of a contour integration along the complete circle at infinity of the $\cos\vartheta'$ - plane, and another contour enclosing the part outside L of this plane. The latter contour can be reduced for $t > \sqrt{\mu_0\varepsilon_1}\;R_{sec}$ to a straight line connecting the only two singularities within it, the singularities being given by (13). The final evaluation results in the following expression if, moreover, we pass to a new integration variable $s = i\sqrt{\varepsilon_1}\cos\vartheta'$:

$$\Pi_{sec} = \frac{M}{4\pi\varepsilon_1} U(t - \sqrt{\mu_0\varepsilon_1}\, R_{sec}) \left[\frac{(\varepsilon_2-\varepsilon_1)}{(\varepsilon_2+\varepsilon_1)} \frac{1}{R_{sec}} + \right.$$

$$\left. + \frac{2i}{\pi} \frac{\varepsilon_1\varepsilon_2}{\varepsilon_2-\varepsilon_1} \int_{-\sqrt{\varepsilon_2-\varepsilon_1}}^{\sqrt{\varepsilon_2-\varepsilon_1}} \frac{s\sqrt{\varepsilon_2-\varepsilon_1-s^2}\, ds}{\left\{(\varepsilon_1+\varepsilon_2)s^2+\varepsilon_1^2\right\}\sqrt{-R_{sec}^2 s^2 + 2it\dfrac{(h+z)}{\sqrt{\mu_0}}s + \dfrac{t^2}{\mu_0} - \varepsilon_1\rho^2}} \right]$$

Adding the primary field we get an expression which is identical with that derived by <u>Van der Pol</u> and <u>Levelt</u> for the special case $h = 0$.

All these expressions only reduce to elementary results if both the transmitter and receiver are on the earth's surface ($z = h = +0$). We then have

$$\Pi = \begin{cases} 0 & \text{for } t < \sqrt{\mu_0\varepsilon_1}\,\rho \\ \dfrac{\varepsilon_2 M}{2\pi\varepsilon_1(\varepsilon_2^2-\varepsilon_1^2)} \left[\dfrac{\varepsilon_2}{\rho} - \dfrac{\sqrt{\mu_0\varepsilon_1\varepsilon_2\varepsilon_{12}}}{\sqrt{t^2-\mu_0\varepsilon_{12}\rho^2}} \right] & \text{for } \sqrt{\mu_0\varepsilon_1}\,\rho < t < \sqrt{\mu_0\varepsilon_2}\,\rho \\ \dfrac{M\varepsilon_2}{2\pi\varepsilon_1(\varepsilon_1+\varepsilon_2)} \dfrac{1}{\rho} & \text{for } t > \sqrt{\mu_0\varepsilon_2}\,\rho \end{cases}$$

$(h = z = +0)$

in which ε_{12} is defined by $1/\varepsilon_{12} = 1/\varepsilon_1 + 1/\varepsilon_2$.

We next consider the pulse solution inside the earth. The results are much more complicated there, which can be explained physically by the refraction associated with the propagation from the transmitter to the receiver. The singularities due to the square root in the denominator of (10) are now to be determined as the zeros of an equation of the fourth degree in $\cos\vartheta'$ or $\cos\vartheta''$, and it is no longer possibile to reduce the integral to one of the elliptic type. The only exception leading to simpler results is that of a vanishing height of either the transmitter or the receiver. It is then again possible to get an elliptic integral, as has been shown by Van der Pol and Levelt. Their result can also be deduced from (10) by a rather tedious analysis. For the sake of completeness we give the final expression for $h=0$, but $z\neq 0$:

$$\Pi = \begin{cases} 0 & \text{if either simultaneously } t<\sqrt{\mu_0\varepsilon_2}\,r \text{ and } \sqrt{\varepsilon_1}\,r>\sqrt{\varepsilon_2}\,\rho, \\ & \text{or simultaneously } t<\sqrt{\mu_0}\left(\sqrt{\varepsilon_1}\rho-\sqrt{\varepsilon_2-\varepsilon_1}\,z\right) \text{ and } \sqrt{\varepsilon_1}\,r<\sqrt{\varepsilon_2}\,\rho \\ -\dfrac{M\varepsilon_2}{\pi^2(\varepsilon_2-\varepsilon_1)}\displaystyle\int^{\sqrt{\varepsilon_2-\varepsilon_1}} \frac{\sqrt{\varepsilon_2-\varepsilon_1-s^2}\; s\,ds}{\left\{(\varepsilon_1+\varepsilon_2)s^2-\varepsilon_2^2\right\}\sqrt{\left(\frac{t}{\sqrt{\mu_0}}+zs\right)^2+\rho^2s^2-\varepsilon_2\rho^2}} & \\ \quad \dfrac{-tz+\rho\sqrt{\mu_0\varepsilon_2 r^2-t^2}}{\sqrt{\mu_0}\,r^2} & \text{if simultaneously } \sqrt{\varepsilon_1}\,r<\sqrt{\varepsilon_2}\,\rho \\ & \text{and } \sqrt{\varepsilon_1}\rho-\sqrt{\varepsilon_2-\varepsilon_1}\,z<\dfrac{t}{\sqrt{\mu_0}}<\sqrt{\varepsilon_2}\,r \\ \dfrac{M}{2\pi}\left[\dfrac{1}{(\varepsilon_1+\varepsilon_2)r}-\dfrac{\varepsilon_2}{\pi(\varepsilon_2-\varepsilon_1)}\displaystyle\int_{-\sqrt{\varepsilon_2-\varepsilon_1}}^{\sqrt{\varepsilon_2-\varepsilon_1}} \frac{s\sqrt{\varepsilon_2-\varepsilon_1-s^2}\;ds}{\left\{(\varepsilon_1+\varepsilon_2)s^2-\varepsilon_2^2\right\}\sqrt{r^2s^2+\frac{2tz}{\sqrt{\mu_0}}s+\frac{t^2}{\mu_0}-\varepsilon_2\rho^2}}\right] & \text{for } t>\sqrt{\mu_0\varepsilon_2}\,r. \end{cases}$$

The quantity r denotes the distance $(x^2 + y^2 + z^2)^{1/2}$ from the transmitter to the point of observation.

The occurrence of two different non-vanishing expressions is connected with the possibility of two different modes of propagation, viz. a direct wave propagating through the earth with the velocity $1/\sqrt{\mu_0 \varepsilon_2}$, and a wave travelling first along the earth's surface with the velocity $1/\sqrt{\mu_0 \varepsilon_1}$, and then refracted into the earth at grazing incidence.

III. The field of a vertical dipole in a homogeneous atmosphere above a spherical homogeneous earth.

IIIa. Derivation of the harmonic series.

When the distances of radiocommunication increased it was realized that the curvature of the earth had to be taken into account. Sommerfeld's problem was extended to that of the diffraction of radio waves around the spherical earth (radius a).

We assume again a vertical dipole situated at $r = b$, $\vartheta = 0$ in a system of spherical coordinates with its origin at the centre of the earth. For a dipole with moment $M e^{-i\omega t}$ the field can be deduced in each of the two media (the outer space and the inner space, labelled 1 and 2 respectively) from a Hertzian vector $\vec{\Pi}$, applying instead of (I.3) the formulas :

$$\vec{E} = i\,\omega\,\mu_o\,\varepsilon_{eff}\,\vec{\Pi} + \frac{i}{\omega}\,\mathrm{grad}\left\{\frac{\partial}{\partial r}(r\Pi)\right\},$$

$$\vec{H} = \varepsilon_{eff}\,\mathrm{curl}\,\vec{\Pi}$$

This vector (which is not determined uniquely by the problem) is assumed here in a radial direction according to

$$\vec{\Pi} = \vec{r}\;\Pi\,, \qquad (1)$$

where $\vec{r}$ is a radial vector of length r. The scalar Π

has to satisfy the following equations corresponding to (I.6):

$$\left.\begin{aligned} (\nabla^2 + k_1^2)\Pi &= -\frac{M}{2\pi\, b^3\, \varepsilon_{eff,1}}\,\delta(r-b)\,\frac{\delta(\vartheta)}{\vartheta}, \quad (r > a) \\ (\nabla^2 + k_2^2)\Pi &= 0\,. \quad (r < a) \end{aligned}\right\} \quad (2)$$

The primary field existing in the absence of the earth can be derived here from the scalar

$$\Pi_{pr} = \frac{M}{4\pi\, b\, \varepsilon_{eff\,1}}\,\frac{e^{ik_1 R}}{R}\,, \qquad (3)$$

R being the distance from the transmitter to the receiver at (r, ϑ, φ). The resulting primary $\vec{E}$, $\vec{H}$ field is the same as that following from (I.7a) when applying (I.3a).

The primary field can now be split into a sum, instead of an integral, of solutions of the homogeneous wave equation that are separated in the two relevant coordinates r and ϑ. In fact, the role of (I.7b) is now taken over by the expansion

$$\Pi_{pr} = \frac{ik_1 M}{4\pi\, b\, \varepsilon_{eff,1}} \sum_{n=0}^{\infty} (2n+1) P_n(\cos\vartheta)\,.\,\zeta_n^{(1)}\left\{k_1 \max(b,r)\right\}.\,\psi_n\left\{k_1 \min(b,r)\right\}, \qquad (4)$$

in which we have introduced the spherical Bessel functions defined by :

$$\zeta_n^{(1)(2)}(z) = \sqrt{\frac{\pi}{2z}}\, H^{(1)(2)}_{n+1/2}(z) , \qquad \psi_n(z) = \sqrt{\frac{\pi}{2z}}\, J_{n+1/2}(z) . \tag{5}$$

The asymptotic expressions for large $k_1 r$ of the Hankel functions show that $\zeta_n^{(1)}(k_1 r)\, P_n(\cos\vartheta)\, e^{-i\omega t}$ and $\zeta_n^{(2)}(k_1 r)\, P_n(\cos\vartheta)\, e^{-i\omega t}$ represent waves which travel away from and towards the origin respectively. This is in accordance with (4) the component waves of which travel only outwards or upwards above the spherical level $r = b$ through the transmitter, and in view of $\psi_n = \frac{1}{2}\zeta_n^{(1)} + \frac{1}{2}\zeta_n^{(2)}$, both outwards or upwards and inwards or downwards below this level.

On the complete analogy of the situation for the flat earth the presence of the spherical earth also produces an additional secondary field Π_{sec} in the outer space, and a field Π in the inner space. These fields are composed of only outwards, and of both outward and inward travelling waves respectively. The former have to satisfy the homogeneous scalar wave equation $(\nabla^2 + k_1^2)\Pi = 0$, the latter the equation $(\nabla^2 + k_2^2)\Pi = 0$. These considerations agree with the following representation for the secondary field :

$$\Pi_{sec} = \frac{i\,k_1 M}{4\pi\, b\, \varepsilon_{eff,1}} \sum_{n=0}^{\infty} (2n+1)\,\zeta_n^{(1)}(k_1 b)\,\psi_n(k_1 a)\,R(n)\,\frac{\zeta_n^{(1)}(k_1 r)}{\zeta_n^{(1)}(k_1 a)}\,P_n(\cos\vartheta)\,.$$

$$(r > a) \qquad (6)$$

On the other hand the waves inside the earth are partly travelling inwards and partly outwards, as may be clear from the property that a chord connecting two pointw on the sphere has a minimum distance to the centre of the earth at its midpoint. However, the total waves inside have to be finite at this centre, which is only possibile for waves of the type $\psi_n(k_2 r)\,P_n(\cos\vartheta)\,e^{-i\omega t}$. The Π scalar of the inner field can be expanded accordingly into such waves.

The boundary conditions expressing the continuity of the horizontal or tangential components of $\vec{E}$ and $\vec{H}$ at the earth's surface reduce to the following conditions in terms of the scalar Π :

$$\frac{\partial}{\partial r}(r\Pi) \text{ and } \varepsilon_{eff}\Pi \text{ or } k^2\Pi \text{ continuous at } r = a \qquad (7)$$

The second condition fixed at once the coefficient entering in the new expansion for the inner field Π when the expansion (6) for the outer field is known. We find :

$$\Pi = \frac{i\,k_1 M}{4\pi b\,\varepsilon_{eff,1}}\,\frac{k_1^2}{k_2^2} \sum_{n=0}^{\infty} (2n+1)\,\zeta_n^{(1)}(k_1 b)\,\psi_n(k_1 a)\left[1 + R(n)\right]\frac{\psi_n(k_2 r)}{\psi_n(k_2 a)}\,P_n(\cos\vartheta)$$

$$(r < a) \qquad (8)$$

The other condition of the continuity of $\partial(r\Pi)/\partial r$ then determines completely the weighting function $\mathcal{R}(n)$. The result reads :

$$\mathcal{R}(n) = \frac{-\frac{1}{x}\frac{d}{dx}\log\left\{x\psi_n(x)\right\}_{x=k_1a} + \frac{1}{x}\frac{d}{dx}\log\left\{x\psi_n(x)\right\}_{x=k_2a}}{\frac{1}{x}\frac{d}{dx}\log\left\{x\zeta_n^{(1)}(x)\right\}_{x=k_1a} - \frac{1}{x}\frac{d}{dx}\log\left\{x\psi_n(x)\right\}_{x=k_2a}} \qquad (9)$$

The substitution of this quantity into (6) and (8) concludes the derivation of the final solution in terms of an expansion in Legendre functions.

IIIb. <u>The Watson transformation. Rainbow terms.</u>

Very unfortunately the convergence of the expansions (6) and (8) is extremely slow in the case of radio waves diffracted by the earth, the number of terms wanted for a reasonable computation being of the order of the large quantity k_1a. The first successful transformation to another rapidly enough converging series has been performed by Watson [8] by a procedure now generally known as the " Watson transformation ". In the present problem this transformation reduces in its simplest form to the following identity :

$$\sum_{n=0}^{\infty}(2n+1)h(n)P_n(\cos\vartheta) = i\int_{0}^{\infty}\frac{dn\ n}{\cos(\pi n)}h(n-\tfrac{1}{2})P_{n-\frac{1}{2}}(-\cos\vartheta) =$$

(10)
$$= -i\int_{-\infty+io}^{\infty+io}\frac{dn\ n}{\cos(\pi n)}h(n-\tfrac{1}{2})P_{n-\frac{1}{2}}(-\cos\vartheta) =$$

$$= i\int_{-\infty+io}^{\infty+io}\frac{dn(n+1/2)}{\sin(\pi n)}h(n)P_n(-\cos\vartheta).$$

The second and last steps only hold if $h(n-1/2)$ is even in n. In this case only the first step leading to an integration around the positive axis in the n-plane can be completed to an integration along and immediately above the total real axis of this plane. It is obvious that the whole procedure presumes the possibility of extending the set of discrete coefficients $h(n)$ to a continuous function $h(n)$ which has no singularities along the real n-axis.

In its original application by Watson $h(n-1/2)$ was not an even function of n. This involved an extra integral along the imaginary axis of the n-plane, which proved to be of no numerical importance in radio practice. However, for mathematical elegance, we prefer a theory [9] which avoids this extra integral by modifying the n dependence of the coefficients $h(n)$ of $(2n+1)P_n(\cos\vartheta)$ in a proper way, such that $h(n-1/2)$ becomes even in n while $h(n)$ remains unchanged for integral n , as it should.

We shall indicate how this can be performed for the outer space $r > a$. An elementary analysis based on a geometric progression (the convergence of which will be assumed on physical grounds) shows that the sum of (4) and (6) can be represented by :

$$\Pi = \frac{ik_1 M}{8\pi b\,\varepsilon_{eff,1}} \sum_{n=0}^{\infty} (2n+1) \sum_{k=-1}^{\infty} h_k(n,r) P_n\{\cos(\vartheta - k\pi - \pi)\} , \qquad (11)$$

in which

$$h_{-1}(n,r) = R_{11}(n) \frac{\zeta_n^{(2)}(k_1 a)}{\zeta_n^{(1)}(k_1 a)} \zeta_n^{(1)}(k_1 b) \zeta_n^{(1)}(k_1 r) + $$

$$+ \zeta_n^{(1)}\{k_1 \max(r,b)\} \zeta_n^{(2)}\{k_1 \min(r,b)\} ,$$

$$h_k(n,r) = R_{12} R_{22}^{k} R_{21} \left\{ \frac{\zeta_n^{(2)}(k_2 a e^{-i\pi})}{\zeta_n^{(2)}(k_2 a)} \right\}^{k+1} \frac{\zeta_n^{(1)}(k_1 b) \zeta_n^{(2)}(k_1 a)}{\zeta_n^{(1)}(k_1 a)} \zeta_n^{(1)}(k_1 r)$$

$$(k \geqslant 0) \qquad (12)$$

Here we have introduced the new quantities :

$$R_{11}(n) = \frac{-\frac{1}{x}\frac{d}{dx}\log\left\{x\,\zeta_n^{(2)}(x)\right\}_{x=k_1a} + \frac{1}{x}\frac{d}{dx}\log\left\{x\,\zeta_n^{(2)}(x)\right\}_{x=k_2a}}{\frac{1}{x}\frac{d}{dx}\log\left\{x\,\zeta_n^{(1)}(x)\right\}_{x=k_1a} - \frac{1}{x}\frac{d}{dx}\log\left\{x\,\zeta_n^{(2)}(x)\right\}_{x=k_2a}} ,$$

$$R_{12}(n) = \frac{k_1^2}{k_2^2}\left\{1 + R_{11}(n)\right\} , \tag{13}$$

while R_{22} and R_{21} are obtained from R_{11} and R_{12} by simultaneously interchanging k_1, k_2 and $\zeta^{(1)}$, $\zeta^{(2)}$.

All these parameters have a clear significance. Thus R_{11} constitutes the reflection coefficient at the outer side of the earth's surface for a travelling wave of the form $\zeta_n^{(2)}(k_1 r)\,P_n(\cos\vartheta)\,e^{-i\omega t}$ which arrives from the outer space. At the boundary $r = a$ it is split into a reflected wave

$$\zeta_n^{(2)}(k_1 a)\,.\,R_{11}(n)\,.\,\frac{\zeta_n^{(1)}(k_1 r)}{\zeta_n^{(1)}(k_1 a)}\,P_n(\cos\vartheta)\,e^{-i\omega t}$$

returning to the outer space, and a wave proportional to $\zeta_n^{(2)}(k_2 r)\,P_n(\cos\vartheta)\,e^{-i\omega t}$ that penetrates by refraction into the earth. Similarly, R_{22} represents a reflection coefficient at the inner side for a wave of the form $\zeta_n^{(1)}(k_2 r)P_n(\cos\vartheta)e^{-i\omega t}$

which approaches the internal earth's surface. The reflected wave generated there, with an amplitude determined by R_{22}, is accompanied by a wave proportional to $\zeta_n^{(1)}(k_1 r) P_n(\cos\vartheta) e^{-i\omega t}$ that escapes by refraction into the outer space.

By inverting the order of summation in (11), which we assume to be justified, we get the splitting :

$$\Pi = \sum_{k=-1}^{\infty} \Pi_k \quad , \tag{14a}$$

where

$$\Pi_k = \frac{i k_1 M}{8\pi b\, \varepsilon_{\mathrm{eff},1}} \sum_{n=0}^{\infty} (2n+1) . h_k(n,r) . P_n\{\cos(\vartheta - k\pi - \pi)\} . \tag{14b}$$

The convergence of these series follows from the asymptotic proportionality for large integral positive n of $|h_k(n,r)|$ to $(a^2/4br)^n/\{n\Pi(n-1/2)\}^2$ for $k \geq 0$, and of $|h_{-1}(n,r)|$ to $(a^2/br)^n/n$. In turn these asymptotic expressions result from the general approximations (to be derived from the power series of the Bessel functions entering in $\zeta_n^{(1)}$ and $\zeta_n^{(2)}$)

$$\zeta_n^{(1)(2)}(\alpha) \sim \mp i \frac{2^n \Pi(n-1/2)}{\sqrt{\pi}\, \alpha^{n+1}} \tag{15}$$

which hold for n tending to infinity in the right half of the complex n-plane.

The occurrence of the factor $R_{12}\ R_{22}^{\ k}\ R_{21}$ in each term of Π_k indicates that Π_k can be interpreted as a contribution produced after a refraction into the earth (depending on R_{12}) of the waves arriving from the transmitter, followed by k reflections against the internal surface of the earth (each of which depends on a single factor R_{22}), and a final refraction (determined by R_{21}) by which the wave returns to the outer space. This geometry of the contributions Π_k can be compared with that of light rays penetrating into a water droplet in the rainbow phenomenon. Here too, after this penetration the energy can escape to the outer space after having suffered a number k of reflections against the internal surface of the droplet. Therefore, the contributions Π_k may be termed " rainbow terms " . The only significant term in the radio case is the first term Π_{-1}, all other contributions becoming extremely small since they depend on the propagation through the highly absorbing earth. On the contrary the rainbow phenomenon depends numerically on the terms Π_1 ("main bow") and Π_2 ("secondary bow"). However, its analytical theory is completely identical with that for the propagation of radio waves.

As mentioned above, the splitting into rainbow terms is suggested mathematically from the wish to apply the Watson transformation in its simple version (10). This can be achieved for any term Π_k individually since for all coefficients $h_k(n-1/2,r)$ proves to be an even function in n , as a conseguence of the properties

$$\zeta^{(1)}_{-n-1/2}{}^{(2)}(\alpha) = e^{\pm i\pi n}\ \zeta^{(1)}_{n-1/2}{}^{(2)}(\alpha) .$$

Therefore, a direct application of (10) yields :

$$\Pi_k = -\frac{k_1 M}{8\pi b\, \varepsilon_{eff,1}} \int_{-\infty+io}^{\infty+io} \frac{dn(n+1/2)}{\sin(\pi n)} h_k(n,r) P_n\left\{\cos(\vartheta - k\pi)\right\}. \quad (16)$$

IIIc. The residue series.

In view of radio practice we restrict ourselves to the main term Π_{-1}. For simplicity we further consider the special case of a transmitter and receiver both on the earth's surface so as to have $r = b = a$. The expression (16) for $k = -1$ then reduces to the following integral if we substitute the value of $h_{-1}(n,r)$ according to (12) :

$$\Pi_{-1} = -\frac{k_1 M}{8\pi b\, \varepsilon_{eff,1}} \int_{-\infty+io}^{\infty+io} \frac{dn(n+1/2)}{\sin(\pi n)} \left\{1+R_{11}(n)\right\} \zeta_n^{(1)}(k_1 a)\, \zeta_n^{(2)}(k_1 a) P_n\left\{\cos(\pi-\vartheta)\right\}.$$

$$(r = b = a).$$

A further application of both (13) and of the Wronskian relation for $\zeta_n^{(1)}$ and $\zeta_n^{(2)}$ leads to the final representation :

$$\Pi_{-1} = \frac{M}{4\pi i k_1^2 a^4 \varepsilon_{eff,1}} \int_{-\infty+io}^{\infty+io} \frac{dn(n+1/2)}{\sin(\pi n)} \frac{P_n\left\{\cos(\pi-\vartheta)\right\}}{N(n)},$$

$$(r = b = a) \quad (17)$$

with

$$N(n) = \frac{1}{x}\cdot\frac{d}{dx}\log\left\{x\zeta_n^{(1)}(x)\right\}_{x=k_1a} - \frac{1}{x}\frac{d}{dx}\log\left\{x\zeta_n^{(2)}(x)\right\}_{x=k_2a} \quad . \tag{18}$$

The integration path of all integrals (16) can be completed to a contour enclosing the upper half $\Im m\, n > 0$ of the n-plane, with its basis just above the real axis. In fact, the contribution to the integral from the upper half $0 < \arg n < \pi$ of the circle at infinity is found to vanish. This can be proved with the aid of (15), a similar expression for n tending to infinity in the left half of the n-plane, and the asymptotic expression :

$$\frac{P_n\left\{\cos(\vartheta-\pi)\right\}}{\sin(\pi n)} \sim -\frac{\sqrt{2i}\; e^{i(n+\frac{1}{2})\vartheta}}{\sqrt{\pi(n+1)\sin\vartheta}} \tag{19}$$

$$(|n|\to\infty\,;\ \Im m\, n>0)\,.$$

According to Cauchy's theorem the contour integrals reduce to the residues at the enclosed singularities. For Π_{-1} these singularities are the first-order poles consisting of the zeros of $N(n)$ with positive imaginary part. The higher--order rainbow terms depend on higher-order poles. In the case of Π_{-1} we get the "residue series"

$$\Pi_{-1} = \frac{M}{2k_1^2a^4\varepsilon_{eff,1}}\sum_{s=0}^{\infty}\frac{(n_s+\frac{1}{2})}{\sin(\pi n_s)}\,\frac{P_{n_s}\left\{\cos(\pi-\vartheta)\right\}}{\left(\frac{\partial N}{\partial n}\right)_{n=n_s}}\quad . \tag{20}$$

This new series is of a sufficiently rapid convergence to compute the diffraction effect of the earth on radio waves of all frequencies when the inhomogeneity of the atmosphere may be neglected. In view of (19) the dependence on the angular distance ϑ from the transmitter to the receiver is given approximately for each term by $\exp\{i(n_s+1/2)\vartheta\}/\sqrt{\sin\vartheta}$, since all eigenvalues n_s prove to be of the order of the large quantity $k_1 a$. Hence, the moduli of the terms are about proportional to $\exp(-\mathfrak{Im}\, n_s . \vartheta)/\sqrt{\sin\vartheta}$ which involves (for $\sin\vartheta$ not too small) a rapid convergence of the residue series, provided that its terms are ordered with respect to increasing values of $\mathfrak{Im}\, n_s$.

In neglecting the rainbow terms our application of the Watson transformation has resulted in the transition from an expansion in modes with integral orders n [the sum of (4) and (6) in the outer space] to another expansion (20). The latter constitutes the special value for $r=b=a$ of the general solution which is obtained by multiplying each term by an additional factor

$$\frac{\zeta^{(1)}_{n_s}(k_1 b)\, \zeta^{(1)}_{n_s}(k_1 r)}{\left\{\zeta^{(1)}_{n_s}(k_1 a)\right\}^2} .$$

The terms thus become proportional to the "radial modes" $\zeta^{(1)}_{n_s}(k_1 r)\, P_{n_s}\{\cos(\pi-\vartheta)\}$ with complex orders ν_s.

The modes of the former type, to be termed "azimuthal modes", are finite and unique throughout space but independent of the boundary conditions at the earth. On the other hand, the "radial modes" are not unique since $P_{\nu_s}(-\cos\vartheta)$ becomes logarithmically infinite along the radius $\vartheta = 0$ for complex ν_s; these latter modes depend strongly on the boundary condition (7) along the earth. Such a boundary condition may be the starting point for determining the "radial modes" straightforwardly. This will be worked out in detail in the more general case of a stratified atmosphere (see next chapter).

The residue series is particularly useful if (for finite heights of the transmitter and receiver above the earth) the point of observation is well beyond the horizon as observed from the transmitter. In the transition region near the horizon the field can also be computed conveniently by investigating the integral (16) and substituting proper approximations in its integrand. This is the essential basis of Fock's [10] many papers on diffraction theory. In the region in front of the horizon the convergence of the residue series becomes slow but can be improved by using another residue series (converging most rapidly for small ϑ) derived by Franz and Deppermann [11]. Finally the factor $1/\sin(\pi n_s)$ in (20) can be expanded into

$$\frac{1}{\sin(\pi n_s)} = -2i \sum_{r=0}^{\infty} e^{i(2r+1)\pi n_s}$$

which leads to a splitting into "creeping waves". In fact, the contribution due to the r-th term can be ascribed to a wave which has "crept" r times around the earth before reaching the receiver.

IIId. Final numerical approximations.

The expansion (20) for $r = b = a$ can be put in a form suitable for numerical computations by using the property that all eigenvalues are of the order of the large quantity $k_1 a$. This justifies us in applying the approximation (19) for the Legendre function, as well as approximations holding for Bessel functions of almost equally large order and argument when evaluating $N(n)$. The complete evaluation [12] of the residue series for Π_{-1} then results in the following expansion :

$$\Pi \sim 2\,\Pi_{pr}\,\sqrt{2\pi i\chi}\,\sum_{s=0}^{\infty} f_s(h_1)\,f_s(h_2)\,\frac{e^{i\tau_s\chi}}{(2\tau_s - 1/\delta^2)}\,,$$

in which $\chi = (k_1 a)^{1/3}\vartheta$ determines the dependence on the distance from the transmitter to the receiver. Further $\tau_s = (n_s - k_1 a)/(k_1 a)^{1/3}$ fixes the eigenvalues according to the equation:

$$\frac{H^{(1)}_{2/3}\left\{\frac{(2\tau_s)^{3/2}}{3}\,e^{-\frac{3}{2}i\pi}\right\}}{H^{(1)}_{1/3}\left\{\frac{(2\tau_s)^{3/2}}{3}\,e^{-\frac{3}{2}i\pi}\right\}} = \frac{e^{-i\pi/6}}{\delta\sqrt{2\tau_s}}$$

in which the parameter

$$\delta = i \frac{k_2^2/k_1^2}{(k_1 a)^{1/3}\sqrt{(k_2^2/k_1^2)-1}}$$

accounts for the influence of the electric properties of the soil. Finally the "height-gain factors" $f_s(h_1)$ and $f_s(h_2)$ determine the influence of the elevations h_1 and h_2 of the transmitter and receiver above the earth. These factors equal unity if $h_1 = h_2 = 0$, and can be evaluated with the aid of further approximations.

Extensive numerical results derived with this theory have been represented in tables and atlasses, for instance those published by the C.C.I.R. [11] which include also the effect of the normal stratification of the atmosphere (see next chapter). For short distances supplementary geometrical-optics approximations prove to be more useful. We finally mention that the expansions degenerate to Sommerfeld's approximation for the flat earth (see chapter I.c) if a tends to infinity.

IV. The field of a vertical dipole in a stratified atmosphere above a spherical homogeneous earth.

IVa. Reduction to an approximative scalar problem.

The effect of the inhomogeneity of the atmosphere can be accounted for in a first approximation by assuming $\varepsilon_{eff,1}$ as a function of the height $h = r - a$ above the earth's surface. For our vertical dipole of current momentum $M e^{-i\omega t}$ (situated at $r = b$, $\vartheta = 0$) Maxwell's equations (I.2) now read, assuming $\mu = \mu_0$ throughout the outer space,

$$\text{curl } \vec{E} - i\,\omega\,\mu_0\,\vec{H} = \vec{0}\,,$$

$$\text{curl } \vec{H} + i\omega\,\varepsilon_{eff,1}(r)\vec{E} = \frac{M}{2\pi b^2}\,\delta(b-r)\,\frac{\delta(\vartheta)}{\vartheta}\vec{u}_r\,, \tag{1}$$

$\vec{u}_r$ representing a unit vector in a radial direction.

It proves to be possible, once again, to solve this set of equations with the aid of a radially directed Hertzian vector as defined by (III.1). In fact, substituting

$$\vec{E} = \sqrt{\varepsilon_{eff,1}(b)}\left[i\omega\,\mu_0\,\sqrt{\varepsilon_{eff,1}(r)}\,\Pi\,\vec{r}\, + \right.$$
$$\left. + \frac{i}{\omega}\,\text{grad}\,\frac{\frac{\partial}{\partial r}\left\{r\sqrt{\varepsilon_{eff,1}(r)}\,\Pi\right\}}{\varepsilon_{eff,1}(r)}\right],$$

$$\vec{H} = \sqrt{\varepsilon_{eff,1}(b)}\;\text{curl}\left\{\sqrt{\varepsilon_{eff,1}(r)}\,\Pi\,\vec{r}\right\}, \qquad (r > a)$$

we find that (1) is satisfied provided that the scalar quantity Π satisfies the scalar wave equation :

$$(\nabla^2 + k^2_{1,eff}(r))\,\Pi = -\frac{M}{\varepsilon_{eff,1}(b)\,2\pi b^3}\,\delta(r-b)\,\frac{\delta(\vartheta)}{\vartheta}, \qquad (2)$$

in which we have introduced a wave number $k_{1,eff}(r)$ defined by :

$$k^2_{1,eff}(r) = \omega^2\mu_0 .\ \varepsilon_{eff,1}(r) - \sqrt{\varepsilon_{eff,1}(r)}\,\frac{d^2}{dr^2}\,\frac{1}{\sqrt{\varepsilon_{eff,1}(r)}}\ .$$

The wave equation (2) for the atmosphere has a non constant coefficient $k_{1,eff}(r)$ whereas the other wave equation $(\nabla^2 + k_2^2)\,\Pi = 0$ with constant k_2 still holds inside the earth. However, our problem may be simplified by accounting for the influence of the earth with the aid of an approximative boundary condition, which is made plausible as follows.

In radio practice the transmitter and receiver are near the earth's surface, not too close to each other. Therefore, the waves propagate mainly in the horizontal direction along this surface. The dominating contribution from the three-dimensional Fourier spectrum of the field thus originates from waves which travel just above the earth, and which are of the type $\exp\ i(k_1 x - \omega t)$ if we neglect for a moment the earth's curvature. After penetration into the earth, by refraction at grazing incidence, these waves become proportional to :

$$e^{i(k_1 x - \sqrt{k_2^2 - k_1^2}\, z - \omega t)}$$

The inner field just below the surface then satisfies approximately the relation :

$$\frac{\partial \Pi_2}{\partial_z} = -i \sqrt{k_2^2 - k_1^2}\, \Pi_2 \,, \qquad \text{or also}$$

$$\frac{\partial}{\partial_r}(r\Pi_2) = -i \sqrt{k_2^2 - k_1^2} \,.\, r\, \Pi_2 \qquad \text{for } r = a - o \;.$$

Further, the boundary conditions (III.7) still hold approximately, so that the latter relation can also be expressed as follows in terms of the field Π just above the earth's surface :

$$\frac{\partial}{\partial r}(r\Pi) = -i \frac{k_1^2}{k_2^2} \sqrt{k_2^2 - k_1^2} \,.\, r\, \Pi \qquad \text{at } r = a + o \;,$$

or abbreviated :

$$\frac{\partial}{\partial r}(r\Pi) = \gamma \,.\, r\, \Pi \qquad \text{at } r = a. \qquad (3)$$

This boundary condition of the Leontovich type reduces the present problem to one of a single medium only, that of the atmosphere. Of course, this simplification could also have been introduced in the preceding theory of the homogeneous atmosphere, as has been worked out by Wait [13]. However, it is then impossible to study the interesting theory of the rainbow terms. Our present problem now amounts to solving the scalar wave equation (2) with a variable coefficient k_1 , in the space $r > a$, and under the assumption of the Leontovich boundary condition (3) at $r = a$, and the radiation condition at infinity.

In contrast to the preceding analysis we shall now derive directly the expansion in radial modes, a method which could of course also have been applied for the homogeneous atmosphere.

IVb. The expansion of the field in radial modes.

This expansion is of the form :

$$\Pi = \sum_s c_s \cdot f_{n_s}(r) \cdot P_{n_s}\left\{\cos(\pi - \vartheta)\right\} \quad , \qquad (4)$$

in which the eigenvalues n_s have to be determined from the boundary conditions, and next the coefficients c_s from the conditions imposed by the transmitting dipole. Substitution into the wave equation (2) yields :

$$\left\{\nabla^2 + k^2_{1,\mathrm{eff}}(r)\right\} \sum_0^{\infty} c_s \cdot f_{n_s}(r) P_{n_s}\left\{\cos(\pi - \vartheta)\right\} =$$

$$= -\frac{M}{\varepsilon_{\mathrm{eff},1}(b) 2\pi b^3} \delta(r-b) \frac{\delta(\vartheta)}{\vartheta} . \qquad (5)$$

The homogeneous wave equation $\left\{\nabla^2+k^2_{1,\mathrm{eff}}(r)\right\}\Pi=0$ is satisfied for the relevant boundary conditions by any modes

$$f_n(r)\,P_n\left\{\cos(\pi-\vartheta)\right\} \quad ,$$

provided that $f_n(r)$ satisfies the " height-gain differential equation "

$$\left\{\frac{d^2}{dr^2}+k^2_{1,\mathrm{eff}}(r)-\frac{n(n+1)}{r^2}\right\}\left\{r\,f_n(r)\right\}=0 \tag{6}$$

and also the Leontovich boundary condition $\frac{\partial}{\partial r}(r\,f_n)=\gamma\, r f_n$ at $r=a$, as well as the radiation condition at infinity.

The resulting generally complex eigenvalues n_s, however, involve infinite values of the Legendre function $P_{n_s}\left\{\cos(\pi-\vartheta)\right\}$ along the axis $\vartheta=0$. The approximation

$$P_{n_s}\left\{\cos(\pi-\vartheta)\right\}\sim\frac{2}{\pi}\sin(\pi n_s)\log\frac{\vartheta}{2}$$

in the vicinity of this axis implies that the above modes satisfy the inhomogeneous equation :

$$\left\{\nabla^2+k^2_{1,\mathrm{eff}}(r)\right\}\left[f_{n_s}(r)P_{n_s}\left\{\cos(\pi-\vartheta)\right\}\right]=\frac{4}{\pi}\sin(\pi n_s)\frac{f_{n_s}(r)}{r^2}\,\frac{\delta(\vartheta)}{\vartheta}. \tag{7}$$

This relation can be checked by integrating it over an infinitesimal cone around the axis $\vartheta=0$, while applying Gauss's divergence theorem.

With the aid of (7) we deduce from (5) :

$$\sum_{0}^{\infty} c_s . \sin(\pi n_s) . f_{n_s}(r) = - \frac{M}{8\,b\,\varepsilon_{eff,1}(b)} \delta(r-b). \qquad (8)$$

From the well-known theory [14] of Sturm-Liouville differential equations we infer the complete orthogonality of the infinite set of eigenfunctions $f_{n_s}(r)$ which are characterized as the solutions of (6) that satify the homogeneous boundary condition (3) at $r = a$ (together with the radiation condition at infinity). Such a discrete set of eigen functions could not be assumed in the rigorous two-media problem. In general the latter also involves a continuous spectrum of eigenvalues, though the contribution from the latter may be small as has been shown for special conditions by Friedman [15].

In view of the integration interval $a < r < \infty$ the expansion with respect to the above eigenfunctions reads for an arbitrary function $\varphi(r)$:

$$\varphi(r) = \sum_{s=0}^{\infty} f_{n_s}(r) \frac{\int_a^\infty \varphi(r) f_{n_s}(r)\,dr}{\int_a^\infty f_{n_s}^2(r)\,dr}$$

In particular we have :

$$\delta(r-b) = \sum_{s=0}^{\infty} \frac{f_{n_s}(b).\,f_{n_s}(r)}{\int_a^{\infty} f_{n_s}^2(r)\,dr}$$

The coefficient of a special function $f_{n_s}(r)$ in (8) yields an equation for c_s. Substituting its result into (4) we obtain :

$$\Pi = -\frac{M}{8b\varepsilon_{eff,1}(b)} \sum_{s=0}^{\infty} \frac{f_{n_s}(b)\,f_{n_s}(r)}{\int_a^{\infty} f_{n_s}^2(r)\,dr} \frac{P_{n_s}\left\{\cos(\pi-\vartheta)\right\}}{\sin(\pi n_s)}. \quad (9)$$

The integral in the denominator can be evaluated as follows. We apply the operator

$$\frac{\partial}{\partial r}(r f_n).1 - r f_n . \frac{\partial}{\partial n}$$

to (6). This leads to the following relation for all solutions of (6) :

$$(2n+1)\,f_n^2(r) = \frac{d}{dr}\left[r^2 f_n^2 . \frac{\partial}{\partial n}\left\{ \frac{\frac{\partial}{\partial r}(r f_n)}{r f_n} \right\}\right].$$

A further integration over the interval $a < r < \infty$ yields, assuming that the expression between braces tends to zero for $r \to \infty$ (as in the case of atmospheric absorption , however small it may be) :

$$\int_a^{\infty} f_n^2(r)\,dr = -\frac{a^2 f_n^2(a)}{(2n+1)} \cdot \frac{\partial}{\partial n}\left\{\frac{\frac{\partial}{\partial r}(r f_n)}{r f_n}\right\}_{r=a}$$

Substitution into (9) results in the final expansion :

$$\Pi = \frac{M}{8a^2 b\varepsilon_{eff,1}(b)} \sum_{s=0}^{\infty} \frac{(2n_s+1)}{\frac{\partial}{\partial n}\left\{\frac{\frac{\partial}{\partial r}(r f_n)}{r f_n}\right\}_{\substack{r=a\\ n=n_s}}} \frac{f_{n_s}(b) f_{n_s}(r)}{f_{n_s}^2(a)} \frac{P_{n_s}\{\cos(\pi-\vartheta)\}}{\sin(\pi n_s)}. \tag{10}$$

In view of Cauchy's theorem and the boundary condition (3) for the eigenfunctions this expansion constitutes the " residue series " of the following contour integral in the simplest case of transmitter and receiver both on the ground $(r = b = a)$:

$$\Pi = \frac{M}{16\pi i a^3 \varepsilon_{eff,1}(a)} \int_L dn \frac{(2n+1)}{\left\{\frac{\frac{d}{dr}(rf_n)}{r f_n} - \gamma\right\}_{r=a}} \frac{P_n\{\cos(\pi-\vartheta)\}}{\sin(\pi n)}. \tag{11}$$

The contour L in the complex n-plane should enclose all eigenvalues n_s and no other singularities of the integrand. As in the case of a homogeneous atmosphere L may be a line enclosing the upper half of the n-plane , leaving the real axis outside, or simply a line parallel and just above the axis (from $-\infty + io$ to $\infty + io$).

IVc. Splitting into contributions depending on different numbers of atmospheric reflections (multi-hop contributions).

A consequence of the stratification of the atmosphere is the partial downward reflection of any wave sent upwards. The reflection effect always exists even if the ray trajectories corresponding to Snell's law (see next paragraph) escape into space; in the latter case, however, the numerical value of the reflection coefficient becomes very small . A reflection coefficient $T_n(r_0)$ which accounts for the reflection effects in the atmosphere above the level $r = r_0$ may be introduced as follows.

We consider an arbitrary mode $\Pi_n = f_n(r) P_n\{\cos(\pi - \vartheta)\}$ which satisfies the height-gain differential equation (6) as well as the radiation condition at $r = \infty$. In order to restrict ourselves to the effect of the stratification above $r = r_0$ we assume a homogeneous space with constant wave number k_0 below this level, but the wave number should be continuous at $r = r_0$; hence $k_0 = k_{1,eff}(r_0)$. The combination of the above mode with the corresponding solution in the homogeneous space is represented by :

$$\Pi_n = \begin{cases} f_n(r)\, P_n\left\{\cos(\pi - \vartheta)\right\} & \text{for } r > r_0 \quad (12a) \\ \left\{\alpha\, \zeta_n^{(1)}(k_0 r) + \beta\, \zeta_n^{(2)}(k_0 r)\right\} P_n\left\{\cos(\pi - \vartheta)\right\} & \text{for } r < r_0 \quad (12b) \end{cases}$$

We assume once again boundary conditions requiring the

continuity of $\partial/\partial r\ (r\Pi_n)$ and $k^2\Pi_n$ at the transition level $r = r_o$. This leads to two equations from which we can solve the values of α and β. In view of the time factor $e^{-i\omega t}$ the first term of (12b) represents a rising wave, the second term a descending wave. The reflection coefficient $T_n(r_o)$ is reasonably defined as the ratio of both waves at the lower boundary $r = r_o$ of the stratification. Hence

$$T_n(r_o) = \frac{\beta \zeta_n^{(2)}(k_o r_o)}{\alpha \zeta_n^{(1)}(k_o r_o)} .$$

Substitution of the values of α and β derived along the lines indicated here results in the expression :

$$T_n(r_o) = \frac{k_o \cdot \frac{d}{dx} \log\left\{x \zeta_n^{(1)}(x)\right\}_{x=k_o r_o} - \frac{d}{dr}\left\{\log(r f_n)\right\}_{r=r_o}}{-k_o \cdot \frac{d}{dx} \log\left\{x \zeta_n^{(2)}(x)\right\}_{x=k_o r_o} + \frac{d}{dr}\left\{\log(r f_n)\right\}_{r=r_o}} . \qquad (13)$$

We can introduce a corresponding reflection coefficient R_n for the earth's surface. In fact, if the space above the earth were homogeneous (wave number k_o) a normalized descending wave

$$\zeta_n^{(2)}(k_o r)\, P_n\left\{\cos(\pi - \vartheta)\right\} e^{-i\omega t}$$

approaching the surface $r=a$ would be reflected there and thus produce a rising wave proportional to $\zeta_n^{(1)}(k_o r)P_n\{\cos(\pi-\vartheta)\}$. The reflection coefficient now constitutes the ratio of both waves at $r=a$. The total wave function in the outer space then becomes :

$$\left\{\zeta_n^{(2)}(k_o r)+R_n\,\zeta_n^{(2)}(k_o a)\,\frac{\zeta_n^{(1)}(k_o r)}{\zeta_n^{(1)}(k_o a)}\right\}P_n\{\cos(\pi-\vartheta)\}\;e^{-i\omega t}$$

This function has to satisfy Leontovich's boundary condition (3) which determines the reflection coefficient R_n. The result reads :

$$R_n=\frac{k_o\cdot\frac{d}{dx}\log\left\{x\,\zeta_n^{(2)}(x)\right\}_{x=k_oa}-\gamma}{-k_o\cdot\frac{d}{dx}\log\left\{x\,\zeta_n^{(1)}(x)\right\}_{x=k_oa}+\gamma}\,. \tag{14}$$

For the sake of simplicity we again consider the case of both transmitter and receiver on the earth's surface $(r=b=a)$. The contour integral (11) then reduced to :

$$\Pi=\frac{M}{16\pi i a^3\varepsilon_{eff,1}(a)}\int_L dn\,\frac{(2n+1)}{\left\{\frac{\frac{d}{dr}(rf_n)}{rf_n}\right\}_{r=a}-\gamma}\,\frac{P_n\{\cos(\pi-\vartheta)\}}{\sin(\pi n)}. \tag{15}$$

We further express the quantity $\frac{d}{dr}\left\{\log(rf_n)\right\}_{r=a}$ in terms of $T_n(a)$, and similary γ in terms of R_n , applying (13) and (14) respectively. The following identity results, also using the Wronskian of $\zeta_n^{(1)}(x)$ and $\zeta_n^{(2)}(x)$

$$\left\{\frac{\frac{d}{dr}(r f_n)}{r f_n}\right\}_{r=a} - \gamma = \frac{2i\left\{1 - T_n(a) R_n\right\}}{k_o a^2 \zeta_n^{(1)}(k_o a)\zeta_n^{(2)}(k_o a)\left\{1+T_n(a)\right\}(1+R_n)} .$$

We replace the denominator of (15) by this expression, and obtain

$$\Pi = -\frac{k_o M}{32\pi a\, \varepsilon_{eff,1}(a)} \int_L dn \frac{(2n+1)(1+T_n)(1+R_n)}{(1-T_n R_n)} \zeta_n^{(1)}(k_o a)\zeta_n^{(2)}(k_o a) \frac{P_n\left\{\cos(\pi-\vartheta)\right\}}{\sin(n\pi)} , \tag{16}$$

omitting henceforth the argument a of T_n .

This representation shows at once two characteristic features of the solution, viz :

(a) the boundary condition (3) for the eingenfunctions, which is equivalent to the vanishing of the denominator of (15) and (16), can also be expressed by

$$1 - T_{n_s} . R_{n_s} = 0 .$$

This indicates the resonance property that each eigenfunction remains unchanged after a simultaneous reflection both in the atmosphere and at the earth's surface;

(b) assuming $|T_n R_n| < 1$ along the contour L we get from an expansion of the denominator into a geometric progression a new series :

$$\Pi = \sum_{j=0}^{\infty} \Pi_j ,$$

in which

$$\Pi_0 = -\frac{k_0 M}{32\pi a\, \varepsilon_{eff,1}(a)} \int_L dn(2n+1)(1+R_n)\zeta_n^{(1)}(k_0 a)\zeta_n^{(2)}(k_0 a)\frac{P_n\{\cos(\pi-\vartheta)\}}{\sin(\pi n)},$$

$$\Pi_j = -\frac{k_0 M}{32\pi a\, \varepsilon_{eff,1}(a)} \int_L dn(2n+1)(1+R_n)^2 R_n^{j-1} T_n^j \zeta_n^{(1)}(k_0 a)\zeta_n^{(2)}(k_0 a)\frac{P_n\{\cos(\pi-\vartheta)\}}{\sin(\pi n)}$$

$$(j \geqslant 1) \qquad (17)$$

The first term Π_0 represents the effect of a homogeneous atmosphere $(T=0)$. Obviously, any Π_j corresponds to the contribution that has suffered j successive reflections in the atmosphere and $j-1$ intermediate reflections against the earth during its propagation from the transmitter to the receiver. In the case of atmospheric reflections by the ionosphere the term " j-hop transmission " is used.

IVd. Derivation of the geometrical-optics properties.

The propagation along geometric-optical trajectories should be recognizable from the general expression (17). The atmospheric properties are included in the reflection coefficient T_n, an approximation of which can be obtained by applying the W.K.B. method to the height-gain differential equation (6). For the general differential equation

$$\frac{d^2 y}{d r^2} + k^2(r)\, y = 0$$

this approximation reads

$$y(r) \sim \frac{A e^{i \int_{r_0}^{r} k(s) ds} + B e^{-i \int_{r_0}^{r} k(s) ds}}{\sqrt{k(r)}}, \tag{18}$$

and therefore breaks down if r approaches a level r_0 for which $k(r_0) = 0$. Such a situation just occurs when ray trajectories reach a highest point in the atmosphere, as in the case of ionospheric reflections. We then have $k^2(r) < 0$ immediately above $r = r_0$, and $k^2(r) > 0$ for $r < r_0$. In both regions the W.K.B. method is applicable up to the vicinity of the transition region $r \sim r_0$. The necessary connection between the W.K.B. solutions at both sides of it can be obtained by assuming a linear dependence of $k^2(r)$ on r in the transition region itself. The analysis has been worked out by Kramers [16)] for quantum-mechanical applications, and the complete problem has been studied in detail by Langer [17]. The extension of the W.K.B. solution thus arrived at can be represented as follows if the solution should behave as a wave $e^{ik_0 r}$ moving outwards to the limiting space $r \rightarrow \infty$ with a constant wave number $k(r) \rightarrow k_0$:

$$y(r) \sim \frac{C}{\sqrt{k(r)}} \sqrt{\int_r^{r_0} k(s)ds} \, . \, H_{\frac{1}{3}}^{(1)} \left\{ \int_r^{r_0} k(s)ds \, . \, e^{-i\pi} \right\} .$$

For $r > r_0$ the integral is to be defined here as

$$e^{i\pi} \int_{r_0}^{r} k(s)\, ds \, .$$

The two mutually connected ordinary W.K.B. solutions in the regions well above and well below the level $r = r_0$ are obtained from the asymptotic approximation of the Hankel function entering in the latter expression. These individual W.K.B. approximative read explicitly :

$$y(r) \sim \begin{cases} iC \sqrt{\dfrac{2}{\pi k(r)}} \, e^{-\int_{r_0}^{r} |k(s)|\, ds - i\frac{5}{12}\pi} & \text{for } r \gg r_0 \quad (19a) \\[2ex] 2C \sqrt{\dfrac{2}{\pi k(r)}} \, e^{-\frac{i\pi}{6}} \cos\left\{ \int_r^{r_0} k(s)ds - \dfrac{\pi}{4} \right\} & \text{for } r \ll r_0 . \quad (19b) \end{cases}$$

In the application to the height-gain differential equation (6) the transition level r_o is found for a special n value from the equation :

$$k^2_{1,eff}(r_o) - \frac{n(n+1)}{r_o^2} = 0 , \qquad (20)$$

and the W.K.B. function below this level is given by :

$$r f_n(r) \sim$$

$$\sim 2C \sqrt{\frac{2}{\pi \sqrt{k^2_{1,eff}(r) - n(n+1)/r^2}}}\, e^{-\frac{i\pi}{6}} \cos\left\{ \int_r^{r_{o,n}} \sqrt{k^2_{1,eff}(s) - \frac{n(n+1)}{s^2}}\, ds - \frac{\pi}{4} \right\}. \qquad (21)$$

According to (19a) the wave in question is attenuated exponentially above $r = r_o$, but it might become sinusoidal again at a higher altitude if k^2 passes through another zero. In the region $r < r_o$ we split the cosine in (21) into its two exponentials so as to obtain :

$$r f_n(r) \sim C \sqrt{\frac{2}{\pi \sqrt{k^2_{1,eff}(r) - n(n+1)/r^2}}} \times$$

$$\times \left\{ e^{\frac{i\pi}{12} - i \int_r^{r_{o,n}} \sqrt{k^2_{1,eff}(s) - \frac{n(n+1)}{s^2}}\, ds} + e^{-i\frac{5\pi}{12} + i \int_r^{r_{o,n}} \sqrt{k^2_{1,eff}(s) - \frac{n(n+1)}{s^2}}\, ds} \right\} \qquad (22)$$

Let the medium become homogeneous, with $k_{1,\mathrm{eff}}=k_0$, below a special level. The phase of the first contribution in (22) increases there with r, that of the second one decreases. In view of our time factor $e^{-i\omega t}$ the first contribution represents a rising wave; the second contribution can be interpreted as a descending wave that is due to the reflection of the former in the transition region $r \sim r_{o,n}$. The atmospheric reflection coefficient $T_n(r)$, defined above, constitutes the ratio of both waves at a special r level. Taking $r=a$ we get from the ratio of the two waves in (22):

$$T_n(a) \sim -i\, e^{2i \int_a^{r_{o,n}} \sqrt{k^2_{1,\mathrm{eff}}(s)-n(n+1)/s^2}\, ds} \tag{23}$$

We next consider the other factors of the integrand of (17). The product $\zeta_n^{(1)}(k_o a)\, \zeta_n^{(2)}(k_o a)$ can be approximated by observing that the functions $f_n(r)=\zeta_n^{(1)(2)}(k_o r)$ constitute the solutions of (6) for a homogeneous space with $k_{1,\mathrm{eff}}(r)=k_o$. The corresponding W.K.B. approximation of the type (18) reads:

$$\zeta_n^{(1)(2)}(k_o r) \sim \mp i\, \frac{e^{\pm i \left\{ \int_{\sqrt{n(n+1)}}^{k_o r} \sqrt{1-n(n+1)/s^2}\, ds + \frac{\pi}{2}\sqrt{n(n+1)} \right\}}}{\sqrt{k_o r}\, \sqrt[4]{k_o^2 r^2 - n(n+1)}}$$

This approximation breaks down near the zero of the denominator, but the expression

$$\zeta_n^{(1)}(k_o a)\,\zeta_n^{(2)}(k_o a) \sim \frac{1}{k_o a\sqrt{k_o^2 a^2 - n(n+1)}}$$

deduced from it is still integrable there with respect to n .

In the last factor of (17) we can substitute (III.19) if we take for L a path just above the real axis of the n-plane. We apply (23) for the factor $\{T_n(a)\}^j$ but we leave unchanged the other factors $1+T_n$, $1+R_n$, R_n^j which are not proportional to an exponential function in n . The mentioned substitutions result in :

$$\Pi_j \sim \frac{\sqrt{2i}\;\;M\,(-i)^j}{32\pi^{3/2}a^2\,\varepsilon_{eff,1}(a)\sqrt{\sin\vartheta}} \times$$

$$\times \int_{-\infty+io}^{\infty+io} dn\,\frac{(2n+1)(1+R_n)^2 R_n^{j-1}\, e^{i\left\{2j\int_a^{r_{o,n}}\sqrt{k_{1,eff}^2(s)-n(n+1)/s^2}\,ds+(n+\frac{1}{2})\vartheta\right\}}}{\sqrt{n+1}\;\sqrt{k_o^2a^2-n(n+1)}} . \qquad (24)$$

We are now able to apply a saddlepoint approximation. The exponential, $e^{i\phi(n)}$ say, has a saddlepoint at the n value n_o for which $\partial\phi/\partial n = 0$. With the aid of (20) we derive :

$$\frac{\partial \phi}{\partial n} = \vartheta - (2n+1)\, j \int_a^{r_{o,n}} \frac{ds}{s\sqrt{k_{1,eff}^2(s)s^2 - n(n+1)}} \qquad (25)$$

A geometrical interpretation of this new integral is arrived at as follows. Snell's law for a trajectory in a concentrically stratified medium with a refractive index $n(r)$ reads $r\, n(r) \sin \tau(r) = \text{constant}$, where $\tau(r)$ is the angle at any point of the trajectory between its tangent and the radius directed towards the centre of symmetry. Let us assume $n(r)$ proportional to the local wave number $k_{1,eff}(r)$ so that the corresponding ray trajectories will satisfy :

$$k_{1_{eff}}(r) \cdot r \cdot \sin \tau(r) = \sqrt{n(n+1)} =$$

$$= k_o a \sin \tau_a\,, \text{ say.} \qquad (26)$$

This equation fixes a trajectory that leaves the earth's surface (wave number k_o) with a tangent forming an angle τ_a with the local vertical (passing through the centre of the earth). At its highest level $r = r_o$ (if any) the trajectory becomes horizontal so as to have there $\tau = \pi/2$ and $k_{1,eff}(r_o) \cdot r_o = \sqrt{n(n+1)}$; in other words the highest altitude is reached at a level given by (20), that is a level at which the W.K.B. approximation breaks down. After this highest point the ray descends to the earth's surface, the descending branch being symmetric to the rising branch when observed from the vertical through the highest point. Therefore, considering the equation $\vartheta = \vartheta(r)$ of the trajectory, we get the following quantity for the angular distance $\vartheta(\tau_a)$ covered between the two contacts with the earth's surface :

$$\vartheta(\tau_a) = 2 \int_{r=a}^{r=r_{o,n}} d\vartheta = 2 \int_{a}^{r_{o,n}} dr \frac{d\vartheta}{dr} = 2 \int_{a}^{r_{o,n}} \frac{dr}{r} \tan \tau(r) .$$

We can express $\tau(r)$ in terms of $\sin \tau(r)$ given by (26) , and obtain :

$$\vartheta(\tau_a) = 2 \sqrt{n(n+1)} \int_{a}^{r_{o,n}} \frac{ds}{s\sqrt{k_{1,eff}^2(s)s^2 - n(n+1)}} . \qquad (27)$$

Therefore, we may replace (25) by

$$\frac{\partial\phi}{\partial n} = \vartheta - j \frac{(2n+1)}{\sqrt{n(n+1)}} \frac{\vartheta(\tau_a)}{2} \sim \vartheta - j\vartheta(\tau_a) , \qquad (28)$$

since the relevant n values always prove to be very large.

The saddlepoint value n_o is thus characterized by

$$\vartheta = j\vartheta(\tau_a) .$$

Hence $n_0 \sim \sqrt{n_0(n_0+1)} = k_0 a \sin \tau_{a,0}$ determines the starting direction $\tau_{a,0}$ at the earth of the trajectory that arrives at the receiver (an angular distance ϑ away) just after j hops. The connection of Π_j with j-hop transmission is confirmed here by the fact that the dominating contribution to Π_j originates from the vicinity of a saddlepoint which depends on the starting direction of the trajectory that overbridges the distance to the receiver in exactly j hops.

The saddlepoint being known, we pass to the evaluation of the corresponding approximation. Its so-called second order approximation is fixed in our case by the relation :

$$\int_{-\infty}^{\infty} dn\, A(n)\, e^{i\phi(n)} \sim A(n_0)\, e^{i\phi(n_0)} \int_{-\infty}^{\infty} dn\, e^{\frac{i}{2}\left(\frac{\partial^2\phi}{\partial n^2}\right)_{n=n_0}(n-n_0)^2} =$$

$$= e^{\pm i\frac{\pi}{4}} \sqrt{\frac{2\pi}{\left|\left(\frac{\partial^2\phi}{\partial n^2}\right)_{n=n_0}\right|}}\; A(n_0)\, e^{i\phi(n_0)} \qquad (29)$$

in which the upper (lower) sign holds according as $(\partial^2\phi/\partial n^2)_{n_0} > (<)\ 0$. With the aid of (28) we get , applying (29) to (24) (also neglecting 1 with respect to n_0)

$$\Pi_j \sim \frac{e^{i(1\mp 1)\pi/4}(-i)^j M}{8\pi a^2 \,\varepsilon_{eff,1}(a)\sqrt{\sin\vartheta}} \; \frac{\sqrt{n_o}\,(1+R_{n_o})^2 R_{n_o}^{j-1}}{\sqrt{k_o^2 a^2 - n_o^2}} \times$$

$$\times \frac{e^{i\phi(n_o)}}{\sqrt{j\left|\partial\vartheta(\tau_a)/\partial n\right|_{n=n_o}}} \qquad \text{for } \left(\frac{\partial\vartheta}{\partial n}\right)_{n=n_o} \gtrless 0 .$$

We next replace, according to (26), n_o by its approximative value $k_o a \sin\tau_{a,o}$. This leads to the following physical representation :

$$\Pi_j \sim \frac{e^{i(1\mp 1)\pi/4} M}{8\pi a\,\varepsilon_{eff,1}(a)} \; \frac{e^{i\phi(n_o)}}{\hat{D}} . \alpha_j \left\{(1+R_n)^2 R_n^{j-1}\right\}_{n=k_o a \sin\tau_{a,o}} (-i)^j , \tag{30}$$

in which

$$\alpha_j = \frac{\hat{D}\sqrt{\left|\left(\frac{d\tau_a}{d\vartheta}\right)_{\tau_a=\tau_{a,o}}\right| \tan\tau_{a,o}}}{a\sqrt{j\sin\vartheta}} \tag{31}$$

while $\hat{D}$ is to be defined as the total length of the j-hop path connecting the transmitter T with the receiver R .

The various factors entering in (30) after the constant factor in front can be interpreted as follows.

The first factor $\exp\left\{ i\,\phi(n_0)\right\}$ determines the phase of the field as expected by geometrical optics. In fact, $\phi\,(n_0)$ proves to be connected with the time of propagation which is given, in view of Snell's law (26), by

$$t_{prop} = \frac{1}{\omega}\int_T^R k_{eff}\,ds = \frac{2j}{\omega}\int_a^{r_{o,n}} k_{eff}\,\frac{ds}{dr}\,dr = \frac{2j}{\omega}\int_a^{r_{o,n}} k_{eff}\,\sec\zeta\,dr =$$

$$= \frac{2j}{\omega}\int_a^{r_{o,n}} \frac{k_{eff}^2\,r}{\sqrt{k_{eff}^2\,r^2 - n(n+1)}}\,dr \quad ,$$

or also by

$$t_{prop} = \frac{2j}{\omega}\left\{\int_a^{r_{o,n}} \frac{\sqrt{k_{eff}^2 . r^2 - n(n+1)}}{r}\,dr + n(n+1)\int_a^{r_{o,n}} \frac{dr}{r\sqrt{k_{eff}^2 . r^2 - n(n+1)}}\right\} .$$

This can further be reduced, with the aid of (27) and the relation $\vartheta = j\,\vartheta\,(\zeta_{a,o})$ [following from the vanishing of (28) at the saddlepoint] , to

$$t_{prop} \sim \frac{1}{\omega}\left\{2j\int_a^{r_{o,n}} \sqrt{k_{eff}^2 - n(n+1)/r^2}\,dr + (n+\tfrac{1}{2})\vartheta\right\} = \frac{1}{\omega}\,\phi(n_0) .$$

The phase $\phi(n_o)$ thus equals ωt_{prop} as it should.

The factor $1/\hat{D}$ accounts for the ordinary inverse-distance decrease of the field in free space. The next factor α_j, given by (31), may be termed " divergence factor " . It characterizes the extra divergence of the ray trajectories that is due to the inhomogeneity of the stratified atmosphere. More exactly, this factor proves to be equal to the square root $\sqrt{dO'/dO}$ of the cross-section dO' of a small conical pencil of rays propagating along straight lines over a distance $\hat{D}$, divided by the actual cross-section dO for these same rays when propagating along this distance through the stratified atmosphere. The proportionality of the field to $\alpha/\hat{D}$ amounts to a proportionality to the quare root of the inverse local cross-sections of a small pencil. This can also be interpreted such that the energy-current density (which is proportional to the square of the field amplitude) changes in inverse proportion with the cross-section. In other words the energy emitted at the source dipole within a small pencil of rays does not escape from this pencil during the propagation.

One factor $1+R_n$ in (30) is to be ascribed to the fact that each ray leaving the transmitter situated just above the earth's surface is accompanied by a ray reflected there against this surface; the amplitude of the latter ray is R times that of the unreflected ray. The other factor $1+R_n$ of $(1+R_n)^2$ is connected in a similar way to the receiver; in fact, the latter is reached by combinations of two rays, one of which arrives directly from the atmosphere, while the other has been reflected against the earth just before the arrival.

Obviously, the next factor R_n^{j-1} accounts for the decrease of the amplitude at each of the $j-1$ intermediate reflections against the earth which alternate with the j reflections in the atmosphere. Finally the factor $(-i)^j$ indicates a phase shift $-\pi/2$ upon each atmospheric reflection. Such a phase shift is characteristic of reflections which can be considered as due to gradual refractions in a continuously changing medium. The remaining effect of the ionospheric reflections is completely contained in the difference of $\phi(n_o)$ from the corresponding phase in a homogeneous space. Therefore, these reflections are merely associated with phase shifts, and not with a change of the modulus of the amplitude. In other words we have to do here with total reflections, as could be proved with the aid of the W.K.B. approximation.

In other cases geometric-optical reflection in the higher atmosphere is impossibile. This is then expressed by the lack of a real zero (with respect to r) of the coefficient in the height-gain differential equation. Formally a complex zero may then exist if the profile for $\varepsilon(r)$ can be continued analytically for complex values of r. The W.K.B. approximation then still constitutes a fair approximation when depending on a zero with a small imaginary part. Such cases correspond to a modulus of (23) smaller than unity; the reflection is therefore partial.

Summarizing, we have found that all properties of geometrical optics can be derived by applying the W.K.B. method together with a saddlepoint approximation to the rigorous solution. The saddlepoint in question fixes the relevant ray trajectories. Such an approximation is better according as the exponent changes effectively at a shorter distance from the saddlepoint. This occurs in particular for high ω and explains geometrical-optics approximations as limits for $\omega \to \infty$. These general considerations apply to all wave equations. Thus the geometric-optical ray propagation in front of the horizon of the transmitter can also be derived (in the case of a homogeneous atmosphere) from a saddlepoint approximation for the quantity Π_{-1} considered in the preceding chapter.

REFERENCES

1) A. Sommerfeld, Ann. Physik 28, 665 (1909); 81, 1135 (1962).

2) H. Weyl, Ann. Physik 60, 481 (1919).

3) C.L. Pekeris and Z. Alterman, J. Appl. Phys. 28, 1317 (1957).

4) Balth, van der Pol and A.H.M. Levelt, Ned. Akad. Wetensch. Proc. A 63 = Indag. Math. 22, 254 (1960).

5) A.T. De Hoop and H.J. Frankena, Appl. Sc. Res. B 8, 369 (1960)

6) Balth. Van der Pol, Trans. I.R.E. AP 4, 288 (1956).

7) H. Brémmer, The pulse solution connected with the Sommerfeld problem for a dipole in the interface between two dielectrics, Proc. Symposium Electromagnetic waves, Madison (Wisc.), April 1961.

8) G.N. Watson, Proc. Roy. Soc. London A 95, 83 (1918).

9) Balth. Van der Pol and H. Bremmer, Phil. Mag. 24, 825 (1937); H.Bremmer, Terrestrial Radio Waves, New York 1949, chapter III.

10) V.A. Fock, J. Phys. U.S.S.R. 9, 256 (1945); J. Theor. Exp. Phys. 15, 480 (1945).

11) W. Franz and K. Deppermann, Ann. Phys. 10, 361 (1952) and 14, 253 (1954).

12) Balth. Van der Pol and H. Bremmer, Phil. Mag. 25, 817 (1938); H.Bremmer, Terrestrial Radio Waves, New York 1949, chapter III.

13) J. Wait, Journ. Res. Nat. Bur. Stand. 56, 237 (1956).

14) R. Courant und D. Hilbert, Methoden der Mathematischen Physik I, Berlin 1924, chapter V.

15) B. Friedman, Comm. Pure Appl. Math. 4, 317 (1951).

16) H.A. Kramers, Z. Phys. 39, 828 (1926).

17) R.E. Langer, Phys. Rev. 51, 669 (1937).

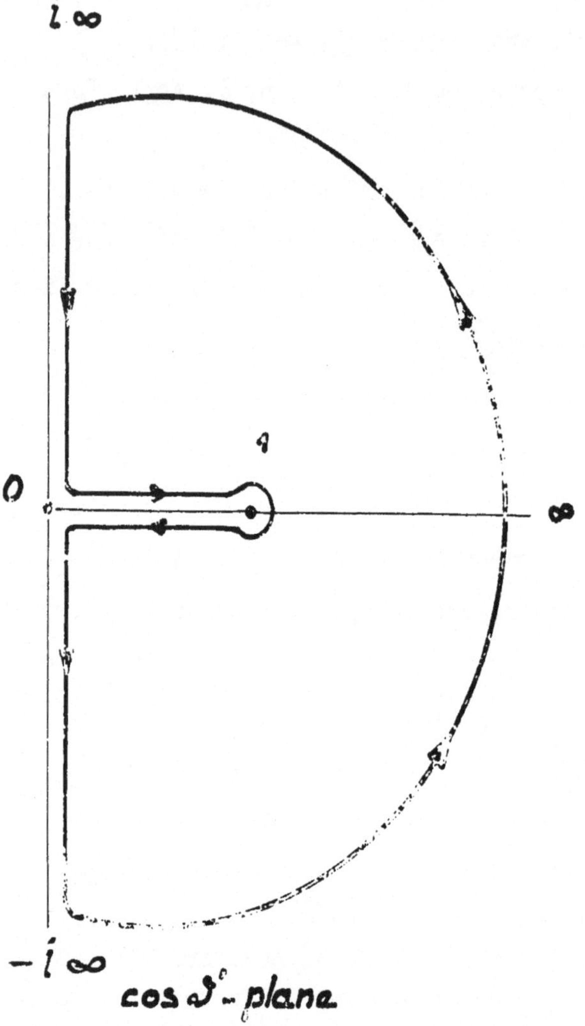

Fig. 1

CENTRO INTERNAZIONALE MATEMATICO ESTIVO

(C.I.M.E.)

L.B. FELSEN

Department of Electrophysics, Polytechnic Institute of Brooklyn[1)]

ASYMPTOTIC EVALUATION OF INTEGRALS[2)]

1) These lectures were presented while the author was on a years' leave of absence with the Office of Naval Research, London Branch.

2) This chapter is based on Chapter IV, "Modal Analysis and Synthesis of Electromagnetic Fields", by L.B.Felsen and N.Marcuvitz, Research Report R-776-59, PIB-705, Microwave Research Institute, Polytechnic Institute of Brooklyn, Oct.1959.

ASYMPTOTIC EVALUATION OF INTEGRALS

by L.B.FELSEN

Department of Electrophysics,
Polytechnic Institute of Brooklyn

1. INTRODUCTORY REMARKS

Radiation and diffraction problems in open regions (with infinite cross section) are usually solved in terms of integral representations for the fields which cannot be evaluated in closed form. In many applications, however, the integrands contain a large parameter, to be called Ω, in terms of which one may obtain an approximation to the integrals. While such an evaluation can be treated for rather general functional dependences of the integrand on Ω, it will suffice within the present context to consider integrals of the following type:[1,2]

$$I(\Omega) = \int_{\bar{P}_z} f(z)\, e^{\Omega q(z)}\, dz , \qquad (1)$$

where f and q are analytic functions of the complex variable z along the path of integration $\bar{P}_z$, whose endpoints lie at infinity, and where the large parameter Ω is assumed to be positive.*

---

* If $\Omega = |\Omega| \exp(i \arg \Omega)$ is complex, the phase term is included in the definition of $q(z)$. Alternatively, it may be more convenient to obtain an asymptotic evaluation if $I(\Omega)$ for real values of Ω and then continue Ω analytically into a range of permitted complex values.

Suppose that $\operatorname{Re} q(z)$ has at the point z_s on $\bar{P}_z$ a maximum value so that $\operatorname{Re} q(z) < \operatorname{Re} q(z_s)$ on the remainder of the path. Since Ω is very large, it follows that $A = \left|\exp\left[\Omega\, q(z)\right]\right|$ likewise has a maximum at z_s and decreases very rapidly away from z_s. It is then suggestive to approximate $I(\Omega)$ by its contribution from the vicinity of z_s only since the contribution from the remainder of the path will be exponentially small in comparison. If $f(z)$ is regular and slowly varying in the vicinity of z_s, this function may be approximated there by $f(z_s)$ and taken outside the integrands in (1), thereby leaving in the integrand only the exponential. Integration of the latter can be effected approximately by expanding $q(z)$ in a power series about z_s and retaining only the first few terms; by a more basic procedure (see Sec.II), the integral is compared with a "canonical" one having similar properties. This, in rough outline, constitutes the basis for an asymptotic approximation of $I(\Omega)$ for large values of Ω; details of the evaluation are given in subsequent sections. In general, A will not have the abovedescribed behavior along the given contour $\bar{P}_z$, but rather along some other path P_z. One then attempts to deform $\bar{P}_z$ into P_z in order to apply the preceding argument. In such a path deformation, proper account must be taken of any interfering singularities of $f(z)$ (such as poles or branch points) in the complex z-plane.

To gain a physical insight into the disposition of the pertinent contours in the complex plane, let us decompose q and z into its real and imaginary parts

$$q(z) = u(x,y) + iv(x,y)\,, \quad z = x + iy\,, \qquad (2)$$

where u, v, x, y are real. The three dimensional plot of u or v vs. x and y resembles a mountain relief map with infinite peaks, bottomless valleys, and passes, arising from the well-known fact that neither u nor v can have absolute maximum or minimum values in the complex plane. Although it is possible to have stationary points with

$$\partial u/\partial x = \partial u/\partial y = \partial v/\partial x = \partial v/\partial y = 0 , \qquad (3)$$

it follows from the Cauchy Riemann equations $\partial u/\partial x = \partial v/\partial y$, $\partial u/\partial y = -\partial v/\partial x$, that $\partial^2 u/\partial x^2 = -\partial^2 u/\partial y^2$, $\partial^2 v/\partial x^2 = -\partial^2 v/\partial y^2$. Hence, if the curvature at a simple stationary point on the surface u(x,y) or v(x,y) is positive along the x direction, it is negative along the (perpendicular) y direction, thereby characterizing the stationary points as "saddle points" (see Fig.1). Equations (3) imply that

$$\frac{dq}{dz} = 0 \text{ at a saddle point } z = z_s . \qquad (4)$$

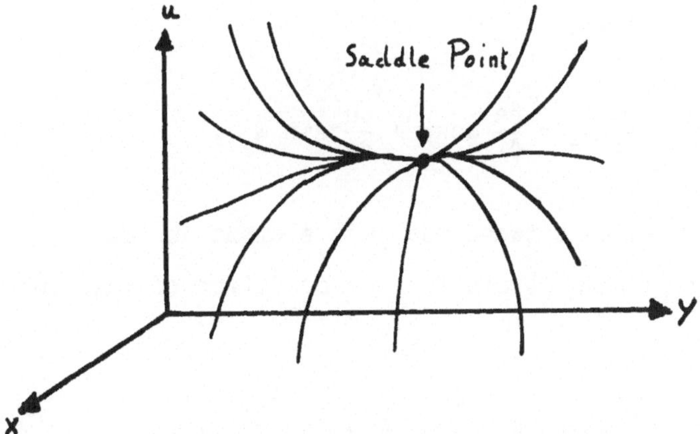

Fig.1 - Relief of the function u in the vicinity of a first-order saddle point.

Saddle points are classified according to the number of derivatives of $q(z)$ which vanish at z_s. If the lowest non--vanishing derivatives at z_s is $q^{(n)}(z_s)$, the saddle point is said to be of order (n-1). It is evident from Fig.1 that the flatness of the saddle increases with the order of the saddle point.

From the preceding remarks concerning the behavior of $A = \exp(\Omega u)$, the most desirable path P_z is the "steepest descent path" (SDP) along which u decreases most rapidly away from its maximum value. There exists a whole family of such curves, for different values of u_{max}, which connects two valley regions, as noted from Fig.1. To obtain as good an approximation to $I(\Omega)$ as possible, one selects that path which is characterized by a minimum value of u_{max}, since both $I(\Omega)$ and its asymptotic approximation -- and hence their difference, the error -- will be proportional to $\exp[\Omega u_{max}]$. The desired contour therefore passes through the saddle point z_s. To ascertain the progress of the SDP through z_s, one seeks that path in the z-plane along which du/ds, the rate of change of u, is a maximum. Since

$$\frac{du}{ds} = \frac{\partial u}{\partial x}\frac{\partial x}{\partial s} + \frac{\partial u}{\partial y}\frac{\partial y}{\partial s} = \frac{\partial u}{\partial x}\cos\alpha + \frac{\partial u}{\partial y}\sin\alpha , \tag{5}$$

where α is the angle between the path element ds and the positive x-axis, one obtains for the condition of maximum rate of change

$$\frac{\partial^2 u}{\partial\alpha\,\partial s} = 0 = -\frac{\partial u}{\partial x}\sin\alpha + \frac{\partial u}{\partial y}\cos\alpha = -\frac{dv}{ds} , \tag{6}$$

wherein the last equality is a consequence of the Cauchy-Riemann equations. Hence, u changes most rapidly along a path on which $\mathrm{Im}q \equiv v = \text{constant}$, i.e., along a " constant phase " path for the function $\exp[\Omega\, q(z)]$. While the constant phase condition characterizes both the steepest ascent and the steepest descent paths, only the latter is of interest herein.

For a first-order saddle point, $q''(z_s) \neq 0$, and one may represent

$$q(z) = q(z_s) + \frac{q''(z_s)}{2!}(z - z_s)^2 + \ldots \tag{7}$$

Upon substituting (7) into $\exp[\Omega\, q(z)]$ and defining the angle ψ as $\psi = \arg(z - z_s) + (1/2)\arg q''(z_s)$, one finds that $|\exp[\Omega\, q(z)]|$ decreases most rapidly when $\psi = \pm\pi/2$, increases most rapidly when $\psi = 0, \pi$, and is constant when $\psi = \pm\pi/4$, $\pm 3\pi/4$ (on the latter contours, the " level " is constant but the phase changes most rapidly). Since only the lowest order term in $(z - z_s)$ has been retained in (7), this simple disposition of the terrain into two valley regions $(\pi/4 < \psi < 3\pi/4,\ -\pi/4 > \psi > -3\pi/4)$ and two mountain regions $(-\pi/4 < \psi < \pi/4,\ 3\pi/4 < \psi < 5\pi/4)$ of equal angular width, holds only in the immediate vicinity of the saddle point.

For a second order saddle point, $q'(z_s) = 0 = q''(z_s)$ whence

$$q(z) = q(z_s) + \frac{q^{(3)}(z_s)}{3!}(z - z_s)^3 + \ldots \tag{8}$$

In this case there are three mountain and three valley regions emanating from the saddle point, each having an angular width of 60°. Similar considerations apply to saddle point of higher order.

To effect an approximate evaluation of $I(\Omega)$ for large values of Ω by the procedure sketched in the biginning of this section, one seeks to deform the given path $\bar{P}_z$ into a steepest descent path through the saddle point z_s. It may sometimes be necessary to pass over more than one saddle point in order to connect the endpoints of $\bar{P}_z$ to steepest descent paths; under these circumstances, one includes the contribution from each saddle point (unless their levels differ appreciably, in which instance only those giving the largest contribution need be retained). It should be pointed out that, in view of the localization of the dominant contribution to $I(\Omega)$ to the vicinity of the saddle point, it is sufficient to follow the SDP near z_s only; for example, between points z_1 and z_2 for which $\operatorname{Re} q(z_{1,2}) < \operatorname{Re} q(z_s)$. If on the remainder of the path $\operatorname{Re} q(z) \lesssim \operatorname{Re} q(z_{1,2})$, the contribution therefrom is proportional to $\exp\left[\Omega\, q(z_{1,2}) - \Omega\, q(z_s)\right]$, i.e., exponentially small as Ω becomes very large. By the same considerations, one can effect an approximate evaluation of a finite integral between the limits z_1 and z_2, provided that the path can be deformed through a saddle point z_s and that $\operatorname{Re} q(z) < < \operatorname{Re} q(z_s)$ along the entire path.

2. TRANFORMATION OF THE GIVEN INTEGRAL INTO A CANONICAL FORM

From the discussion in Sec. I it is noted that the asymptotic evaluation of the integral $I(\Omega)$ (as $\Omega \to \infty$) by the method of saddle points is intimately connected with the properties of the analytic function $q(z)$, in particular with the nature and disposition of its stationary points. In this section, it will be assumed that the given integration path has been deformed into a steepest (or, less stringently, a rapid) descent path through one or more of the pertinent saddle points of the function $q(z)$, and attention is given to the actual evaluation of $I(\Omega)$. In most cases of interest, the relevant stationary points are isolated, and the asymptotic evaluation of $I(\Omega)$ is effected by treating separately the contribution from each. However, there are occasions when two (or more) saddle points are clustered in close proximity and can no longer be considered in isolation; instead, the combined configuration must be taken into account. Moreover, the analytic properties of $f(z)$ in the integrand of (1) influence the manner of evaluation of the integral. If $f(z)$ possesses singularities near the pertinent saddle points, their effect must likewise be considered.

Since the major contribution arises from the vicinity of each isolated configuration of stationary points, it is desirable to transform the integral in (1) into a " canonical " form wherein the function $q(z)$ is replaced by another function which characterizes the saddle point arrangement at z_s in the simplest manner. Such a function, having the same number and order of saddle points as $q(z)$ (i.e., zeros of $q'(z)$), is a polynomial. The transformation will be phrased in terms of a new variable s and the polynomial denoted by $\tau(s)$:

$$q(z) = \tau(s) \quad ; \tag{9}$$

moreover, it will be convenient to have the point z_s in the complex z-plane correspond to $s = 0$ in the complex s-plane. Upon changing variables from z to s , one obtains from (1)

$$I(\Omega) = \int_P G(s)\, e^{\Omega\tau(s)}\, ds \quad , \tag{10a}$$

where

$$G(s) = f(z)\frac{dz}{ds} \quad , \quad \text{and} \quad \frac{dz}{ds} = \frac{\tau'(s)}{q'(z)} \quad . \tag{10b}$$

P is a steepest descent contour leading to infinity away from $s = 0$ (i.e., $\text{Re}\,\tau(s) < \text{Re}\,\tau(0)$ on P), so that the exponential term decays very rapidly away from the origin. Hence, if G(s) is regular and slowly varying near $s = 0$, one may approximate $I(\Omega)$ in (10a) by its asymptotic representation

$$I(\Omega) \sim G(0) \int_P e^{\Omega\,\tau(s)}\, ds \, , \qquad \Omega \to \infty .^{*} \tag{11}$$

---

* If a cluster of saddle points exists near $s = 0$, G(0) is generally replaced by a suitable measure of G(s) near $s = 0$.[1]

In (11), $I(\Omega)$ is expressed in terms of a canonical integral which can be reduced to known functions for certain $\tau(s)$. (11) represents only a lowest order approximation to $I(\Omega)$; higher order terms, smaller by inverse (generally fractional) powers of Ω, can also be obtained (see reference 1).

If it is assumed that $f(z)$ is regular near z_s, the regularity of $G(s)$ near $s=0$ implies a like behavior for the mapping derivative (dz/ds). Thus, $\tau'(s)$ must be chosen so as to possess at the points s_s zeros of the same order as those of $q'(z)$ at z_s, where the saddle points s_s in the s-plane correspond to z_s in the z-plane. From these remarks one deduces for an isolated first-order saddle point at z_s the transformation

$$q(z) = \tau(s) = q(z_s) - s^2 , \qquad (12)$$

while for an isolated Mth order saddle point,

$$q(z) = \tau(s) = q(z_s) - s^{M+1} . \qquad (13)$$

If two first-order saddle points at z_1 and z_2 are located so near one another that they must be treated jointly, one employs[3]

$$q(z) = \tau(s) = a_0 + \sigma s - \frac{s^3}{3} , \qquad (14)$$

where a_0 and σ are constants and the factor 1/3 in the s^3 term has been included for convenience. (The s^2-term in the polynomial is not required). The two distinct first-order zeros of $\tau'(s)$, $s_{1,2} = \pm\sqrt{\sigma}$, correspond to $z_{1,2}$ and from $q(z_{1,2}) = \tau(\pm\sqrt{\sigma})$, one finds

$$a_0 = \tau(0) = \frac{1}{2}\left[q(z_1) + q(z_2)\right] \quad , \quad \frac{2}{3}\sigma^{3/2} = \frac{1}{2}\left[q(z_1) - q(z_2)\right]. \qquad (14a)$$

For more general saddle point configurations, the polynomial $\tau(s)$ takes on a more complicated form. However, the above-listed examples cover most of the cases arising in connection with diffraction problems.

The canonical integral in (11) can be evaluated for each of the special forms in (12)-(14). For (12) and (13), the integral is expressed in terms of the gamma function, while (14) leads to Airy functions.**

So far, it has been assumed that $f(z)$ has no singularities near the pertinent saddle point(s). If such singularities do exist, their effect has to be considered explicitly[4]. For example, if $G(s)$ has at $s = b$ (b small) a pole of order N, then one may represent

$$G(s) = \frac{a_{-N}}{(s-b)^N} + \frac{a_{-(N-1)}}{(s-b)^{N-1}} + \ldots + \frac{a_{-1}}{(s-b)} + T(s) , \qquad (15)$$

where $T(s)$ is regular at $s = b$ and at $s = 0$. The contribution to $I(\Omega)$ from $T(s)$ can be treated as before. Inclusion of the

---

* Conditions on $|\sigma|$, which allow the two saddle points to be treated as isolated, are discussed in Sec. IV.

** For three equally spaced, colinear saddle points, the canonical integral (11) is expressible in terms of parabolic cylinder functions.

remaining terms in (15) requires evaluation of a new class of canonical integrals

$$A_N(\Omega) = \int_P (s-b)^{-N} e^{\Omega\tau(s)} ds, \quad N=1,2... \tag{16}$$

For a first-order saddle point, with $\tau(s)$ given by (12), the integrals in (16) can be evaluated in terms of the error function. If $G(s)$ also has a branch point singularity at $s=0$ so that the expression on the right-hand side of (15) is multiplied by s^{β}, $\beta > -1$, then with $\tau(s)$ as in (12), the integrals (16) can be expressed in terms of Whitaker's confluent hypergeometric function[11].

3. INTEGRANDS WITH ISOLATED SADDLE POINTS

A. First-order Saddle Points

1. First-order Approximation

We now consider in detail the first-order asymptotic approximation of the integral in (1), as well as its complete asymptotic expansion, for the case where $q(z)$ has one relevant first-order saddle point at z_s and $f(z)$ has no singularities near z_s. The pertinent change of variable from the z-plane to the s-plane is that given in (12) and the steepest descent path P in the s-plane, along which Im $\tau(s)$ = constant, is clearly the real s-axis. Thus, the integral in (10a) is in this case

$$I(\Omega) = e^{\Omega q(z_s)} \int_{-\infty}^{+\infty} G(s)\, e^{-\Omega s^2}\, ds\,, \tag{17}$$

with G(s) given by (10b), and

$$\frac{dz}{ds} = \frac{-2s}{q'(z)}\,. \tag{17a}$$

Since G(s) is assumed regular near $s = 0$, it can be expanded into a power series

$$G(s) = G(0) + G'(0)\,s + G''(0)\left(\frac{s^2}{2!}\right) + \ldots + G^{(n)}(0)\left(\frac{s^n}{n!}\right) + \ldots, \tag{18}$$

which converges uniformly inside a circle with finite radius r certered at $s = 0$, r being the distance to the nearest singularity of G(s). Upon applying L'Hopital's rule to the indeterminate form for (dz/ds) in (17a) when $s = 0$, i.e., $z = z_s$, one evaluates the first coefficient of the expansion as

$$G(0) = f(z_s)\left(\frac{dz}{ds}\right)_{s=0},\quad \left(\frac{dz}{ds}\right)_{s=0} = \pm\sqrt{\frac{-2}{q''(z_s)}}\,, \tag{18a}$$

where $q''(z_s) \neq 0$ at the first-order saddle point. The choice of sign of $(dz/ds)_{s=0}$ depends upon the direction of integration along the steepest descent path through z_s in the z-plane. (Since ds is positive along the path of integration and, in particular, at $s = 0$, arg(dz/ds) at $s = 0$ is equal to arg(dz) at z_s along the steepest descent path.) If G(s) is approximated by G(0) only, one obtains as the first-order asymptotic approximation of $I(\Omega)$ in (17) :

$$I(\Omega) \sim G(0)\, e^{\Omega q(z_s)} \int_{-\infty}^{\infty} e^{-\Omega s^2}\, ds \,. \qquad (19)$$

Since the integral in (19) is equal to $\sqrt{\pi/\Omega}$, one obtains from (19) and (18a)

$$I(\Omega) \sim \pm \sqrt{\frac{-2\pi}{\Omega q''(z_s)}}\, f(z_s)\, e^{\Omega q(z_s)}, \quad \Omega \to \infty. \qquad (20)$$

2. Complete Asymptotic Expansion

A complete asymptotic expansion for $I(\Omega)$ in (17) is obtained upon substituting for $G(s)$ the power series expansion in (18) and integrating term by term:+

$$I(\Omega) \sim e^{\Omega q(z_s)} \sum_{n=0}^{\infty} \frac{G^{(n)}(0)}{n!} I_n(\Omega) \,. \qquad (21)$$

+ This manipulation is not rigorously justifiable since the radius of convergence r of the power series expansion is generally finite so that the series representation for $G(s)$ cannot be employed over the infinite range in s . However, the error incurred by this procedure (as $\Omega \to \infty$) is exponentially small, thereby implying the " asymptotic " validity of the resulting expansion (see ref. 1 and 2; also G.N.Watson, Proc. London Math. Soc. (2), 17 (1918)),

The integral $I_n(\Omega)$ can be evaluated in terms of the gamma function $\Gamma(z)$

$$I_n(\Omega) = \int_{-\infty}^{+\infty} s^n e^{-\Omega s^2} ds = \frac{\Gamma(\frac{1+n}{2})}{\Omega^{(1+n)/2}} , \text{ n even} \tag{22a}$$

$$= 0 , \text{ n odd} , \tag{22b}$$

where, in view of the symmetrical integration interval, (22b) results from the fact that the integrand is an odd function of s . The values of $\Gamma(n+\frac{1}{2})$, $n=0,1,2,\ldots$, are readily inferred from the recursion formula

$$\Gamma(z+1) = z\,\Gamma(z) , \quad \Gamma(\tfrac{1}{2}) = \sqrt{\pi} . \tag{23}$$

Alternatively, one may express $I_n(\Omega)$ as

$$I_n(\Omega) = (-1)^{n/2} \frac{d^{n/2}}{d\,\Omega^{n/2}} \int_{-\infty}^{\infty} e^{-\Omega s^2} ds \tag{24a}$$

$$= \left(-\frac{d}{d\Omega}\right)^{n/2} \sqrt{\frac{\pi}{\Omega}} , \text{ n even} . \tag{24b}$$

(24b) gives rise to a recursion relation between I_n and I_{n+2} , evident from (22a)

$$I_{n+2}(\Omega) = \left(-\frac{d}{d\Omega}\right) I_n(\Omega) , \quad n=0,2,4\ldots \tag{25}$$

Thus, the complete asymptotic expansion for $I(\Omega)$, as $\Omega \to \infty$, is given by :

$$I(\Omega) \sim \frac{e^{\Omega q(z_s)}}{\sqrt{\Omega}} \sum_{n=0}^{\infty} \frac{G^{(2n)}(0)}{(2n)!} \frac{\Gamma(n+1/2)}{\Omega^n} , \tag{26}$$

or, alternatively,

$$I(\Omega) \sim e^{\Omega q(z_s)} \sum_{n=0}^{\infty} \frac{G^{(2n)}(0)}{(2n)!} \left(-\frac{d}{d\Omega}\right)^n \sqrt{\frac{\pi}{\Omega}} . \tag{27}$$

(27) can be written in a convenient operator notation as

$$I(\Omega) \sim e^{\Omega q(z_s)} G_e\left(\sqrt{-\frac{d}{d\Omega}}\right) \sqrt{\frac{\pi}{\Omega}} , \tag{27a}$$

where the representation of the even function $G_e(x)$ in terms of a power series about $x = 0$ is as follows :

$$G_e(x) = \sum_{n=0}^{\infty} \frac{G_e^{(2n)}(0)}{(2n)!} x^{2n} , \tag{27b}$$

whence the series representation for (27a) is that in (27). That Eqs. (26) or (27) indeed constitute the asymptotic expansion of $I(\Omega)$ as $\Omega \to \infty$ follows from the recognition that the ratio between successive terms of the series, i.e., between the $(N+1)$th and Nth terms , approaches zero as

$\Omega \to \infty$, for any N.*

As a general comment on asymptotic expansions it should be pointed out that a formal asymptotic series as in (27) may diverge for a fixed value of Ω, as $N \to \infty$. The asymptotic expansion may nevertheless be employed to compute the numerical value of $I(\Omega)$ for a given Ω since the first terms in the expansion decrease in magnitude. In general, the series should be broken off at the smallest contributing term, for a given value of Ω.

One notes that the lowest order approximation to $I(\Omega)$ arising from the $n = 0$ term in (26) or (27) is given by (20) as required. For the evaluation of the higher order terms one must know the higher order derivatives of $G(s) = f(z)\,dz/ds$, evaluated at $s = 0$. Formally, the derivatives

---

* A function $I(\Omega)$ is said to have an asymptotic expansion

$$I(\Omega) \sim \sum_{n=0}^{\infty} a_n f_n(\Omega), \quad \text{as} \quad \Omega \to \infty,$$

if, for any N and for $\arg \Omega$ in a given integral,

$$\frac{I(\Omega) - I_N(\Omega)}{f_N(\Omega)} \to 0 \quad \text{as} \quad \Omega \to \infty,$$

where $I_N(\Omega) = \sum_{n=0}^{N} a_n f_n(\Omega)$. For details on general properties of asymptotic expansions see references 2.

$(d^n z/ds^n)$ can be obtained by successive differentiation of (10b) and evaluation of the resulting indeterminate form at $s=0$. An alternative procedure is to expand $q(z)$ in (12) in a power series about $z=z_s$ to obtain the dependence of s as a function of $(z-z_s)$. To obtain the required behavior of $(z-z_s)$ as a function of s, this power series must be inverted. (See ref. 10 for inversion of series).

Although it has been assumed throughout that Ω is real, it is evident that $I_n(\Omega)$ as defined in (22a) can be continued analytically into the range $|\arg \Omega| < \pi/2$, since the integral converges as well in this extended range of Ω for which $\mathrm{Re}(\Omega s^2) > 0$. This process of analytic continuation is frequently more convenient that the inclusion of the factor $\exp(i \arg \Omega)$ in $q(z)$, if Ω is complex (as mentioned in the beginning of this Chapter). Thus, the asymptotic expansions in (26) or (27) are valid also for complex Ω, in the range $|\arg \Omega| < \pi/2$.

<u>Example</u>

Consider the integral

$$I_1(\Omega, \alpha, \beta) = \int_{\bar{P}_z} \frac{e^{i\Omega \cos(z-\alpha)}}{z-\beta}\, dz, \quad 0 \leqslant \alpha < \frac{\pi}{2}, \tag{28}$$

taken over the path $\bar{P}_z$ shown in Fig.2. This integral is typical of those which arise in several of the excitation and diffraction problems studied in this course. Since

$$\mathrm{Im} \cos(z-\alpha) = -\sin(x-\alpha)\sinh y, \tag{29}$$

and Ω is assumed positive, one notes that the exponential term decays in the regions $y > 0$, $-\pi < (x-\alpha) < 0$, and $y < 0$, $0 < (x-\alpha) < \pi$, which are shown shaded in Fig.2. The integrand contains a simple pole at $z = \beta$, where β is arbitrary. Upon comparison with (1) one notes that

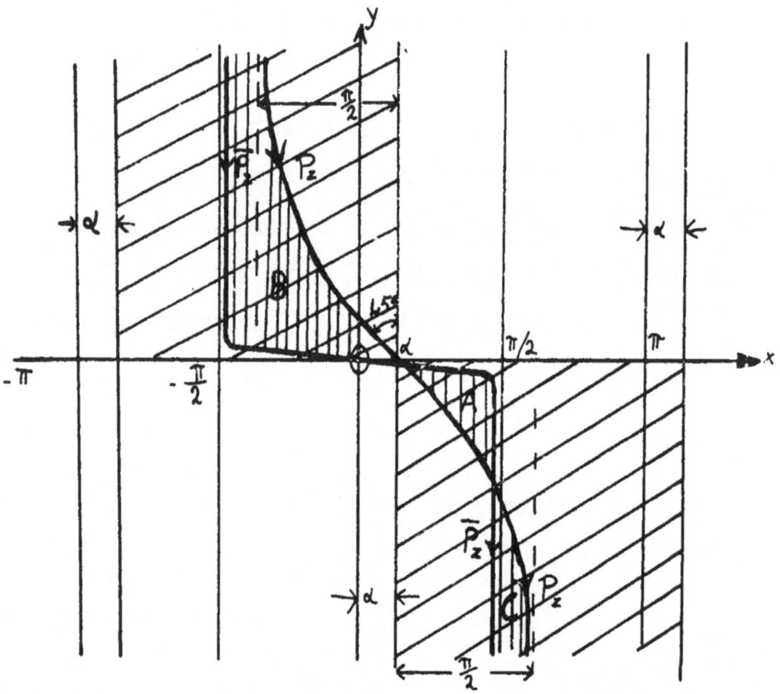

Fig.2 - Contours of integration in z-plane (z=x+iy)

$$q(z) = i\cos(z-\alpha), \quad f(z) = \left(\frac{1}{z-\beta}\right). \tag{30}$$

The pertinent saddle point z_s in the z-plane is located at

$$\frac{d}{dz} q(z) = - i \sin(z - \alpha) = 0 \text{, i.e. at } z_s = \alpha . \qquad (30a)$$

The steepest descent path P_z through the saddle point is defined by

$$\mathrm{Im}\, q(z) = \mathrm{Im}\, q(z_s) = i . \qquad (30b)$$

The change of variable to the s-plane is then effected via (12) by

$$i \cos(z - \alpha) = i - s^2 , \qquad (31a)$$

or

$$s = \pm \sqrt{2}\, e^{i\pi/4} \sin\left(\frac{z-\alpha}{2}\right) . \qquad (31b)$$

Although we could proceed immediately to the s-plane via the transformation (31), we study first the nature of the steepest descent path P_z in the z-plane. Along P_z, s is real, and the slope of P_z at $z = z_s = \alpha$ is inferred from (18a) or (31b) to be

$$\left.\frac{dz}{ds}\right|_{s=0} = \pm \sqrt{2}\, e^{-i\pi/4} . \qquad (32)$$

Thus, P_z makes an angle of (-45°) with the x axis at $z = \alpha$. The direction of integration along $\bar{P}_z$ in Fig.2 suggests a choice of the indicated direction along P_z, so that $\arg(z-z_s) = -\pi/4$ along P_z near $z = z_s$, and the positive sign is chosen in Eqs. (32) and (31). The complete steepest descent path can be plotted readily from Eq. (30b) which requires that $\mathrm{Im}\left[i \cos(z-\alpha)\right] = i$ along P_z, or

$$x - \alpha = \cos^{-1}(\text{sech } y) \quad \text{along } P_z \; . \qquad (33)$$

(33) yields the x-coordinate of any point on P_z for an assumed value of y. The resulting path is shown in Fig.2 and has as its asymptotes the lines $x = \alpha \pm \pi/2$.

The transformation of the path $\bar{P}_z$ from the z-plane to the s-plane is accomplished via Eq. (31b), where it is recalled that the positive sign is chosen. The resulting contour $\bar{P}$ is shown in Fig.3 and has as its asymptotes in the lower half of the s-plane the lines $\arg s = -\alpha/2$ and $\arg s = \pi + (\alpha/2)$. Corresponding regions in the z- and s-plane, where the exponential term decays in magnitude, are shown shaded with slanted lines. Since $(ds/dz) = 0$ at $(z-\alpha) = \pm\pi$, the transformation in (31b) gives rise to first-order branch point singularities at $s = \pm s_b = \pm\sqrt{2}\exp(i\pi/4)$ in the s-plane. These branch points, and the associated choice of branch cuts, are also shown in Fig.3. The steepest descent path P in the s-plane, corresponding to the path P_z in the z-plane, extends along the real s-axis.

Since all paths considered begin and terminate in a shaded region of the z- or s-plane it is evident that the contours $\bar{P}_z$ and $\bar{P}$ can be deformed at infinity into the contours P_z and P, respectively. Concerning the deformation of the paths in the remainder of the z- or s-planes, attention must be given to the location of the pole singularity at $z = \beta$ in (28). If $z = \beta$ (or $s = s_\beta = \sqrt{2}\, e^{i\pi/4} \sin\frac{\beta-\alpha}{2}$) is situated in any of the vertically shaded regions A, B, C in Fig.2 (or Fig.3), the residue at the pole must be taken into account in

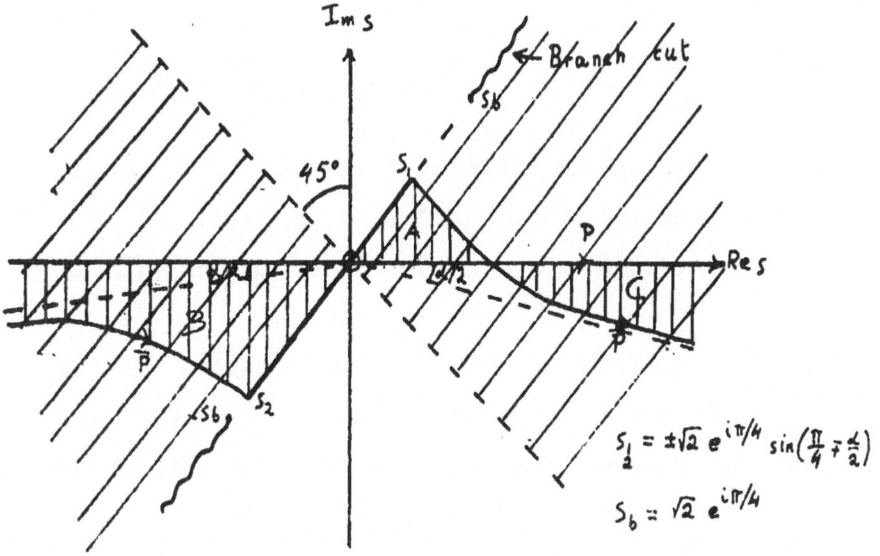

Fig.3 - Contours of integration in s-plane

the contour deformation. Thus,

$$I_1(\Omega, \alpha, \beta) = 2\pi i \; e^{i\Omega\cos(\beta-\alpha)}\,\varepsilon(\beta) + e^{i\Omega}\int_{-\infty}^{\infty} G(s)\, e^{-\Omega s^2}\, ds, \quad (34)$$

where

$$\varepsilon(\beta) = \begin{cases} +1, & \text{if } \beta \text{ (or } s_\beta) \text{ lies in regions B or C} \\ -1, & \text{if } \beta \text{ (or } s_\beta) \text{ lies in region A} \\ 0, & \text{if } \beta \text{ (or } s_\beta) \text{ lies outside regions A,B,C.} \end{cases} \quad (34a)$$

In view of the change of variable in (31), G(s) is given by

$$G(s) = \frac{1}{z-\beta}\frac{dz}{ds}, \quad \frac{dz}{ds} = \frac{-2is}{\sin(z-\alpha)} = \frac{-2is}{\sqrt{1-\cos^2(z-\alpha)}}. \quad (34b)$$

The simple form of (dz/ds) permits a direct determination of the complete series expansion by the binomial theorem

$$\frac{dz}{ds} = \sqrt{2}\, e^{-i\pi/4} \left(1+\frac{is^2}{2}\right)^{-1/2}$$

$$= \sqrt{2}\, e^{-i\pi/4}\left(1-\frac{is^2}{4}-\frac{3}{32}\, s^4+\ldots\right) , \qquad (34c)$$

which converges in the interior of a circle of radius $|s| = \sqrt{2}$ passing through the branch point singularities. Thus, the range of convergence of the power series expansion for $G(s)$ is $|s| < \sqrt{2}$ if $|s_\beta| > \sqrt{2}$ and is $|s| < |s_\beta|$ if $|s_\beta| < \sqrt{2}$, i.e., within a circle of finite radius, provided $|s_\beta| > 0$. The first two terms in the asymptotic expansion of $I_1(\Omega, \alpha, \beta)$ in (28) (as $\Omega \to \infty$) are therefore given via (34) and (26) by

$$I_1(\Omega, \alpha, \beta) \sim 2\pi i\, e^{i\Omega\cos(\beta-\alpha)}\varepsilon(\beta) + \frac{e^{i(\Omega-\frac{\pi}{4})}}{\alpha-\beta}\sqrt{\frac{2\pi}{\Omega}}\left\{1-\frac{i}{4\Omega}\left[\frac{4}{(\alpha-\beta)^2}+\frac{1}{2}\right]+\ldots\right\}, \qquad (35)$$

Concerning the residue contribution from the pole to Eq.(35), one notes that the magnitude of the exponential term behaves like $\exp\left[-\Omega\left|\sin(\beta_r-\alpha)\sinh\beta_i\right|\right]$, where β_r and β_i are the real and imaginary parts of β, respectively. If $\beta_i \neq 0$ and $\beta_r \neq \alpha$, the pole contribution is exponentially small and can be neglected in comparison with the remaining terms. On the other hand, if $\beta_i \to 0$ or if $\beta_r \to \alpha$, the residue contribution may be the dominant one since its magnitude then remains constant as $\Omega \to \infty$.

Although the asymptotic representation in (35) remains valid (as $\Omega \to \infty$) for any $\beta \neq \alpha$, the proximity of the pole near the saddle point ($\beta \approx \alpha$) is seen to influence the accuracy of any numerical evaluation for large fixed values of Ω . In particular, if the pole approaches the saddle point ($(\alpha - \beta) \to 0$) , the utility of the representation in (35) ceases since the error incurred by use of even the first term becomes large. (This implies that the radius of convergence of the series representation for G(s) shrinks to zero). One observes in this connection that the quantity $\sqrt{\Omega}\ |\alpha-\beta|$ plays a special role in assessing whether the pole is near enough to the saddle point to invalidate (35). If for $\Omega \gg 1$, one also has $\sqrt{\Omega}\ |\alpha-\beta| \gg 1$, the pole can be considered far from the saddle point and (35) applies; on the other hand, if $\sqrt{\Omega}\ |\alpha-\beta| \lesssim 1$, with $\Omega \gg 1$, the terms in the asymptotic series are no longer small and the validity of the expansion is in question. The dependence of the asymptotic expansion of $I_1(\Omega,\alpha,\beta)$ on the magnitude of $\sqrt{\Omega}\ |\alpha-\beta|$ is studied in detail in Section V, wherein is treated the evaluation of integrals whose integrands contain a pole near a saddle point.

B. Saddle Points of Higher Order

If q(z) in (1) has one pertinent M-th order saddle point at z_s , i.e., $q^{(n)}(z_s) = 0$, $n = 1 \ldots M$, $q^{(M+1)}(z_s) \neq 0$, the appropriate transformation to the s-plane is given by (13):

$$q(z) = \tau(s) = q(z_s) - s^{M+1} \quad . \tag{36}$$

The requirement $\operatorname{Im} q(z) = \text{constant}$, i.e., s^{M+1} positive along the steepest descent paths, leads to the following possible contours in the s-plane ,

$$\arg s = 2\,\mu\pi/(M+1)\ ,\ \mu = 0,1,2\ldots M\ . \qquad (37)$$

Thus, the (M+1) steepest descent paths originating at the saddle point z_s in the z-plane map in the s-plane into the (M+1) straight lines defined in Eq.(37), which originate at $s=0$ and extend to $|s| = \infty$. Since s^{M+1} is positive along any of the paths defined in Eq.(37), we may consider without loss of generality the case $\mu = 0$. Let the subscript $_0$ denote the contribution to $I(\Omega)$ in Eq.(1) from a steepest descent path which maps into the positive real s-axis. Then the integrals $I_n(\Omega)$ in (21) are given in view of (36) by:

$$I_n(\Omega) = \int_0^{\infty} s^n\, e^{-\Omega s^{M+1}}\, ds = \qquad (38a)$$

$$= \Gamma\left(\frac{1+n}{M+1}\right) \frac{\Omega^{-(1+n)/(M+1)}}{M+1}\ , \qquad (38b)$$

where the integral in (38a) has been evaluated in terms of the gamma function. The asymptotic nature of the resulting expansion in (21) as $\Omega \to \infty$ follows by the same considerations as before.

Upon expanding $q(z)$ in (36) in a power series about the point z_s, one obtains directly

$$\left.\frac{dz}{ds}\right|_{s=0} = \rho\left[\frac{-(M+1)!}{q^{(M+1)}(z_s)}\right]^{\frac{1}{M+1}}\ ,\quad \rho = e^{i2\pi\mu/(M+1)}\ , \qquad (39)$$

where ρ is the appropriate (M+1)-th root of unity whose choice is determined by the given steepest descent path (see (37)), and where the principal value of the (M+1)-th root is taken in the remaining term. From (21) one obtains the following first-order approximation to the integral $\bar{I}_0(\Omega)$ ($\mu = 0$):

$$\bar{I}_0(\Omega) \equiv \int_0^\infty G(s)\, e^{\Omega\tau(s)} ds \sim \left[\frac{-(M+1)!}{q^{(M+1)}(z_s)}\right]^{\frac{1}{M+1}} f(z_s)\, e^{\Omega q(z_s)} \frac{\Gamma(\frac{1}{M+1})}{(M+1)\Omega^{\frac{1}{M+1}}}, \quad (40)$$

where it is recalled that $G(s) = f(z)(dz/ds)$. When $M = 1$, one recovers from Eq.(40) the previously derived result in (20), save for a factor of 2 which arises since the interval of integration in (19) extends from $s = -\infty$ to $s = \infty$.

4. INTEGRANDS WITH TWO ADJACENT FIRST-ORDER SADDLE POINTS

If $q(z)$ in (1) has two pertinent first-order saddle points which approach each other very closely, the numerical evaluation of the integral along a path through one saddle point is markedly influenced by the presence of the other. In this case one seeks to approximate $I(\Omega)$ by a known standard integral whose integrand also contains two saddle points[3]. An appropriate class of standard integrals are the Airy integrals $Ai(\sigma)$ and $Bi(\sigma)$[8] which are defined as follows:

$$Ai(\sigma) = \frac{1}{2\pi i} \int_{L_{32}} e^{\sigma s - s^3/3}\, ds, \quad (41a)$$

$$Bi(\sigma) = \frac{1}{2\pi} \int_{L_{21}+L_{31}} e^{\sigma s - s^3/3} ds , \qquad (41b)$$

where L_{ij}, the contours of integration in the complex s-plane, are shown in Fig.4. Each contour begins and ends at $|s| = \infty$ in a shaded region, $\text{Re}\, s^3 > 0$, wherein convergence of the integrals is assured. The exponent in the integrands of Eqs.(41) manifestly has two saddle points located at $s_{1,2} = \pm\sqrt{\sigma}$. Thus, σ is a measure of the distance separating the two saddle points.

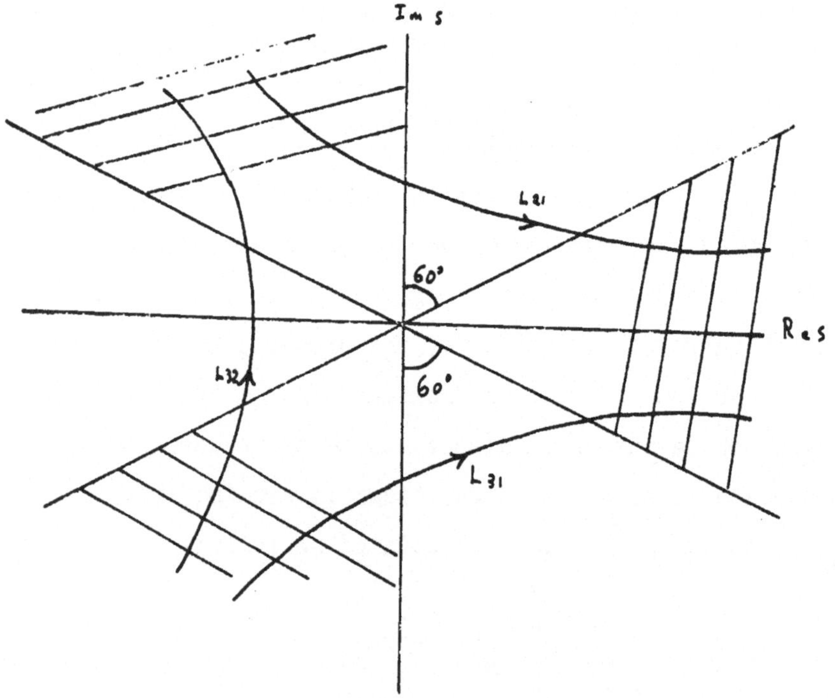

Fig.4 - Airy integral paths.

Let us return to a consideration of the integral in (1), when the derivative $q'(z)$ has two simple neighboring zeros at $z_s = z_{1,2}$ and $f(z)$ has no singularities near $z_{1,2}$. As discussed in Sec. I, we wish to transform the region containing the two saddle points $z_{1,2}$ into the vicinity of the origin in the complex s-plane. The required transformation has been given in Eq.(14), with the constant a_0 and σ defined in (14a). (Special attention must be given to the choice of the proper branch of $(\sigma^{3/2})^{2/3}$, as required for the evaluation of σ from (14a)).

Upon substituting (14) into (11) one obtains the first-order asymptotic approximation for $I(\Omega)$ in Eq.(1) as $\Omega \to \infty$, valid for small values of σ :

$$I(\Omega) \sim \Omega^{-1/3} G(0)\, e^{a_0 \Omega}\, C(\sigma \Omega^{2/3}) \; , \tag{42}$$

with a_0 and σ defined as in (14a) and

$$G(0) = \left[f(z) \left(\frac{dz}{ds}\right) \right]_{s=0} \; , \tag{42a}$$

$$C(\rho) = \int_P e^{\rho t - t^3/3}\, dt \; . \tag{42b}$$

See footnote associated with Eq.(11). For arbitrary σ , G(0) in (42) must be replaced by $(1/2)\left[G(\sqrt{\sigma}) + G(-\sqrt{\sigma})\right]$ (see ref.1,3).

The integral in (42b) is readily identified in terms of the Airy integrals in (41), for any specified allowable path P in Fig.4. Since $\int_{L_{21}} - \int_{L_{31}} + \int_{L_{32}} = 0$ one notes that

$$\mathcal{O}(\rho) = \begin{cases} 2\pi i \, Ai(\rho) & , \text{ if } P = L_{32} \\ \pi \left[Bi(\rho) - i \, Ai(\rho) \right] & , \text{ if } P = L_{21} \, . \\ \pi \left[Bi(\rho) + i \, Ai(\rho) \right] & , \text{ if } P = L_{31} \end{cases} \tag{43}$$

Eq.(42) is valid for small values of σ, i.e., for $z_1 \approx z_2$. In this case, since f(z) and (dz/ds) are assumed to be regular and slowly varying functions in the vicinity of $s = 0$, one may write approximately

$$f(z)\Big|_{s=0} \approx f(z_1) \approx f(z_2) \quad , \tag{44a}$$

and

$$\left(\frac{dz}{ds}\right)_{s=0} \approx \left(\frac{dz}{ds}\right)_{s=\sqrt{\sigma}} \approx \left(\frac{dz}{ds}\right)_{s=-\sqrt{\sigma}} . \tag{44b}$$

From (14), one finds readily that (see also (18a))

$$\left(\frac{dz}{ds}\right)_{s=\sqrt{\sigma}} = \pm \, \sigma^{1/4} \sqrt{\frac{-2}{q''(z_1)}} \, , \quad \left(\frac{dz}{ds}\right)_{s=-\sqrt{\sigma}} = \pm \, \sigma^{1/4} \sqrt{\frac{2}{q''(z_2)}} \, , \tag{44c}$$

where the choice of sign depends as in (18a) on the direction of integration along the path. The derivative $(dz/ds)_{s=0}$ must have a unique value, implying that $q''(z_2) \not\approx -q''(z_1)$ for $z_1 \approx z_2$, i.e., $\sigma \approx 0$; it then follows that $q''(z_2) = q''(z_1) = 0$ if $z_1 = z_2$.

If the two first-order saddle points $z_{1,2}$ coincide, the point $z_1 = z_2$ is a second-order saddle point since both $q'(z_s)$ and $q''(z_s)$ vanish. Thus, in (14), $\sigma = 0$, $\tau'(0) = \tau''(0) = 0$, and the origin in the s-plane is likewise a second order saddle point. To evaluate the indeterminate form as $\sigma \to 0$, write (44c) as

$$\left(\frac{dz}{ds}\right)^2_{s=\sqrt{\sigma}} = \left.\frac{-2s}{q''(z)}\right|_{\substack{s=\sqrt{\sigma}\\ z=z_1}} \tag{45}$$

As $\sigma \to 0$, $q''(z_1) \to 0$. Hence, by L'Hopital's rule,

$$\left(\frac{dz}{ds}\right)^2_{s=0} = \left.\frac{-2}{q^{(3)}(z_1)\,\frac{dz}{ds}}\right|_{\substack{s=0\\ z=z_1}} \tag{46}$$

whence

$$\left(\frac{dz}{ds}\right)_{s=0} = \pm\left[\frac{-2}{q^{(3)}(z_s)}\right]^{1/3} \quad \text{when} \quad z_1 = z_2 = z_s \tag{47}$$

(47) contains an ambiguity regarding the proper choice of the cube root. Its resolution depends on the integration path in the problem under consideration and is demonstrated in an example at the end of this section. The limiting form of (42) for $\sigma = 0$ is now given by

$$I(\Omega) \sim \pm \left[\frac{-2}{\Omega\, q^{(3)}(z_s)} \right]^{1/3} f(z_s)\, e^{\Omega q(z_s)}\, C(0) \,, \tag{48}$$

where in view of (38) and (42b)

$$C(0) = \begin{cases} i\, 3^{-1/6}\, \Gamma(\frac{1}{3}) \,, & \text{if } P = L_{32} \\ e^{-i\pi/6}\, 3^{-1/6}\, \Gamma(\frac{1}{3}), & \text{if } P = L_{21} \\ e^{i\pi/6}\, 3^{-1/6}\, \Gamma(\frac{1}{3}), & \text{if } P = L_{31} \end{cases} \,. \tag{48a}$$

One verifies readily that (48) agrees with the second-order saddle point result in (40) (with $M = 2$) provided that the path P for C(0) is taken from $s = 0$ to $s = \infty$ only. In this case, the value for C(0) as obtained from (38) is $3^{-2/3}\,\Gamma(1/3)$; The expressions in (48a) are given in terms of this result as follows :

$$\int_{L_{32}} e^{-s^3/3}\, ds = \int_{\infty e^{-i2\pi/3}}^{0} + \int_{0}^{\infty e^{i2\pi/3}} = (e^{i2\pi/3} - e^{-i2\pi/3}) \int_0^{\infty} , \tag{49}$$

etc.

When $\sigma\,\Omega^{2/3}$ in (42) becomes large one may use the asymptotic representation for the Airy-type function $C(\sigma\Omega^{2/3})$. Since $\Omega \to \infty$, one notes that $\sigma\,\Omega^{2/3} \to \infty$ if σ is proportional to $\Omega^{-2/3+\alpha}$, where $\alpha > 0$ but may be small.

As will be seen below, the use of the asymptotic representation for $C(\sigma\Omega^{2/3})$ reduces (42) **approximately** to the result obtained when each saddle point is treated separately. Thus, the double saddle point procedure is required only when $\sigma = 0(\Omega^{-2/3})$, i.e., very small. To be specific, let us assume that the path P in Eq.(42) is the same as L_{32} in Fig.4. Then, from (43) and the asymptotic representation **(see Ref.2c)**

$$Ai(\sigma) \sim \frac{1}{2\sqrt{\pi}\,\sigma^{1/4}} e^{-(2/3)\sigma^{3/2}}, \quad \sigma \to \infty, \quad |\arg \sigma| < \frac{\pi}{3}, \tag{50}$$

one obtains

$$C(\sigma\Omega^{2/3}) = 2\pi i\, Ai(\sigma\Omega^{2/3}) \approx \frac{i\sqrt{\pi}}{\sigma^{1/4}\Omega^{1/6}} e^{-\frac{2}{3}\Omega\sigma^{3/2}}, \quad |\arg(\sigma\Omega^{2/3})| < \frac{\pi}{3}, \tag{51}$$

as $\sigma\Omega^{2/3} \to \infty$. Since (51) is valid for small values of σ, i.e., $z_1 \approx z_2$, Eqs.(44) still apply and substitution of (51) into (42) yields for σ small but $\sigma\Omega^{2/3} \to \infty$ as $\Omega \to \infty$,

$$I(\Omega) \sim \pm f(z_2)\sqrt{\frac{-2\pi}{\Omega q''(z_2)}}\, e^{\Omega q(z_2)}, \tag{52}$$

which is identical with the saddle point result in (20). (52) involves the saddle point at $s_2 = -\sqrt{\sigma}$ which is evidently the pertinent one for the contour L_{32} in Fig.4 when Ω is real and $|\arg \sigma| < \pi/3$. To achieve the form in (20), the quantities in Eqs.(44) have been evaluated at z_2 and s_2.

Similar results are obtainable when the original integration path is deformable into any of the other contours in Fig.4.

For the complete asymptotic expansion of $I(\Omega)$ in (1), with $q(z)$ given by (14), the reader is referred to reference 3.

Example

Consider the Hankel function $H_\nu^{(1)}(\Omega)$ defined by the integral[9]

$$H_\nu^{(1)}(\Omega) = \frac{1}{\pi} \int_{\bar{P}_z} e^{\Omega q(z)}\, dz \,, \qquad (\nu, \Omega \text{ positive}), \tag{53}$$

$$q(z) = i\left[\cos z + \left(z - \frac{\pi}{2}\right)\sin\alpha\right], \quad \sin\alpha = \frac{\nu}{\Omega}, \tag{53a}$$

where the path $\bar{P}_z$ is the same as in Fig.2. The saddle points of $q(z)$ are located at

$$q'(z) = -i\left[\sin z - \sin\alpha\right] = 0\,, \tag{54a}$$

which has pertinent solutions at

$$z_1 = \alpha\,, \quad z_2 = \pi - \alpha\,. \tag{54b}$$

We assume that $(\nu/\Omega) \lesssim 1$ so that $0 < \alpha \lesssim \pi/2$, and seek a first-order asymptotic evaluation of the integral in (53) as $\Omega \to \infty$ and $\alpha \to \pi/2$, i.e., when the order and argument of the Hankel function are both large and almost equal.

From (53a) and (54b),

$$q(z_{\frac{1}{2}}) = \pm i \left[\cos\alpha - \left(\frac{\pi}{2} - \alpha\right)\sin\alpha\right] \quad . \tag{55}$$

Since $q(z_{\frac{1}{2}})$ is imaginary, $\sigma^{3/2}$ as defined in (14a) is imaginary and can be satisfied by $\sigma < 0$. Let us introduce

$$\eta = -\sigma > 0 \ , \tag{56a}$$

and choose

$$s_{\frac{1}{2}} = \pm\sqrt{\sigma} = \pm i\sqrt{\eta} \quad , \quad \sqrt{\eta} > 0 \ . \tag{56b}$$

Then from (14)

$$\frac{2}{3}\eta^{3/2} = \cos\alpha - \left(\frac{\pi}{2} - \alpha\right)\sin\alpha \geqslant 0 \ , \tag{57a}$$

$$\tau(0) = 0 \ , \tag{57b}$$

while from (44c)

$$\left(\frac{dz}{ds}\right)_{s_1} = \left(\frac{dz}{ds}\right)_{s_2} = \pm i\,\eta^{1/4}\left(\frac{2}{\cos\alpha}\right)^{1/2} \quad . \tag{57c}$$

Since $f(z) = 1$ in (53) and in view of (57c), one notes that (44a) and (44b) are satisfied exactly in this case. Near $\alpha = \pi/2$, one obtains from (57a) and the requirement that $\sqrt{\eta} \geqslant 0$,

$$\frac{2}{3}\eta^{3/2} \approx \frac{1}{3}\left(\frac{\pi}{2} - \alpha\right)^3, \text{ i.e., } \eta^{1/2} \approx 2^{-1/3}\left(\frac{\pi}{2} - \alpha\right) . \tag{58a}$$

Thus,

$$\left(\frac{dz}{ds}\right)_{s_1} = \left(\frac{dz}{ds}\right)_{s_2} = \pm\ i\ 2^{1/3} \quad , \text{ when } \eta = 0 \tag{58b}$$

(58b) could also have been obtained directly from (47) with $z_1 = z_2 = \pi/2$, with the cube root so chosen that $\sigma^{1/2}$ is imaginary (see (56b)). From (14a) and the power series expansion of $q(z_1)$ about z_2, one finds for $q'(z_2) = q''(z_2) = 0$ that $\sigma^{1/2} \sim (z_1 - z_2)\left[q^{(3)}(z_2)\right]^{1/3}$. Since $q^{(3)}(z_2) = i$ and $(z_1 - z_2)$ is real, this requirement implies that the principal value of the cube root in (47) and (48) is multiplied in this case by $\exp(i\,4\pi/3)$.

To determine the contour of integration in the s-plane, it suffices to consider the transformation in (14) as $\sigma \to 0$ (i.e., $\eta \to 0$) in order to establish the location of the endpoints of the transformed path. Thus, we examine

$$q(z) = i\left[\cos z - \left(\frac{\pi}{2} - z\right)\right] = -\frac{s^3}{3}\ , \quad (\eta = 0)\ , \tag{59}$$

which can be written near $z = \pi/2$ as

$$\frac{i}{6}\left(\frac{\pi}{2} - z\right)^3 \approx \frac{s^3}{3}\ . \tag{59a}$$

Upon taking the cube root of (59a) one obtains

$$s \approx 2^{-1/3}\left(\frac{\pi}{2} - z\right) e^{i\pi/6}\ e^{i2n\pi/3}\ , \quad n = 0, 1, 2\ , \tag{60}$$

where the last factor expresses the three possible values of the cube root of unity. The branch to be chosen is that for which $\pm(ds/dz)$ evaluated at $z = \pi/2$ assumes the value in (58b). As noted before the proper choice is $n = 2$ so that

$$s \approx i\, 2^{-1/3}\,(z-\frac{\pi}{2})\,, \quad z \approx \frac{\pi}{2}\,. \tag{60a}$$

One observes from (60a) that $\arg s = 0$ when $z = (\pi/2) + iz_i$, $z_i < 0$, and the $\arg s = -\pi/2$ when z is real and $z < \pi/2$. It then follows from (59) and a continuity argument that the entire line $\mathrm{Re}\, z = \pi/2$, $\mathrm{Im}\, z < 0$, maps into the positive real s-axis, while the remainder of the path $\bar{P}_z$ terminates in the shaded region in the third quadrant of Fig.4. Thus, the contour $\bar{P}_z$ is transformed into the path L_{31} in the s-plane as shown in Fig.4. From the direction of integration along $\bar{P}_z$ in the neighborhood of $z = \pi/2$ one notes that the choice of the minus sign in (57c) and (58b) is appropriate.

The first-order asymptotic representation (as $\Omega \to \infty$) for $H_\nu^{(1)}(\Omega)$ in (53) can now be written down directly from (42) and from (55) and (57). Since $P = L_{31}$, one obtains:

$$H_\nu^{(1)}(\Omega) \sim \left(\frac{2}{\cos\alpha}\right)^{1/2} \frac{\eta^{1/4}}{\Omega^{1/3}} \left[\mathrm{Ai}(-\eta\,\Omega^{2/3}) - i\,\mathrm{Bi}(-\eta\,\Omega^{2/3})\right], \tag{61}$$

where α and η are defined in Eqs.(53a) and (57a), respectively. When $\alpha = \frac{\pi}{2}$, use of (48a) and (58b) yields

$$H_\nu^{(1)}(\Omega) \sim \left(\frac{2}{\Omega}\right)^{1/3} \left[\mathrm{Ai}(0) - i\mathrm{Bi}(0)\right] = \left(\frac{2}{\Omega}\right)^{1/3} \frac{3^{-1/6}}{\pi}\,\Gamma\!\left(\frac{1}{3}\right) e^{-i\pi/3}\,. \tag{62}$$

When α is sufficiently different from $\pi/2$ to yield $\eta\Omega^{2/3}$ large, we may employ the asymptotic expressions for the Airy functions [2c]

$$Ai(-\sigma) \sim \frac{1}{\sqrt{\pi}\ \sigma^{1/4}} \sin\left(\frac{2}{3}\sigma^{3/2} + \frac{\pi}{4}\right) \tag{63a}$$

$$Bi(-\sigma) \sim \frac{1}{\sqrt{\pi}\ \sigma^{1/4}} \cos\left(\frac{2}{3}\sigma^{3/2} + \frac{\pi}{4}\right), \tag{63b}$$

valid as $\sigma \to \infty$, $|\arg\sigma| < \pi/3$, to obtain the Debye formula

$$H_\nu^{(1)}(\Omega) \sim \sqrt{\frac{2}{\pi\,\Omega\cos\alpha}}\; e^{i\Omega\left[\cos\alpha - \left(\frac{\pi}{2}-\alpha\right)\sin\alpha\right] - i\pi/4}. \tag{64}$$

The condition $\eta\,\Omega^{2/3} \gg 1$ required for the validity of (64) can be phrased in view of (58a),

$$\eta \approx 2^{-2/3}\cos^2\alpha = 2^{-2/3}\left[1-\left(\frac{\nu}{\Omega}\right)^2\right] \approx 2^{1/3}\,\frac{\Omega-\nu}{\Omega}, \tag{65a}$$

as

$$(\Omega-\nu) \gg \Omega^{1/3}. \tag{65b}$$

If $(\Omega-\nu) = O(\Omega^{1/3})$, one must employ (61).

5. INTEGRANDS WITH A POLE SINGULARITY NEAR A FIRST-ORDER SADDLE POINT.

A. Simple Pole

If $f(z)$ in the integrand of (1) has a simple pole singularity at $z = z_0$ near a first-order saddle point z_s, $G(s)$ in (10a) will possess correspondingly a simple pole singularity in the vicinity of $s = 0$, at say $s = b$. Suppose that $G(s)\ (s-b) \to a$ as $s \to b$; then $G(s)$ can be represented in the vicinity of $s = 0$ by (see (15))

$$G(s) = \frac{a}{s-b} + T(s) . \tag{66}$$

It will be convenient to employ the identity

$$\frac{a}{s-b} = \frac{as}{s^2 - b^2} + \frac{ab}{s^2 - b^2} , \tag{66a}$$

and to expand

$$T(s) = T(0) + T'(0)\, s + T''(0) \frac{s^2}{2!} + \ldots, \tag{66b}$$

which is regular as $s = b$ and has a radius of convergence uninfluenced by the presence of the pole.

Since the saddle point is of order one, the transformation in (12) applies and the integral in Eq.(1) can be written as:

$$I(\Omega , b) = e^{\Omega q(z_s)} \int_{-\infty}^{+\infty} G(s)\, e^{-\Omega s^2}\, ds , \tag{67}$$

or, via (66a,b) and (27a) as[4]

$$I(\Omega ,b) \sim e^{\Omega q(z_s)} \left[A(\Omega ,b) + T_e \left(\sqrt{-\frac{d}{d\Omega}} \right) \sqrt{\frac{\pi}{\Omega}} \right] . \tag{68}$$

The function $A(\Omega,b)$ is given in terms of the integral

$$A(\Omega, b) = ab \int_{-\infty}^{+\infty} \frac{1}{s^2 - b^2} e^{-\Omega s^2} ds \quad . \tag{69}$$

The integral in (67), with $G(s)$ as in (66), is not defined when b is real. Viewed as a function of b, a study of the analytic properties of $I(\Omega, b)$ as $\mathrm{Im}\, b \to 0$ reveals that the integral is discontinuous across the real b-axis. Suppose that b approaches the real b-axis from the range of positive imaginary b, i.e., $b \to b_r + i\delta$, with b_r, δ real and $\delta = 0^+$. Then the path of integration is indented at $s = b_r + i\delta$ as shown in Fig.5(a). Similarly, when $b \to b_r - i\delta$,

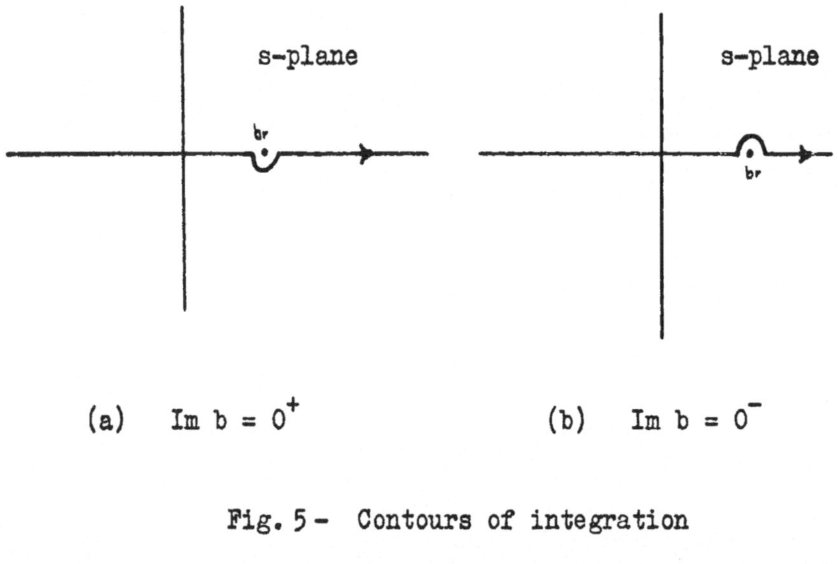

(a) $\mathrm{Im}\, b = 0^+$ (b) $\mathrm{Im}\, b = 0^-$

Fig. 5 - Contours of integration

the appropriate path is that in Fig.5(b). Upon constructing the difference $I(\Omega, b_r + i\delta) - I(\Omega, b_r - i\delta)$ to exhibit the discontinuity in $I(\Omega, b)$ across the real b-axis, one notes that the contributions to the integrals from the straight portions of the paths in Fig.5 cancel, and there remains a small circular contour enclosing in the positive sense the pole at $s = b_r$. Since $T(s)$ in (66) is regular inside this circle, its contribution vanishes and one obtains from the residue at $s = b$

$$I(\Omega, b_r + i\delta) - I(\Omega, b_r - i\delta) = 2\pi i a\, e^{-\Omega b_r^2} e^{\Omega q(z_s)}, \quad \delta \to 0^+. \tag{70}$$

It is easily verified that $A(\Omega, b)$ satisfies the differential equation

$$\left(\frac{d}{d\Omega} + b^2\right) A(\Omega, b) = -ab\sqrt{\frac{\pi}{\Omega}} \quad . \tag{71}$$

To solve, substitute

$$A(\Omega, b) = e^{-\Omega b^2} B(\Omega, b) \tag{72}$$

into (71) whence

$$\frac{dB}{d\Omega} = -ab\; e^{\Omega b^2} \sqrt{\frac{\pi}{\Omega}} \quad . \tag{73}$$

Upon integrating (73) ove Ω between the limits $\bar{\Omega}$ and ∞, one obtains

$$B(\bar{\Omega}, b) = ab\sqrt{\pi} \int_{\bar{\Omega}}^{\infty} e^{\Omega b^2} \Omega^{-1/2}\, d\Omega \quad , \tag{74}$$

where it has been assumed for the moment that $b^2 < 0$ * so that in view of (72), $B(\infty, b) = 0$. A change of variable in (74) from Ω to $(-x^2/b^2)$, or $x = \mp ib\sqrt{\Omega}$, yields

$$B(\Omega, b) = 2a\sqrt{\pi}\,\frac{b}{(\mp i b)}\,Q\left[(\mp ib)\sqrt{\Omega}\right], \tag{75}$$

where $Q(y)$ is the " error function complement "

$$Q(y) = \int_y^\infty e^{-x^2}\,dx \quad . \tag{75a}$$

The ambiguity in sign introduced into (75) by the change of variable is resolved by the previously imposed requirement $B(\infty, b) = 0$ for $b^2 < 0$. Since the integral in (75a) will vanish if the lower limit approaches infinity along the positive real axis, we require for $b^2 < 0$ a choice of sign such that

$$(\mp ib) > 0 \ , \tag{75b}$$

i.e., the minus sign when $b = i|b|$ and the plus sign when $b = -i|b|$.

The validity of (75) can be extended by analytic continuation to values other than $b^2 < 0$. Since the expressions

---

* More generally $\mathrm{Re}\, b^2 < 0$.

for B in (75) and (72) (with (69)) represent identical functions of b^2 when $b^2 < 0$, and since (69) clearly remains valid for all except positive and zero values of b^2, the expression in (75) can likewise be continued analytically to all values of b^2 except $b^2 \geqslant 0$. This continuation from pure imaginary values of b to complex values must be consistent with condition (75b). The analytic properties of the error function complement imply a choice of sign in (75) according to the more general condition $\mathrm{Re}(\mp ib) > 0$, i.e., the $\left\{\begin{matrix}\text{minus}\\ \text{plus}\end{matrix}\right\}$ sign applies when $\mathrm{Im}\, b \gtrless 0$.

From Eqs.(68), (72), and (75), we can now write down the asymptotic expansion of the integral in Eq.(67) for $\Omega \gg 1$ and for arbitrary values of b :[4]

$$I(\Omega,b) \sim e^{\Omega q(z_s)} \left\{ \pm i2a\sqrt{\pi}\, e^{-\Omega b^2} Q(\mp ib\sqrt{\Omega}) + T_e\left(\sqrt{-\frac{d}{d\Omega}}\right)\sqrt{\frac{\pi}{\Omega}} \right\}, \quad \mathrm{Im}\, b \gtrless 0 . \tag{76}$$

The function $e^{-\Omega b^2} Q(\mp ib\sqrt{\Omega})$ is tabulated for real and complex values of $b\sqrt{\Omega}$.[5] Thus, the asymptotic expansion of an integral whose integrand contains a simple pole near a saddle point has the same form as that for an integrand without a pole, except for an additional term involving the error function complement Q.

It is of interest to verify from (76) the previously noted expression for the discontinuity in the value of $I(\Omega,b)$ when Im b changes from positive to negative values. As before, we define

$$b_{\frac{1}{2}} = b_r \pm i\delta, \quad b_r, \delta \text{ real}, \quad \delta \to 0^+ . \tag{77}$$

Since $T_e\left(\sqrt{-\frac{d}{d\,\Omega}}\right)\sqrt{\frac{\pi}{\Omega}}$ is continuous for all b, the jump $\left[I(\Omega,b_1)-I(\Omega,b_2)\right]$ in the value of $I(\Omega, b)$ is given by

$$I(\Omega,b)\Big|_{b=b_2}^{b=b_1}=2ia\sqrt{\pi}\,e^{-\Omega b_r^2}\left[Q(-ib_r\sqrt{\Omega})+Q(ib_r\sqrt{\Omega})\right]e^{\Omega q(z_s)}. \quad (78)$$

To treat the sum $\left[Q(i\,\alpha)+Q(-i\,\alpha)\right]$, α real, we choose for $Q(\pm i\,\alpha)$ in (75a) a path of integration from $\pm i\,\alpha$ to 0 and then from 0 to ∞ along the real axis. Thus,

$$Q(i\alpha)+Q(-i\,\alpha)=\int_{-i\alpha}^{0}e^{-x^2}dx+\int_{i\alpha}^{0}e^{-x^2}dx+2\int_{0}^{\infty}e^{-x^2}dx. \quad (79)$$

The first two terms on the right-hand side of Eq.(79) cancel while the third is equal to $\sqrt{\pi}$. Thus, (78) reduces to the previous result in (70).

When b is large enough so that $(b\sqrt{\Omega})$ is likewise large, one may employ an asymptotic expansion for $Q(\mp ib\sqrt{\Omega})$ in (76). This expansion is obtained directly from the representation in (74) by repeated integrations by parts:

$$B(\Omega,b)=\pm\,2ia\sqrt{\pi}\,Q(\pm\,ib\sqrt{\Omega})\sim-\frac{a}{b}e^{\Omega b^2}\sqrt{\frac{\pi}{\Omega}}\left[1+\frac{1}{2b^2\Omega}+0\left(\frac{1}{b^4\Omega^2}\right)\right]. \quad (80)$$

The first-order asymptotic representation for $I(\Omega,b)$ in (76) is then given by

$$I(\Omega,b) \sim e^{\Omega q(z_s)} \sqrt{\frac{\pi}{\Omega}}\, G(0), \quad |b|\sqrt{\Omega} \gg 1 , \tag{81}$$

where

$$G(0) = -\frac{a}{b} + T(0) \tag{81a}$$

In this instance, the pole is situated " far " from the origin in the s-plane and the expression in (81) is identical with that obtained in (20). Just how large $|b|\sqrt{\Omega}$ has to be before (80) can be employed to within a given accuracy can be assessed by comparing (80) with the exact expression (75) whose values for a given b are found from numerical tables. A detailed comparison is made in (91) et seq. for the special case $\arg(\pm b) = \pi/4$.

When $|b| \to 0$, $Q(\mp ib\sqrt{\Omega}) \to \frac{1}{2}\sqrt{\pi}$.

Example

We return now to the evaluation of the integral in (28) for the case when the pole at $z = \beta$ is situated near the saddle point $z_s = \alpha$. The representation of $I_1(\Omega, \alpha, \beta)$ in (34) still applies. However, G(s) in Eq.(34b) should now be represented as in (66), with the pole contribution exhibited separately. To determine the behavior of $(z-\beta)^{-1}$ as a function of s , first expand s in Eq.(31b) in a power series about $z = \beta$ (note: the plus sign applies in (31b)),

$$s = \sqrt{2}\, e^{+i\pi/4}\left[\sin\left(\frac{\beta-\alpha}{2}\right) + \frac{1}{2}\left(\cos\frac{\beta-\alpha}{2}\right)(z-\beta) - \frac{1}{8}\left(\sin\frac{\beta-\alpha}{2}\right)(z-\beta)^2 + \dots\right], \tag{82}$$

and invert the series[10] to obtain

$$z-\beta=\left[\frac{\sqrt{2}}{e^{i\pi/4}\cos\frac{\beta-\alpha}{2}}\right](s-b)+\left[\frac{\tan\frac{\beta-\alpha}{2}}{2i\cos^2\left(\frac{\beta-\alpha}{2}\right)}\right](s-b)^2+\ldots, \qquad (83a)$$

where

$$b=\sqrt{2}\;e^{i\pi/4}\sin\frac{\beta-\alpha}{2}\;. \qquad (83b)$$

Thus,

$$\frac{1}{z-\beta}=\frac{e^{i\pi/4}\cos\left(\frac{\beta-\alpha}{2}\right)}{\sqrt{2}}\;\frac{1}{s-b}+\text{(terms finite at } s=b\text{)}, \qquad (84)$$

and "a" in (66) is given by

$$a=\lim_{s\to b}\Big[G(s)(s-b)\Big]=\frac{e^{i\pi/4}\cos\left(\frac{\beta-\alpha}{2}\right)}{\sqrt{2}}\left(\frac{dz}{ds}\right)_{s=b}=1\;, \qquad (85)$$

with the value of (dz/ds) at $s=b$ obtained from (83a). By (76) the asymptotic expansion (as $\Omega\to\infty$) of the integral in (28) or (34) can now be written down directly as

$$I_1(\Omega,\alpha,\beta)\sim 2\pi i e^{i\Omega\cos(\alpha-\beta)}\varepsilon(\beta)+e^{i\Omega}\left\{\pm i2\sqrt{\pi}\,e^{-\Omega b^2}Q(\mp ib\sqrt{\Omega})+T_e\left(\sqrt{-\frac{d}{d\Omega}}\right)\sqrt{\frac{\pi}{\Omega}}\right\}$$

$$\operatorname{Im} b \gtrless 0\,, \qquad (86)$$

where $\varepsilon(\beta)$ is defined in (34a), b is given in (83b), and

$$T(s) = \frac{1}{z-\beta}\frac{dz}{ds} - \frac{1}{s-b}\,, \tag{87a}$$

$$T(0) = \frac{1}{\alpha-\beta}\sqrt{2}\,e^{-i\pi/4} + \frac{e^{-i\pi/4}}{\sqrt{2}\,\sin\frac{\beta-\alpha}{2}}\,. \tag{87b}$$

It is of interest to note that the expression in (86) is a continuous function of b although various terms therein are discontinuously represented. This is verified upon an inspection of Fig.5, (34a) and (70). If the pole at $s=b$ crosses the real axis in Fig.5, the term inside the braces in (86) experiences a jump as in (70). However, the first term on the right-hand side of (86) also changes discontinuously under these conditions and compensates exactly for the first-mentioned discontinuity. For values of b such that $|b|\sqrt{\Omega} \gg 1$, (86) reduces via (80) to (35).

The special case where β is real is of particular importance in various diffraction problems. In this instance, $\sin\left[(\beta-\alpha)/2\right]$ in (83b) is real and for $|\beta-\alpha|<\pi$,

$$\mp ib\sqrt{\Omega} = (1-i)\xi\,, \quad \xi = \sqrt{\Omega}\,\sin\left|\frac{\beta-\alpha}{2}\right|, \ \text{Im}\, b \gtrless 0\,. \tag{88}$$

Moreover, $\text{Im}\, b \gtrless 0$ if $(\beta-\alpha) \lessgtr 0$. Thus, the following term in (86) can be rewritten as

$$\pm i\,2\sqrt{\pi}\,e^{-\Omega b^2} Q(\mp ib\sqrt{\Omega}) = 2i\sqrt{\pi}\,\text{sgn}(\beta-\alpha)\,e^{-2i\xi^2} Q\left[(1-i)\xi\right]\,, \tag{89}$$

where

$$\operatorname{sgn}(\beta - \alpha) = \pm 1, \quad (\beta - \alpha) \gtrless 0 \quad . \tag{89a}$$

The function

$$e^{i\pi/4} Q\left[(1-i)\xi\right] = e^{i\pi/4} \int_{(1-i)\xi}^{\infty} e^{-x^2} dx = e^{i\pi/4}\left[\frac{\sqrt{\pi}}{2} - \int_0^{(1-i)\xi} e^{-x^2} dx\right] \tag{90a}$$

can be expressed via the change of variable

$$x = \frac{\sqrt{\pi}}{2} (1-i)\, t \tag{90b}$$

in terms of the well-tabulated Fresnel integrals C(x) and S(x) as

$$e^{i\pi/4} Q\left[(1-i)\xi\right] = \frac{1}{2}\sqrt{\pi}\, e^{i\pi/4} - \sqrt{\frac{\pi}{2}}\left[C\left(\frac{2\xi}{\sqrt{\pi}}\right) + i\, S\left(\frac{2\xi}{\sqrt{\pi}}\right)\right], \tag{90c}$$

where

$$C(x) = \int_0^x \cos\left[\left(\frac{\pi}{2}\right) t^2\right] dt\, , \quad S(x) = \int_0^x \sin\left[\left(\frac{\pi}{2}\right) t^2\right] dt \tag{90d}$$

To provide an estimate of how large ξ has to be before the asymptotic representation for $Q\left[(1-i)\xi\right]$ in (80) can be employed, the function

$$F(\xi) = \frac{2}{\sqrt{\pi}}\, e^{-i2\xi^2} Q\left[(1-i)\xi\right], \quad \xi \geqslant 0\, , \tag{91}$$

has been plotted. For $\xi \to 0$, $F(\xi) \to 1$, while for $\xi \gg 1$ one has from (80)

$$F(\xi) \sim \frac{e^{i\pi/4}}{\xi\sqrt{2\pi}} \quad , \qquad \xi \gg 1 \; . \tag{91a}$$

Upon comparing the graphs of (91) and (91a) as shown in Fig. (6a), one notes that the first-order asymptotic formula in (91a) holds with very good accuracy when $\xi \geqslant 3$. In terms of this estimate one finds from (88) that the " transition region ", inside which the simple asymptotic representation in (35) fails to apply, is given approximately by $|\beta - \alpha| \leqslant 6\,\Omega^{-1/2}$. An analogous estimate can be found for the case of a double pole singularity, in which case one requires values for the derivative of $F(\xi)$ (see (95)). The function[7]

$$\bar{F}(\xi) = 1 - 2\sqrt{2}\,\xi\, e^{-i(2\xi^2+\pi/4)}\, Q\left[(1-i)\xi\right] \quad , \tag{92}$$

which occurs in this connection, as well as its asymptotic approximation

$$\bar{F}(\xi) = \frac{i}{4\xi^2} \quad , \qquad \xi \gg 1 \quad , \tag{92a}$$

is plotted in Fig.(6b).

B. Multiple Pole

If $G(s)$ has a pole of order N at $s = b$, one employs the representation in (15). To infer the asymptotic expansion of $I(\Omega, b)$ in (67) in this case, one must investigate integrals of the form

$$A_{-N}(\Omega, b) = \int_{-\infty}^{+\infty} \frac{1}{(s-b)^N}\, e^{-\Omega s^2}\, ds \; . \tag{93}$$

PLOT OF $F = |F|\, e^{i\chi} = \frac{2}{\sqrt{\pi}}\, e^{-i2\xi^2} \int_{(1-i)\xi}^{\infty} e^{-y^2}\, dy$

FOR LARGE ξ : $F \sim \frac{e^{i\pi/4}}{2\pi\xi} + O\left(\frac{1}{\xi^3}\right)$

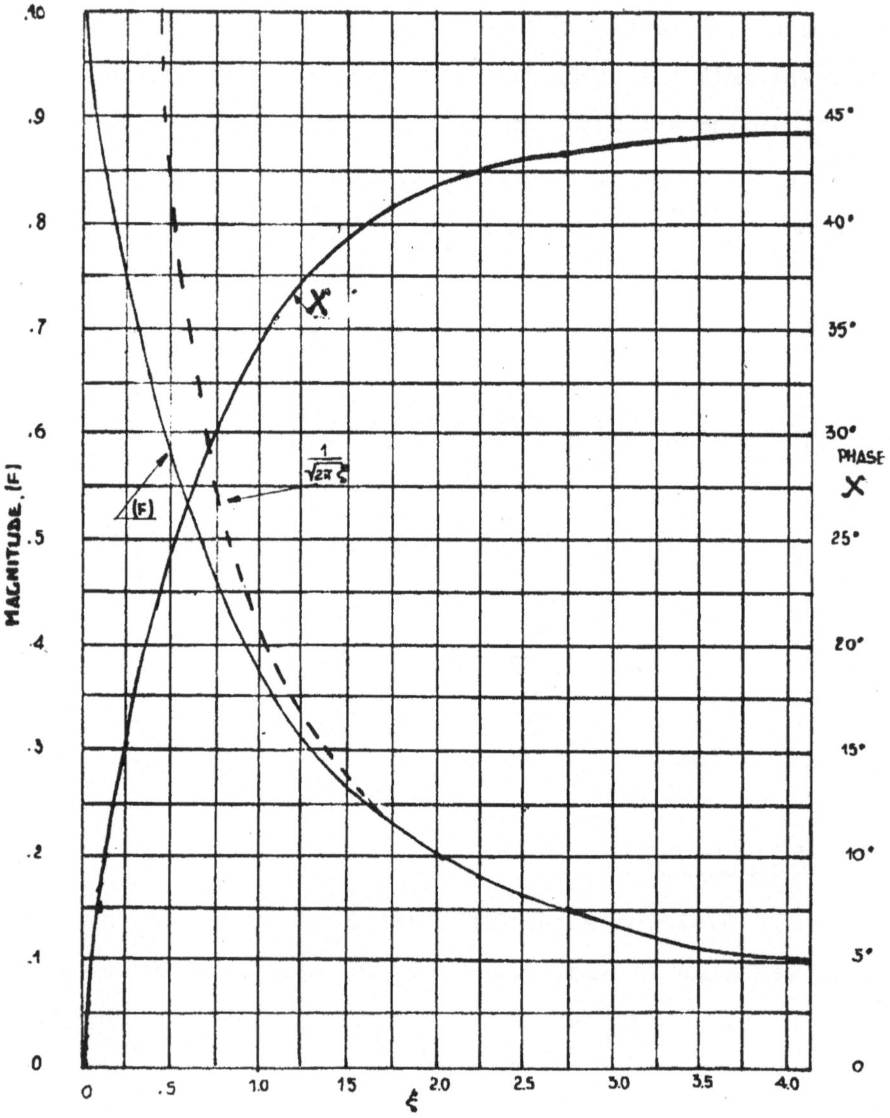

Fig. 6(a)

L.B.Felsen

PLOT OF $\bar{F} = |\bar{F}|e^{i\zeta} = 1-2\sqrt{2}\,\xi e^{-i(2\xi^2+\pi/4)}\int_{(1-i)\xi}^{\infty} e^{-y^2}dy$

FOR LARGE ξ : $\bar{F} \sim \frac{i}{4\xi^2} + O(\frac{1}{\xi^4})$

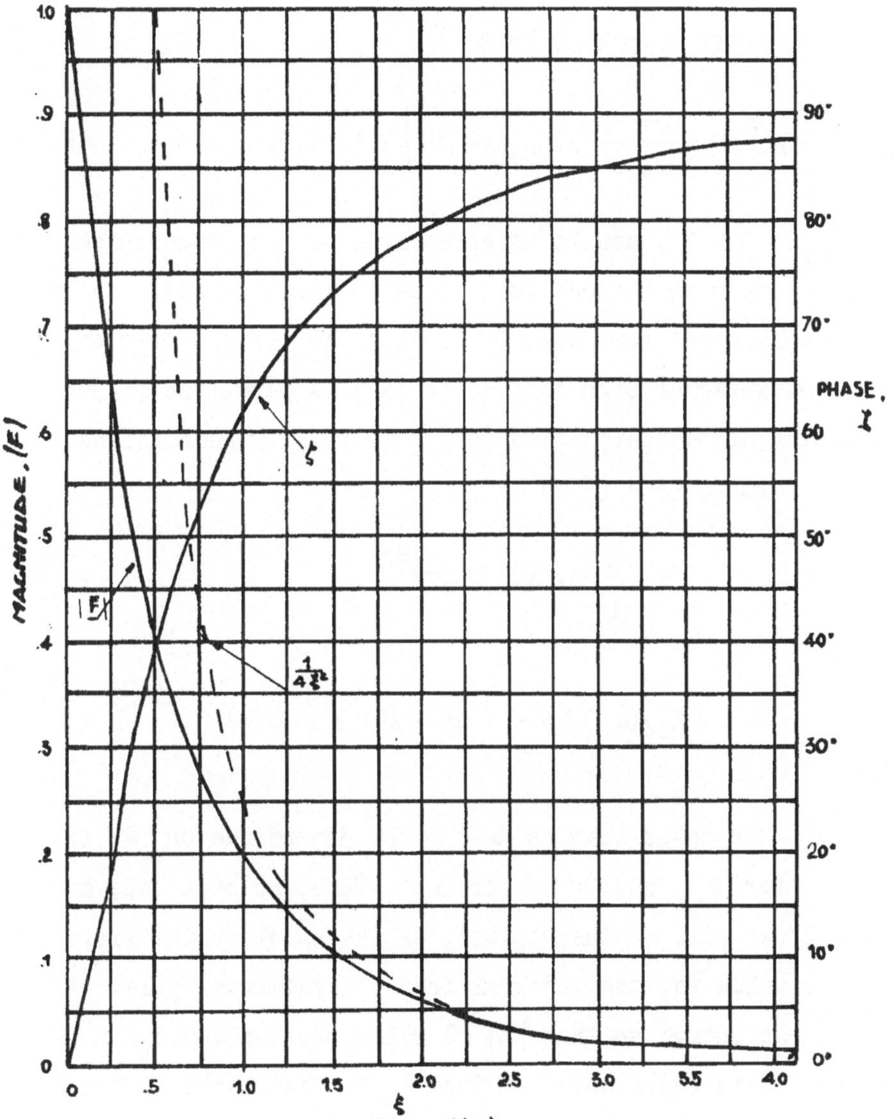

Fig. 6(b)

For $N=1$ (with $a_{-1}=1$), the result has been obtained in (75) (with (72):

$$A_{-1}(\Omega,b) = \pm 2i\sqrt{\pi}\, e^{-\Omega b^2} Q(\mp ib\sqrt{\Omega}\,), \quad \operatorname{Im} b \gtrless 0. \tag{94}$$

Since one notes from (93) that

$$A_{-N}(\Omega,b) = \frac{1}{N-1}\frac{d}{db}\left[A_{-N+1}(\Omega,b)\right], \quad N=2,3\ldots, \tag{95}$$

all A_{-N}, $N \geqslant 2$, can be inferred from A_{-1} by successive differentiation with respect to b (the integral in (93) is uniformly convergent for $\operatorname{Im} b \neq 0$ so that the differentiation under the integral sign implied in (95) is permitted). The evaluation of the derivatives of Q is readily accomplished via the formula (see (75a))

$$\frac{d}{dy} Q(y) = -e^{-y^2}, \tag{96}$$

so that

$$A_{-2}(\Omega,b) = -2\sqrt{\Omega\pi} - 2b\Omega A_{-1}(\Omega,b) \tag{97}$$

etc.

One notes that as $b \to 0$ the dependence on Ω is $O(\sqrt{\Omega}\,)$ for A_{-2} and $O(1)$ for A_{-1}. Thus, as expected, a second-order pole in the vicinity of the saddle point yields a larger value for the integral than a first-order pole. If b is large enough so that $|b|\sqrt{\Omega} \gg 1$, one can employ in (97) the asymptotic representation from (80). The result is found to be

$$A_{-2}(\Omega,b) \sim \sqrt{\frac{\pi}{\Omega}} \cdot \frac{1}{b^2} , \quad \sqrt{\Omega}|b| \gg 1 , \qquad (98)$$

which agrees with that obtained from (93) by a direct asymptotic evaluation.

Radiation from a Vertical Dipole above a Lossy Plane Earth

As an illustration of the direct application of Eq. (86) to a problem which has received much attention in this course, consider the evaluation of the field radiated by a vertical electric dipole located near a plane interface between a lossy and a lossless half-space. Let ε_1, μ and ε_2, μ be the dielectric constant and permeability of the lossless and lossy regions, respectively, where ε_1 and μ are positive real while ε_2 is complex ($0 < \arg \varepsilon_2 < \pi/2$, for an assumed time dependence $\exp(-i\omega t)$). The electromagnetic fields in both regions can be inferred from a single scalar Hertz potential function, Π. If the lossless medium (region I) occupies the half-space $z < 0$ and the source (having a momentum M) is located at $\underline{r}' = (\rho', z')$, $\rho' = 0$, $z' < 0$, where (ρ, z) are cylindrical coordinates, the Hertz potential Π_1 in region I is given by* :

$$\Pi_1 = \frac{M e^{ik_1|\underline{r}-\underline{r}'|}}{4\pi\varepsilon_1|\underline{r}-\underline{r}'|} + \Pi_s , \quad k_{1,2} = \omega\sqrt{\mu\varepsilon_{1,2}} . \tag{A-1}$$

$$\Pi_s = -\frac{Mi}{8\pi\varepsilon_1} \int_{\infty e^{i\pi}}^{\infty} \xi H_0^{(1)}(\xi\rho) \frac{e^{i\varkappa_1|z+z'|}}{\varkappa_1} \frac{\varkappa_2 - \varepsilon\varkappa_1}{\varkappa_2 + \varepsilon\varkappa_1} d\xi , \tag{A-2}$$

---

* See chapter by H.Bremmer, Eqs.(7a) and (12). Also, the chapter " Alternative Green's Function Representations for a Grounded Dielectric Slab " by L.B.Felsen, Eqs.(19a) and (24), with $\vec{Z}$ in Eq.(16) given by $\vec{Z} = \kappa_{x1}/\omega\varepsilon_1$ (to switch to the $\exp(-i\omega t)$ time dependence above, let $j \to -i$) .

where $\varkappa_{1,2} = \sqrt{k_{1,2}^2 - \xi^2}$, $\varepsilon = \varepsilon_2/\varepsilon_1$, and $\mathrm{Im}\,\varkappa_{1,2} \geqslant 0$ ($\mathrm{Re}\,\varkappa_1 > 0$ when ξ is real and $-k_1 < \xi < k_1$).

If ρ is assumed to be very large, one may replace the Hankel function by its asymptotic approximation $H_0^{(1)}(\xi\rho) \sim (2/\pi\xi\rho)^{1/2} \exp(i\xi\rho - i\pi/4)$, provided that $|\xi\rho| \gg 1$. The latter condition can be met by distorting the integration path away from $\xi = 0$. Upon changing variables to $w = \sin^{-1}(\xi/k_1)$, and introducing the spherical coordinates $\hat{r}, \theta$ via $|z+z'| = \hat{r}\cos\theta$, $\rho = \hat{r}\sin\theta$, $0 < \theta < \pi/2$, one obtains

$$\Pi_s \sim -\frac{Me^{i\pi/4}}{8\pi\varepsilon_1}\sqrt{\frac{2k_1}{\pi\hat{r}\sin\theta}}\int_{\bar{P}}\sqrt{\sin w}\;\Gamma(k_1\sin w)\,e^{ik_1\hat{r}\cos(w-\theta)}\,dw, \tag{A-3}$$

where

$$\Gamma(k_1\sin w) = \frac{\sqrt{\varepsilon - \sin^2 w} - \varepsilon\cos w}{\sqrt{\varepsilon - \sin^2 w} + \varepsilon\cos w} \tag{A-3a}$$

and we have defined $\varkappa_1 = +k_1 \cos w$. ($\sqrt{\varepsilon - \sin^2 w} \equiv \sqrt{\varepsilon}$ when $w=0$). The integration path $\bar{P}$ in the w-plane is identical with $\bar{P}_z$ in the complex z-plane as shown in Fig.2 (except that the path avoids the origin). Since the letter z denotes a space coordinate here, w has been chosen as the complex integration variable. To effect an asymptotic evaluation for large values of $\hat{r}$, the path $\bar{P}$ is deformed into the steepest descent path P through the saddle point at $w = \theta$ as in Fig.2.

The singularities of the integrand in (A-3) are the branch points at $\sin w_b = \pm\sqrt{\varepsilon}$, and the poles w_p at which the denominator in (A-3a) vanishes. If $|\varepsilon| \gg 1$ and $\arg \varepsilon \approx \pi/2$, appropriate to a highly dissipative medium, any branch cut integrals arising from the crossing of a branch point during the deformation of $\bar{P}$ into P can be neglected since their contribution is proportional to

$$e^{ik_1 \hat{r}\cos(w_b - \theta)} = e^{ik_1 \hat{r}\left[\cos\theta\sqrt{1-\varepsilon} + \sin\theta\sqrt{\varepsilon}\right]},$$

where the square roots have positive imaginary parts. Hence, the branch cut integrals are exponentially small. The nearest pole singularity is located at

$$w_p \approx \frac{\pi}{2} + \frac{e^{-i\pi/4}}{\sqrt{|\varepsilon|}}, \quad |\varepsilon| \gg 1, \quad \arg\varepsilon \approx \pi/2,$$

and will not be near the steepest descent path except when $\theta = \pi/2$, i.e., when the source and observation points are located on the interface. Therefore, the presence of the pole can be ignored in the asymptotic evaluation of (A-3) except in the vicinity of $\theta = \pi/2$.

To obtain an asymptotic approximation of (A-3) valid for all observation angles in the interval $0 < \theta \leq \pi/2$, it is convenient to write Γ as

$$\Gamma(k_1 \sin w) = -1 + \hat{\Gamma}(k_1 \sin w), \quad \hat{\Gamma} = \frac{2\sqrt{\varepsilon - \sin^2 w}}{\sqrt{\varepsilon - \sin^2 w} + \varepsilon\cos w}, \qquad \text{(A-4)}$$

wherein the numerator of $\hat{\Gamma}$, unlike that of Γ, is a slowly varying function of w in the entire interval $0 < w \leq \pi/2$, w real. The contribution to Π_s from the (-1) term in (A-4) is readily shown to be equal to $(M/4\pi\varepsilon_1\hat{r})\exp(ik_1\hat{r})$ (see (19)). Hence,

$$\Pi_s \sim \frac{Me^{ik_1\hat{r}}}{4\pi\varepsilon_1\hat{r}} - \frac{Me^{i\pi/4}}{8\pi\varepsilon_1}\sqrt{\frac{2k_1}{\pi\hat{r}\sin\theta}}\int_{\bar{P}} F(w)\,\frac{e^{ik_1\hat{r}\cos(w-\theta)}}{w-w_p}\,dw\,, \tag{A-5}$$

where

$$F(w) = \sqrt{\sin w}\;\hat{\Gamma}(k_1\sin w)(w-w_p)\,. \tag{A-5a}$$

Since $(w-w_p)\hat{\Gamma}$ and $\sqrt{\sin w}$ are slowly varying in the interval $0 < w \leq \pi/2$, w real, F(w) may be approximated (in a first-order evaluation) by its value $F(\theta)$ at the saddle point, and removed from the integral. The remaining integral is identical with that in (28). Its asymptotic approximation is given in (86), whence

$$\Pi_s \sim \frac{Me^{ik_1\hat{r}}}{4\pi\varepsilon_1\hat{r}} - \frac{Me^{i\pi/4}}{8\pi\varepsilon_1}\sqrt{\frac{2k_1}{\pi\hat{r}\sin\theta}}\,F(\theta)\left\{i2\sqrt{\pi}\,e^{-k_1\hat{r}b^2}Q(-ib\sqrt{k_1\hat{r}}) + T(0)\sqrt{\frac{\pi}{k_1\hat{r}}}\right\}e^{ik_1\hat{r}} \tag{A-6}$$

where

$$b = \sqrt{2}\;e^{i\pi/4}\sin\left(\frac{w_p-\theta}{2}\right),\quad Q(y) = \int_y^\infty e^{-x^2}\,dx\,, \tag{A-6a}$$

$$T(0) = \frac{\sqrt{2}\, e^{-i\pi/4}}{\theta - w_p} + \frac{e^{-i\pi/4}}{\sqrt{2}\,\sin(\frac{w_p-\theta}{2})} \,. \qquad \text{(A-6b)}$$

While (A-6) is valid for $k_1\hat{r} \gg 1$ and for all θ in the interval $0 < \theta \lesssim \pi/2$, its present form need be retained only when $b \to 0$, i.e., $\theta \to \pi/2$; for other observation angles, one may employ the simpler formula (35). When $\theta \to \pi/2$, the second term inside the braces in (A-6) can be neglected in comparison with the first term. In particular, one finds for $\theta = \pi/2$ $(z = z' = 0,\ \hat{r} = \rho)$,

$$\Pi_s \sim \frac{M e^{ik_1\hat{r}}}{4\pi\epsilon_1\hat{r}}\left[1 + i4\sqrt{\rho}\ e^{-\rho} Q(-i\sqrt{\rho})\right], \quad k_1\hat{r} \gg 1,\ |\epsilon| \gg 1,\ \arg\epsilon \approx \frac{\pi}{2}, \qquad \text{(A-7)}$$

where $\rho = k_1\hat{r}/2\,|\epsilon|$ is Sommerfeld's numerical distance. This result agrees with that derived by Prof.Bremmer* by a different method.

---

* See chapter by H.Bremmer, Sec. I.C.

L.B.Felsen

REFERENCES

1. L.B.Felsen and N.Marcuvitz, " Modal Analysis and Synthesis of Electromagnetic Fields", Microwave Research Institute, Polytechnic Institute of Brooklyn, Report R-776-59, PIB-705, Oct. 1959.

2. For general information on the asymptotic evaluation of integrals, see:
   a) H.Jeffreys and B.Jeffreys, " Methods of Mathematical Physics'', Cambridge University Press, 1946, Chapter 17.
   b) E.T.Copson,'' The asymptotic Expansion of a Function Defined by a Definite Integral or a Contour Integral " , Admiralty Computing Service, London. (1946)
   c) A.Erdelyi, " Asymptotic Expansions" , Dover Publishing Co., 1956, Chapter 2.
   d) N.G. deBruijn, " Asymptotic Methods in Analysis", Interscience Pub. Co., New York, 1958, Chapters 4-6.
   e) K.O.Friedrichs, " Special Topics in Analysis", Notes prepared by K.O.Friedrichs and H.Kranzer, New York University, 1953-1954, Part B, pp. 1-67.

3. C.Chester, B.Friedman and F.Ursell, " An Extension of the Method of Steepest Descents", Proc. of the Cambridge Phil. Soc., Vol. 53, 1957, pp. 599-611.

4. B.L. Van der Waerden, " On the Method of Saddle Points", Applied Sci. Research, B2, pp. 33-45. (For a related method of evaluating steepest descent integrals in the vicinity of a pole, see: W.Pauli, Phys. Rev. Vol. 54, 1938; H.Ott, Annalen d. Physik (Lpz), Vol. 43, 1943; P.C.Clemmow, J.Mech. App. Math., Vol. 3, 1950, and Proc. Roy. Soc. London, Sec. A, Vol. 205, 1951).

5. P.C.Clemmow and C.M.Munford, " A Table of $\frac{1}{2}\sqrt{\pi}\, e^{i\frac{1}{2}\rho^2\pi}\int_{\rho}^{\infty} e^{-\frac{1}{2}i\pi\lambda^2}\, d\lambda$ for Complex Values of ρ ", Phil. Transaction of the Royal Society of London, Vol. A, 1952, pp. 189-211.

6. Cf. E.Jahnke and F.Emde, " Tabels of Functions" , Dover Publications, New York, 1945, p. 35-40.

7. F.Horner, " A Table of a Function Used in Radio-Propagation Theory", Proc. of the Institution of El. Engineers, Vol. 102, Part C, No. 1, March 1955, pp. 134-137.

8. J.C.P. Miller , " The Airy Integral", British Association Mathematical Tables, Cambridge University Press, 1946.

9. A.Sómmerfeld, " Partial Differential Equations in Physics", Academic Press, New York, 1949, p. 89,

10. For a procedure utilizing Cauchy's theorem, see E.T. Copson, " Theory of Functions of a Complex Variable " , Oxford University Press, London, 1935, Sec. 6.23.

11. F.Oberhettinger, " On a Modification of Watson's Lemma ", J. Research Natl. Bureau of Standards, 63B, July-Sept. 1959.

CENTRO INTERNAZIONALE MATEMATICO ESTIVO

(C.I.M.E.)

L. B. FELSEN

Department of Electrophysics, Polytechnic Institute of Brooklyn[1)]

ALTERNATIVE GREEN'S FUNCTION REPRESENTATATIONS FOR A GROUNDED DIELECTRIC SLAB

[1)] These lectures were presented while the author was on a years' leave of absence with the Office of Naval Research, London Branch.

ALTERNATIVE GREEN'S FUNCTION REPRESENTATIONS
FOR A GROUNDED DIELECTRIC SLAB

by L. B. FELSEN

Department of Electrophysics,
Polytechnic Institute of Brooklyn

1. FORMULATION OF THE PROBLEM

Consider the configuration shown in Fig.1, wherein a grounded dielectric slab of width d is excited by an electric current element of strength J directed perpendicular to the interface. The media in the slab and exterior regions are characte-

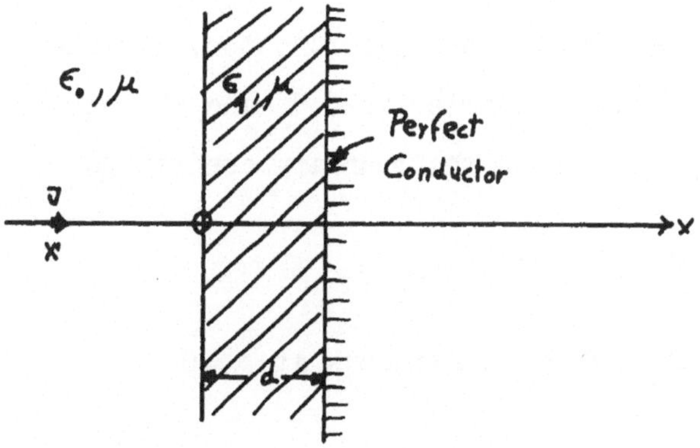

Fig. 1 - Physical configuration

rized by dielectric constants ε_1 and ε_0, respectively; the permeability μ is assumed to be the same for both media. The interface coincides with the $x = 0$ plane, and the current element is located at $\underline{r}' \equiv (x', y', z') = (x', 0, 0)$, $x' < 0$. The electromagnetic fields for the configuration in Fig.1 satisfy the inhomogeneous Maxwell field equations

$$\nabla \times \underline{E}_{\alpha}(\underline{r}) = -j\omega\mu\, \underline{H}_{\alpha}(\underline{r}) \quad , \qquad \alpha = 0,1 \tag{1a}$$

$$\nabla \times \underline{H}_{\alpha}(\underline{r}) = j\omega\varepsilon_{\alpha}\, \underline{E}_{\alpha}(\underline{r}) + \hat{\underline{J}}(\underline{r}) \quad , \tag{1b}$$

where the subscripts $_0$ and $_1$ refer to the regions $x < 0$ and $0 < x < d$, respectively, and an underline denotes a vector quantity. A time dependence $\exp(+j\omega t)$ is assumed throughout. The current density $\hat{\underline{J}}(\underline{r})$ is given in the present instance by

$$\hat{\underline{J}}(\underline{r}) = \underline{x}_0\, J\, \delta(\underline{r} - \underline{r}') \quad , \tag{1c}$$

where $\underline{x}_0$ is a unit vector in the positive x-direction, and $\delta(x)$ denotes the Dirac delta function. For the unique specification of the electric field $\underline{E}$ and the magnetic field $\underline{H}$ in (1a,b), we append the boundary conditions

$$\underline{E}_t = 0 \quad \text{at} \quad x = d \tag{2a}$$

$$\underline{E}_t \text{ and } \underline{H}_t \text{ continuous at } x = 0 \tag{2b}$$

$$\text{radiation condition at } r \to \infty \;. \tag{2c}$$

The subscript $_t$ denotes the vector components transverse to x.

The dielectric slab can support two independent sets of waves: TE waves $(E_x = 0)$ and TM waves $(H_x = 0)$ with respect to the x- direction. Since the x-directed current element excites only TM waves and the slab does not couple the TM and TE waves, the radiation problem can be described in terms of TM waves only. In this instance it is possible to derive the electromagnetic fields in (1) from a single scalar Green's function $G(\underline{r}, \underline{r}')$ via the relation

$$\underline{E}_\alpha(\underline{r}) = \frac{J}{j\omega\,\varepsilon_\alpha} \nabla \times \nabla \times \left[\underline{x}_0\, G_\alpha(\underline{r}, \underline{r}')\right] , \qquad (3a)$$

$$\underline{H}_\alpha(\underline{r}) = J \;\nabla \times \left[\underline{x}_0\, G_\alpha(\underline{r}, \underline{r}')\right] , \qquad (3b)$$

with G defined by the inhomogeneous wave equation

$$(\nabla^2 + k_\alpha^2)\, G_\alpha(\underline{r}, \underline{r}') = -\delta(\underline{r} - \underline{r}') = -\frac{\delta(\rho)\delta(x-x')}{2\pi\rho} , \quad k_\alpha = \omega\sqrt{\mu\varepsilon_\alpha} , \qquad (4)$$

subject to the boundary conditions

$$(\partial G_1/\partial x) = 0 \quad \text{at} \quad x = d \qquad (4a)$$

$$G \text{ and } (1/\varepsilon)(\partial G/\partial x) \text{ continuous at } x = 0 \qquad (4b)$$

$$\text{radiation condition at } r \to \infty \qquad (4c)$$

The Green's function G is related to the familiar Hertz potential function Π as follows[1] :

$$G_\alpha = j\,\omega\varepsilon_\alpha \Pi_\alpha / J \qquad . \qquad (5)$$

In view of the axial symmetry of the dipole field and of the slab configuration, the electromagnetic fields in Fig.1 will be rotationally symmetric about the x-axis. In terms of a cylindrical (ρ, Θ, x) coordinate system, where ρ and Θ are polar coordinates in the plane trasverse to x, this symmetry property implies that $(\partial/\partial \Theta) \equiv 0$.

In the above-mentioned coordinate system the operator $(\nabla^2 + k_\alpha^2)$ has the representation

$$\nabla^2 + k_\alpha^2 \longrightarrow \frac{1}{\rho}\frac{\partial}{\partial\rho}\rho\frac{\partial}{\partial\rho} + \frac{\partial^2}{\partial x^2} + k_\alpha^2 , \tag{6}$$

whence the problem is two-dimensional, in the variables ρ and x. The separable form of $(\nabla^2 + k_\alpha^2)$ and that of the boundary conditions (4a-c), permits the construction of the solution for G_α by the method of separation of variables. By this technique, one seeks a representation for the two--dimensional function G_α in terms of the complete set of eigenfunctions in either the ρ-domain, the x-domain, or both. The radial eigenfunctions $f(\rho, k_\rho)$ annihilate the ρ-dependent part of the operator in (6),

$$\frac{1}{\rho}\frac{d}{d\rho}\rho\frac{d}{d\rho} f(\rho, k_\rho) = -k_\rho^2 f(\rho, k_\rho) , \quad 0 \leqslant \rho < \infty , \tag{7}$$

and satisfy at the endpoints of the radial domain boundary conditions compatible with those in (4a-c). The eigenvalue k_ρ is commonly termed the radial wavenumber.

The axial eigenfunctions $h(x, k_x)$ are defined by the differential equation

$$\frac{d^2}{dx^2} h(x, k_x) = -k_x^2 h(x, k_x) , \quad -\infty < x \leqslant d , \tag{8}$$

where, in view of the discontinuous nature of the medium parameters in the x-domain, the axial wavenumber k_x is expressed as

$$k_x^2 = \begin{cases} k_{xo}^2 & , \quad -\infty < x < 0 \\ k_{x1}^2 = k_{xo}^2 + (k_1^2 - k_o^2) & , \quad 0 < x < d \end{cases} \quad . \tag{8a}$$

The boundary conditions on $h(x, k_x)$ are chosen to be the same as those on G in the x-domain:

$$\frac{dh}{dx} = 0 \quad \text{at} \quad x = d \quad , \tag{8b}$$

$$h \text{ and } \frac{1}{\varepsilon}\frac{dh}{dx} \text{ continuous at } x = 0 \quad . \tag{8c}$$

At the singular endpoint $x = -\infty$, h is assumed to be bounded.

In a radial eigenfunction representation, one expresses G as a superposition of the complete set of eigenfunctions (or modes) $f(\rho, k_\rho)$, with x-dependent amplitude factors which remain to be determined. Depending on the nature of the spectrum of the wavenumber k_ρ, the representation is either discrete (a sum) or continuous (an integral). In an axial eigenfunction representation, the above remarks apply to the mode functions $h(x,k_x)$ with ρ-dependent modal amplitude. Each of these formulations involves a (discrete or continuous) summation over a single parameter. The previously mentioned third possibility, utilizing both $f(\rho, k_\rho)$ and $h(x, k_x)$, results in a more complicated form wherein double summation occur. Since this latter representation is less suitable for deriving approximate expressions permitting the explicit calculation of G, it will receive no further attention herein.

The above-sketched approach may be termed the conventional separation-of-variables procedure and leads to quite dissimilar alternative representation for the two-dimensional Green's function G. Alternative representations are desirable since they generally lend themselves to explicit approximate evaluation in different ranges of the source point, observation point, and medium parameters. In the radial eigenfunction representation (Sec. II), the structure in Fig.1 is viewed as a uniform waveguide along the x-axis having a circular cross--section (of infinite extent), and the x-dependent modal amplitudes describe the reflection and refraction properties of the dielectric slab; this formulation is well suited to the study of the fields far from the slab. The axial eigenfunction representation (Sec.III) highlights the guiding properties of the dielectric slab and views the structure in Fig.1 as a radial waveguide wherein cylindrical waves propagate outward in the ρ-direction; it is especially appropriate for an investigation of the surface wave fields near the slab surface $x = 0$. The two representation can be converted one into the other through the intervention of complex function theory as shown in Sec. IV. The ensuing manipulations are simplified by the use of characteristic Green's function (Sec. V) which lead directly to a general contour integral formulation from which either the uniform or the radial waveguide representation can be deduce.

2. AXIAL WAVEGUIDE REPRESENTATION

Let the configuration in Fig.1 be regarded as a uniform waveguide extending along the x-direction. The mode functions for the circular waveguide with infinite cross section are given by $f(\rho, k_\rho) = J_0(k_\rho \rho)$, where $J_0(z)$ is the Bessel function of order zero and argument z (see (7)) and the complete range of the continuous eigenvalues k_ρ covers the real k_ρ-axis between $k_\rho = 0$ and $k_\rho = \infty$.[2] In terms of these mode functions, the axially symmetric Green's function $G_\alpha(\underline{r}, \underline{r}') \equiv G(\rho, x; 0, x')$ is represented as

$$G_\alpha(\underline{r}, \underline{r}') = \int_0^\infty k_\rho J_0(k_\rho \rho) G_{x\alpha}(x, x'; k_\rho) \, dk_\rho \quad , \tag{9}$$

with the modal amplitudes $G_{x\alpha}(x,x'; k_\rho)$ to be determined. While we have preferred to use a waveguide terminology above, Eq.(9) is simply an application of the Fourier-Bessel integral theorem[3] to the representation of the function G_α. Upon noting that

$$\frac{\delta(\rho)}{\rho} = \int_0^\infty k_\rho J_0(k_\rho \rho) \, dk_\rho \quad , \tag{10}$$

substituting (9) and (10) into (4), interchanging the orders of differentiation and integration, and employing (7), one obtains

$$\int_0^\infty k_\rho J_0(k_\rho \rho) \left[\left(\frac{d^2}{dx^2} + k_\alpha^2 - k_\rho^2 \right) G_{x\alpha}(x,x'; k_\rho) + \frac{\delta(x-x')}{2\pi} \right] dk_\rho = 0 \ , \tag{11}$$

By the Fourier-Bessel transform theorem, the expression inside the square brackets in the integral of (11) vanishes identically, whence $G_{x\alpha}$ is a one-dimensional Green's function defined by the equation:

$$\left(\frac{d^2}{dx^2} + k_{x\alpha}^2\right) G_{x\alpha}(x,x';k_\rho) = -\frac{\delta(x-x')}{2\pi}, \quad k_{x\alpha}^2 = k_\alpha^2 - k_\rho^2 . \qquad (12)$$

To assure that G_α in (9) satisfies the necessary boundary conditions, one imposes upon $G_{x\alpha}$ the same requirements as those listed in (4a-c), with G replaced by G_x.

While G_x can be evaluated by a conventional Green's function procedure [4], we will utilize herein a network method. Consider a TM mode transmission line, excited at the point x' by a series voltage generator having a strength v and zero internal impedance. Then the voltage V and current I on the transmission line satisfy the " transmission line equations"[2,5]

$$-\frac{dV}{dx} = jk_x ZI + v\delta(x-x') , \qquad (13a)$$

$$-\frac{dI}{dx} = jk_x YV , \qquad (13b)$$

where k_x is the propagation constant, and Z and Y, the characteristic impedance and admittance, respectively, are given by

$$Z = \frac{1}{Y} = \frac{k_x}{\omega\varepsilon} . \qquad (13c)$$

The positive polarities of V, I and v are shown in Fig. 2.

Upon differentiating (13b) with respect to x and substituting into (13a), one obtains the second-order differential equation for the current I ,

$$\left(\frac{d^2}{dx^2} + k_x^2\right) I = j k_x Y v \delta(x - x') \quad . \tag{14}$$

Comparison of (12) and (14) shows that

$$I = G_x \quad \text{if} \quad v = \frac{j}{2\pi\omega\bar{\varepsilon}} \quad , \tag{15}$$

where $\bar{\varepsilon}$ is the dielectric constant of the medium containing the source point x'. At a junction of two transmission lines characterized by Z_o, k_{xo} and Z_1, k_{x1}, respectively, V and I are continuous, implying via (13b,c) the continuity of I and $(1/\varepsilon)(dI/dx)$; at a short circuit, $dI/dx = 0$. Hence, the boundary conditions on I are the same as those on G_x, thereby completing the equivalence stated in (15). The corresponding

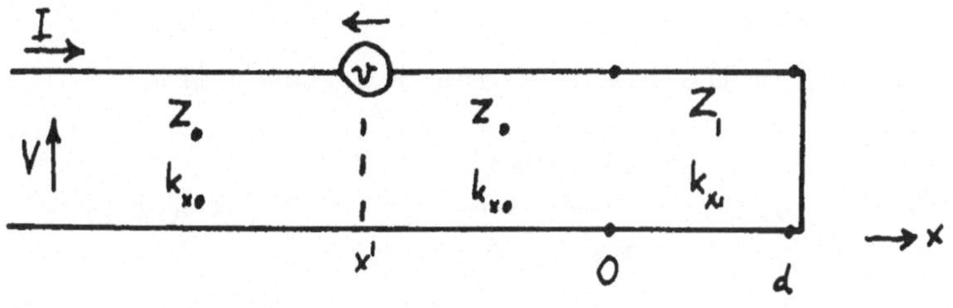

Fig. 2 - Network equivalent (x-transmission formulation).

equivalent network is schematized in Fig.2 where the source point location has been chosen in conformity with Fig. 1 .

However, the network representation applies also when the source point lies in the slab region.

The current at every point of the network in Fig.2 can be determined by a conventional circuit analysis. The reflection coefficient Γ seen looking to the right at $x=0^-$ (i.e., slightly to the left of $x=0$) is given by

$$\Gamma = \frac{\vec{Z} - Z_0}{\vec{Z} + Z_0} , \quad Z_a = \frac{k_{xa}}{\omega \varepsilon_a} , \quad a = 0,1 , \tag{16}$$

where $\vec{Z}$ is the input impedance of the short-circuited transmission line between $x=0$ and $x=d$:

$$\vec{Z} = j\, Z_1 \tan k_{x1} d . \tag{16a}$$

The current to the left of the junction point $x=0$ can then be written in terms of incident and reflected waves as

$$I_0(x,x') = -\frac{Y_0 v}{2}\left[e^{-jk_{xo}|x-x'|} - \Gamma\, e^{+jk_{xo}(x+x')}\right], \quad x,x' \leq 0. \tag{17}$$

In the region to the right of the junction point,

$$I_1(x,x') = I_0(0,x')\,\frac{\cos k_{x1}(d-x)}{\cos k_{x1} d} , \qquad 0 \leq x \leq d , \tag{18}$$

with $I_0(0,x')$ evaluated from (17).

Upon substituting (17) and (18), with (15), into (9), one obtains the axial transmission representation for the Green's function $G(\underline{r},\underline{r}')$. In the region $x<0$, the first term on the right-hand side of (17) constitutes the result for an infinitely extended medium with dielectric constant ε_o. The corresponding contribution to the integral (9) is the free-space Green's function G_{fo} which can be evaluated in the closed form (reference 1, p. 242)

$$G_{fo}(\underline{r}, \underline{r}') = \frac{e^{-jk_o|\underline{r}-\underline{r}'|}}{4\pi\,|\underline{r}-\underline{r}'|} . \tag{19a}$$

The second term in (17) yields the secondary contribution G_{so} due to the presence of the dielectric slab :

$$G_{so}(\underline{r},\underline{r}') = \frac{j}{4\pi}\int_0^\infty k_\rho\, J_o(k_\rho\rho)\,\Gamma(k_\rho)\,\frac{e^{jk_{xo}(x+x')}}{k_{xo}}\,dk_\rho , \tag{19b}$$

where the indirect dependence of Γ on k_ρ has been exhibited explicitly. This decomposition of the total Green's function G_o,

$$G_o = G_{fo} + G_{so} , \quad (x, x' < 0) , \tag{20}$$

into a primary (unperturbed) and secondary part highlights the suitability of this representation for a study of the perturbing effect of the slab, as noted previously.

The propagation constants $k_{x\alpha}$ in (17)-(19) are given via (12) by $k_{x\alpha} = (k_\alpha^2 - k_\rho^2)^{1/2}$, with $\mathrm{Im}\,k_{x\alpha} \leqslant 0.$ (21)

(For a unique specification of the square root, we define $k_{x\alpha} = k_\alpha$ when $k_\rho = 0$). The condition $\mathrm{Im}\,k_{x\alpha} \leqslant 0$ assures the convergence of the integral in (9). The integrand possesses first-order branch points at $k_\rho = \pm k_o$, and simple poles at $k_\rho = \hat{k}_\rho$, where $\hat{k}_\rho$ is determined from the resonance relation

$$\vec{Z}(\hat{k}_\rho) + Z_o(\hat{k}_\rho) = 0 \quad . \tag{22}$$

(No branch points exist at $k_\rho = \pm k_1$ since both I_o and I_1 are even functions of k_{x1}; hence, a power series expansion about $k_{x1} = 0$ contains only integral powers of k_{x1}^2). The nature of the solutions $\hat{k}_\rho$ of (22), and their interpretation as " leaky wave " or "surface wave " poles, has been discussed in another section of this course and will therefore not be repeated here (see also reference 6, Ch.11). If k_o and k_1 are complex (with Im $k_\alpha < 0$, appropriate to a dissipative medium), none of the singularities of the integrand lies on the integration path which extends along the positive real k_ρ-axis. Should k_α be real, it is simplest to determine the disposition of the integration contour by first assigning to k_α a small negative imaginary part which is then allowed to approach zero. In this manner one finds that the path is indented above the singularities, into the first quadrant of the complex k_ρ-plane.

While the " axial waveguide " analysis above leads directly to the representation (19b), it is more convenient for subsequent considerations to have an integration path which begins and ends at infinity in the complex k_ρ-plane. The desired form (see reference 1, p.252) is achieved by writing

$$J_o(k_\rho \rho) = \frac{1}{2}\left[H_o^{(1)}(k_\rho \rho) + H_o^{(2)}(k_\rho \rho)\right] \quad , \tag{23a}$$

introducing into the part of the integrand containing $H_0^{(1)}(k_\rho \rho)$ the change of variable $\bar{k}_\rho = k_\rho \exp(-j\pi)$, and making use of the circuital relation

$$H_0^{(1)}(\bar{k}_\rho e^{j\pi} \rho) = -H_0^{(2)}(\bar{k}_\rho \rho) \quad , \tag{23b}$$

whence

$$G_{so}(\underline{r},\underline{r}') = \frac{j}{8\pi} \int_{\infty e^{-j\pi}}^{\infty} k_\rho H_0^{(2)}(k_\rho \rho) \Gamma(k_\rho) \frac{e^{jk_{xo}(x+x')}}{k_{xo}} d k_\rho \quad . \tag{24}$$

Due to the presence of the Hankel function in the integrand of (24), an additional (logarithmic) branch point singularity has been introduced at $k_\rho = 0$; the associated branch cut is drawn along the negative real k_ρ-axis and the integration path extends along the lower shore of the cut.

The disposition of the path of integration C and of the singularities in the complex k_ρ-plane is shown in Fig.3. Branch cuts have been introduced, thereby assuring the single-valuedness of the integrand in (24) on a multisheeted Riemann surface. Those cuts leading to the branch points at $k_\rho = \pm k_o$ have been selected to make $\operatorname{Im} k_{xo} < 0$ on the entire top sheet, the " permissible " sheet on which the integral converges. Possible surface wave poles of $\Gamma(k_\rho)$ which lie on this sheet have also been exhibited (the leaky wave poles lie on the sheet $\operatorname{Im} k_{xo} > 0$). While it has been assumed that ε_o and ε_1 are real (with $\varepsilon_1 > \varepsilon_o$), the singularities in Fig.3 have been slightly displaced from the real axis for greater clarity.

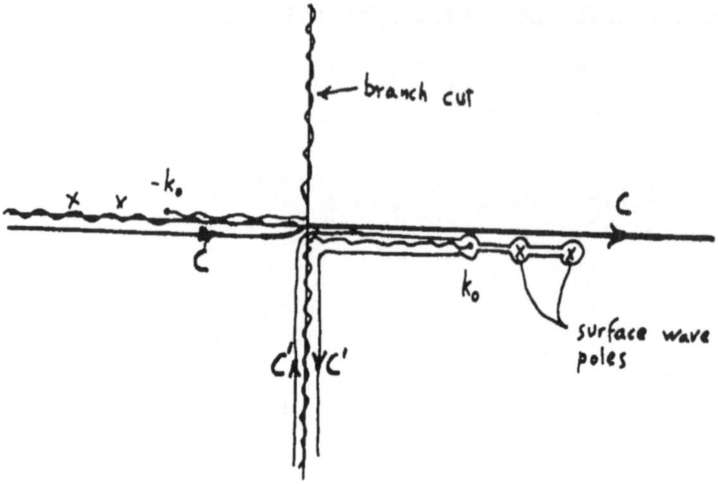

Fig. 3 - Contours of integration and singularities in complex k_ρ-plane.

The branch point singularities at $\pm k_o$ can be removed, and the integral simplified, by transforming from the k_ρ-plane to the w-plane via the change of variable

$$k_\rho = k_o \sin w , \tag{25}$$

whence $k_{xo} = k_o \cos w$. If one also introduces the polar coordinates R , φ relative to the point $x = -x'$,

$$|x+x'| = R \cos \varphi \quad , \quad \rho = R \sin \varphi , \quad 0 < \varphi < \pi/2, \tag{26}$$

one obtains

$$G_{so}(\underline{r},\underline{r}') = \frac{jk_o}{8\pi} \int_{\bar{P}} \sin w \, H_o^{(2)}(k_o R \sin w \sin\varphi) e^{-jk_o R \cos w \cos \varphi} \Gamma(k_o \sin w) dw, \tag{27}$$

with the contour $\bar{P}$ shown in Fig.4. Since the various Riemann

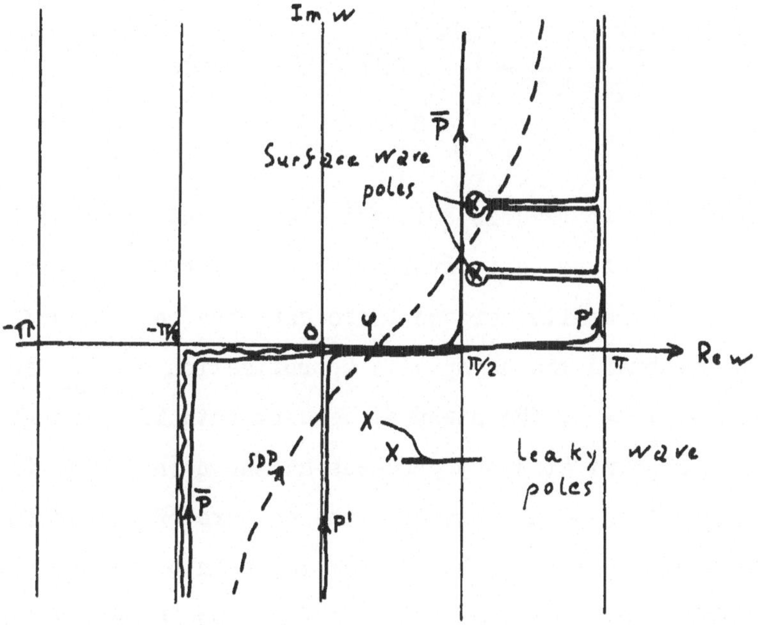

Fig.4 - Contours of integration and singularities in complex w-plane

sheets in the complex k_ρ-plane appear as adjacent strips in the complex w-plane, some pertinent leaky wave poles are also shown. The four quadrants of Fig.3 map into the strips $-\pi < \mathrm{Re}\, w < 0$, $\mathrm{Im}\, w < 0$, and $0 < \mathrm{Re}\, w < \pi$, $\mathrm{Im}\, w > 0$, in Fig. 4, in which $\mathrm{Im} \cos w < 0$.

For an approximate evaluation of the integral (27) for large values of R , one may substitute for the Hankel function the asymptotic formula

$$H_0^{(2)}(x) \sim \sqrt{\frac{2}{\pi x}}\, e^{-j(x-\pi/4)}, \quad |x| \to \infty, \quad -2\pi < \arg x < \pi \qquad (28)$$

to obtain*

$$G_{so}(\underline{r},\underline{r}') \sim \frac{jk_0}{8\pi}\left[\frac{2j}{\pi k_0 R \sin\varphi}\right]^{1/2} \int_{\bar{P}} \sqrt{\sin w}\, \Gamma(k_0 \sin w)\, e^{-jk_0 R\cos(w-\varphi)}\, dw. \qquad (29)$$

The integrand in (29) is quite similar to that treated in the Chapter " Asymptotic Evaluation of Integrals ", Sec.III A and V A ; the results derived there can thus be employed directly to obtain the asymptotic approximation of G_{so} as $(k_0 R\sin\varphi) \to \infty$. The steepest descent path SDP through the saddle point at $w = \varphi$ proceeds as shown in Fig.4. In the deformation of $\bar{P}$ into SDP, one or more of the surface or leaky wave poles may have to be taken into account. If $\varphi \not\approx \pi/2$, i.e., the observation point lies far above the slab surface, any contribution from surface wave poles is exponentially small and can be neglected (the leaky wave contribution to the far field is likewise neglibible). In this instance, the first-order asymptotic approximation of (29) is found to be

$$G_{so}(\underline{r},\underline{r}') \sim -\Gamma(k_0 \sin\varphi)\, \frac{e^{-jk_0 R}}{4\pi R}, \quad R \to \infty, \quad \varphi \not\approx \pi/2. \qquad (30)$$

* The integration path is deformed to avoid the region $w = 0$, whence the argument of the Hankel function can be made to obey the restriction (28).

This result highlights the utility of the axial waveguide representation (19b), et seq., for a determinantion of the reflecting properties of the slab configuration.

3. RADIAL WAVEGUIDE REPRESENTATION

If the configuration in Fig.1 is regarded as a radial waveguige wherein successive cross-sections have the form of circular cylinders centered on the x-axis, one requires for the representation of the Green's function $G(\underline{r},\underline{r}')$ the complete set of (axially symmetric) mode functions in the radial waveguide cross-section, i.e., the eigenfunctions in the x-domain as defined in Eqs.(8). If $\varepsilon_1 > \varepsilon_0$ (ε_1 and ε_0 are assumed real), the mode spectrum supported by the dielectric slab waveguide will contain a discrete (surface wave) and continuous (space wave) part. The completeness relation required for the representation of an arbitrary function in the domain $d > x > -\infty$ is expressed succintly in terms of the representation of the delta function $\delta(x-x')$:

$$\frac{\varepsilon_\alpha}{\varepsilon_0}\delta(x-x') = \sum_\nu h(x,k_{x\alpha}^{(\nu)})\,h^*(x',k_{x\alpha}^{(\nu)}) + \int_0^\infty h(x,k_{x\alpha})h^*(x',k_{x\alpha})\,dk_{xo}\,, \tag{31}$$

where the subscript $\alpha = 0,1$ distinguishes the regions $-\infty < x < 0$ and $0 < x < d$, respectively, and the asterisk denotes the complex conjugate. The orthonormal eigenfunctions in the discrete spectrum are given by:[7,8]

$$h(x,k_{xo}^{(\nu)}) = \frac{e^{t_\nu x/d}}{A_\nu} \quad , \qquad -\infty < x < 0 \ , \tag{32a}$$

$$h(x,k_{x1}^{(\nu)}) = \frac{\cos\left[r_\nu\left(1-\frac{x}{d}\right)\right]}{A_\nu \cos r_\nu} \quad , \qquad 0 < x < d \ , \tag{32b}$$

where we have put $k_{xo}^{(\nu)} = -j t_\nu/d$, $k_{x1}^{(\nu)} = r_\nu/d$. r_ν and t_ν, $\nu = 1,2 \ldots N$, are the positive solutions of the (transcendental) transverse resonance equations (see (22))

$$\frac{\varepsilon_o}{\varepsilon_1} r_\nu \tan r_\nu = t_\nu \quad , \quad r_\nu^2 + t_\nu^2 = (k_1^2 - k_o^2)\, d^2 \ , \tag{32c}$$

and A_ν is the normalization constant, with

$$A_\nu^2 = \frac{d}{2}\left\{\left[1+\left(\frac{t_\nu}{r_\nu}\right)^2\right]\frac{1}{t_\nu} + \left[1+\left(\frac{t_\nu \varepsilon_1}{r_\nu \varepsilon_o}\right)^2\right]\frac{\varepsilon_o}{\varepsilon_1}\right\} \ . \tag{32d}$$

The eigenfunction in the continuous spectrum cover the range of eigenvalues $0 < k_{xo} < \infty$ and are given by:

$$h(x,k_{xo}) = \frac{1}{\sqrt{2\pi}}\left[e^{-jk_{xo}x} - \Gamma e^{jk_{xo}x}\right], \quad -\infty < x < 0 \ , \tag{33a}$$

$$h(x,k_{x1}) = \frac{1}{\sqrt{2\pi}}\left[1-\Gamma\right]\frac{\cos k_{x1}(d-x)}{\cos k_{x1} d}, \quad 0 < x < d \ . \tag{33b}$$

The relation between k_{x1} and k_{xo} is stated in (8a), and the reflection coefficient Γ is given in (16).

Upon multiplying (31) by $(\varepsilon_0/\varepsilon_\alpha)G(\rho,x';\underline{r}'')$ and integrating over x' between the limits $x'=-\infty$ and $x'=d$, one obtains the radial waveguide representation in the form

$$G_\alpha(\underline{r},\underline{r}'') = \sum_\nu h(x,k_{x\alpha}^{(\nu)})\,Q(\rho,x'';k_x^{(\nu)}) + \int_0^\infty h(x,k_{x\alpha})\,Q(\rho,x'';k_{xo})\,dk_{xo}\,, \qquad (34)$$

with the function Q to be determined. Upon substituting (34) and (31) into (4), performing the differentiations, recalling (8), and noting that the eigenfunctions h constitute an orthogonal set, one obtains

$$Q(\rho,x'';k_x) = h^*(x'',k_{x\alpha})\,G_\rho(\rho;k_x)\,, \qquad (35)$$

where G_ρ is the radial Green's function defined by the equation

$$\left(\frac{1}{\rho}\frac{d}{d\rho}\rho\frac{d}{d\rho}+k_\rho^2\right)G_\rho(\rho;k_x) = -\frac{\delta(\rho)}{2\pi\rho}\,, \quad k_\rho^2 = k_\alpha^2 - k_{x\alpha}^2\,. \qquad (36)$$

The solution of (36), satisfying at $\rho\to\infty$ a radiation condition, is

$$G_\rho(\rho;k_x) = -\frac{j}{4}H_0^{(2)}(k_\rho\rho)\,, \quad \mathrm{Im}\,k_\rho \leqslant 0\,. \qquad (37)$$

From a network viewpoint, G_ρ represents the voltage on an infinite radial transmission line excited at $\rho=0$ by a shunt current generator with infinite internal impedance (see reference 2), as shown in Fig.5. The resulting solution (which

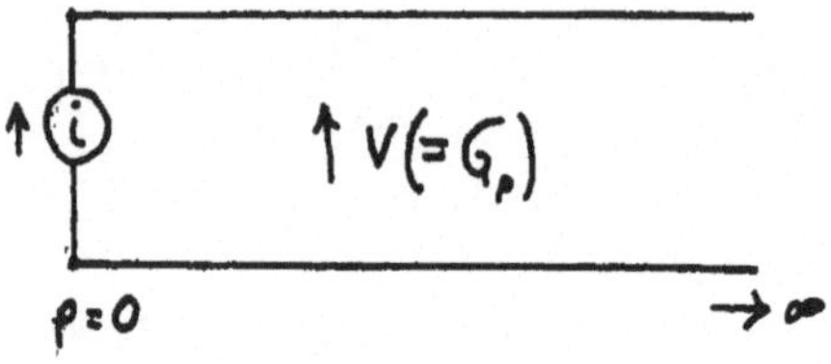

Fig.5 - Network equivalent (ρ-transmission formulation)

satisfies all the requirements in Eqs.(4) is thus given by (only the region $x<0$ will be considered):

$$G(\underline{r},\underline{r}') = -\frac{j}{4}\sum_{\nu}\frac{e^{t_\nu(x+x')/d}}{A_\nu^2} H_0^{(2)}\left(\sqrt{k_o^2+(t_\nu^2/d^2)}\;\rho\right)$$

$$-\frac{j}{4}\int_0 h(x,k_{xo})h^*(x',k_{xo})H_0^{(2)}\left(\sqrt{k_o^2-k_{xo}^2}\;\rho\right)dk_{xo}\quad,\quad x<0\;, \tag{38}$$

where $h(x,k_{xo})$ is listed in (33a), and $\sqrt{k_o^2-k_{xo}^2}$ is positive and negative imaginary for $k_{xo}<k_o$ and $k_{xo}>k_o$, respectively. To avoid the branch point at $k_{xo}=k_o$, the integration path is indented into the first quadrant of the complex k_{xo}-plane; none of the pole singularities of h are located on the positive real k_{xo}-axis.

The radial transmission representation in (38)

exhibits explicitly, and in the most direct manner, the splitting of the radiated field into a surface wave and " space wave " part. It is suitable for an approximate evaluation of $G(\underline{r},\underline{r}')$ when both the source and observation point locations are near the interface. In particular, if $x = x' = 0$, the cylindrical surface wave fields decay with increasing ρ like $(1/\sqrt{\rho})$, owing to the asymptotic properties of the Hankel function. The space wave integral contribution can be estimated from a saddle point evaluation wherein $H_0^{(2)}(k_\rho \rho)$ is approximated by $\sqrt{2/(\pi k_\rho \rho)} \exp\left[-jk_\rho \rho + j\pi/4\right]$; the saddle point of the exponent lies at $k_{xo} = 0$. The lowest order term $(\sim 1/\rho)$ in the asymptotic expansion vanishes since the amplitude factor $h(0,0) = 0$. Hence, the space wave is $O(1/\rho^2)$, i.e., negligible at large distances in comparison with the surface wave field. The above-sketched asymptotic method of evaluation of the space wave integral can also be employed for finite $|x + x'|$, provided that $|x + x'| \ll \rho$.[9]

4. TRANSITION FROM AXIAL TO RADIAL WAVEGUIDE REPRESENTATION

The axial and radial waveguide analyses have led to the quite dissimilar Green's function representations in (19)-(20) and in (38), respectively. Their utility in effecting an approximate explicit evaluation of $G(\underline{r},\underline{r}')$ has been emphasized. The connection between the two representations may be established by the analytic continuation of the integrands into the complex plane and subsequent deformation of contours. It will be convenient to begin with the axial waveguide expression in (19a,b).

Since a deformation of the integration path is contemplated, it is desirable to deal with a contour whose endpoints lie at infinity in the complex k_ρ-plane. The necessary manipulations have been discussed in connection with (24) and lead via (9), (15) and (17) to (note: $x, x' < 0$):

$$G(\underline{r},\underline{r}') = -\frac{j}{8\pi}\int_{\infty e^{-j\pi}}^{\infty} k_\rho H_0^{(2)}(k_\rho \rho)\left[\frac{e^{-jk_{xo}|x-x'|} - \Gamma e^{jk_{xo}(x+x')}}{k_{xo}}\right]dk_\rho . \tag{39}$$

The path of integration C in the complex k_ρ -plane is shown in Fig.3.

The contour is to be closed about the singularities of G_{xo} which are also shown in Fig.3. Since $\operatorname{Im} k_{xo} < 0$ on the entire top sheet of the multisheeted k_{xo}- surface, the exponential terms in the integrand of (39) decay everywhere at infinity in the complex k_ρ-plane. Due to its asymptotic behavior $\sim \exp\left[-jk_\rho \rho\right]$, the Hankel function decays as $k_\rho \to \infty$ in the lower half-plane. Hence, the integrand vanishes exponentially along quarter circles with infinite radius in the third and fourth quadrants, and the contour C can be changed into the path C' in Fig.3, surrounding the singularities of G_{xo}. C' can be resolved into a branch cut integral plus closed contours about the surface wave poles of Γ. Upon evaluating the integrals over the latter contours by the Cauchy residue theorem, one obtains exactly the series contribution in (38). In the branch cut integral, one introduces k_{xo} as a new variable,

$$k_{xo} = \sqrt{k_o^2 - k_\rho^2} \quad , \tag{40}$$

whence k_{xo} is real and runs from $+\infty$ to $-\infty$ as k_ρ moves along the branch cut part of C'. Upon simplifying the k_{xo} integral and noting from (16) that $\Gamma(-k_{xo}) = \Gamma^*(k_{xo})$, one obtains the desired contribution from the continuous spectrum in (38).

The above-described deformation of C into C' in the k_ρ-plane corresponds in the w-plane to a change from $\bar{P}$ to P' as shown in Fig.4.

5. CHARACTERISTIC GREEN'S FUNCTION FORMULATION

The axial and radial waveguide analyses lead to Green's function representations in terms of the radial and axial mode spectra, respectively; however, each of these formulations is of limited utility for the derivation of explicit approximate expressions for the field. Most useful is a contour integral representation (in the complex plane) as in (39), from which either the expression in (19a,b) or that in (38) can be derived by a suitable path deformation. Moreover, the path in easily distorted into the steepest descent path through pertinent saddle points, thereby facilitating an asymptotic evaluation of the integral. The derivation of this more general representation from the special formulations mentioned above is not always straightforward. Thus, the transition from (38) to (39) is not at all trivial, and even the simpler transformation from (19a,b) to (24) or (39) requires intermediate steps which are not immediately obvious. The desired expression can, however, be obtained directly by the characteristic Green's function procedure sketched below.

Instead of the eigenvalue problems in the axial or radial domains, consider the associated one-dimensional Green's functions G_x and G_ρ defined in (12) and (36), respectively. Rather than restricting k_ρ^2 as before*, this parameter is now allowed to range over a suitable domain of complex values. The associated Green's functions may then be termed " characteristic " since one may infer from them by contour integration the complete set of characteristic functions defined in Eqs.(8) and (7), respectively. For example, in the x-domain, the pertinent completeness relation can be phrased in terms of the representation theorem for the delta function

$$\delta(x - x') = -j \oint_C G_{x\alpha}(x,x';k_\rho)\, d k_\rho^2 \quad , \tag{41}$$

where C is a closed contour encircling in the positive sense all of the singularities of $G_{x\alpha}$ in the complex k_ρ^2-plane, as shown in Fig.6 (these singularities have been discussed in connection with Eqs.(21) and (22)).

The two-dimensional function $G_\alpha(\underline{r},\underline{r}')$ can now be represented in terms of the characteristic Green's functions as follows :

---

* In (12) k_ρ^2 is restricted to the spectrum of radial eigenvalues; in (36), $k_{x\alpha}^2$ takes on only the eigenvalues in the x-domain, thereby restricting k_ρ^2 .

$$G_\alpha(\underline{r},\underline{r}') = -j \oint_C G_{x\alpha}(x,x';k_\rho)\, G_\rho(\rho;k_x)\, dk_\rho^2 \quad , \tag{42}$$

where $G_{x\alpha}$ and G_ρ are given in (15) - (18) and in (37), respectively, and the contour C encloses only the singularities of $G_{x\alpha}$ (see Fig.6). To verify that (42) indeed constitutes the desired solution, one notes first that the boundary conditions (4a-c) are satisfied by virtue of the similar boundary conditions on $G_{x\alpha}$ and G_ρ. To check that (42) satisfies the inhomogeneous wave equation, operate on the right-hand side with $(\nabla^2 + k_\alpha^2)$ to obtain (due to the exponentially damped integrand as $|k_\rho^2| \to \infty$, the ∇^2 operator can be taken under the integral sign):

$$(\nabla^2 + k_\alpha^2)\, G_\alpha(\underline{r},\underline{r}') = \tag{43a}$$

$$= -j \oint_C \left(\frac{1}{\rho}\frac{\partial}{\partial\rho}\rho\frac{\partial}{\partial\rho} + \frac{\partial^2}{\partial x^2} + k_\alpha^2\right) G_{x\alpha} G_\rho \; dk_\rho^2 \tag{43b}$$

$$= -j \oint_C \left\{ G_{x\alpha}\left[-k_\rho^2 G_\rho - \frac{\delta(\rho)}{2\pi\rho}\right] + G_\rho\left[k_\rho^2 G_{x\alpha} - \frac{\delta(x-x')}{2\pi}\right]\right\} dk_\rho^2 \tag{43c}$$

$$= -\frac{\delta(\rho)}{2\pi\rho}(-j)\oint_C G_{x\alpha}\, d\, k_\rho^2 = -\frac{\delta(\rho)\,\delta(x-x')}{2\pi\rho} \; . \tag{43d}$$

The transition from (43b) to (43c) is accomplished via (36) and (41); the last term in the integrand of (43c) integrates to zero by the Cauchy theorem since G_ρ has no singularities inside the contour C. Upon changing variables from k_ρ^2 to k_ρ, one obtains from (42) the integral (39), taken over the contour C' in Fig.3. Since $G_{x\alpha}$ and G_ρ decay at $|k_\rho^2| \to \infty$ on the entire top sheet of the k_ρ^2-plane in Fig.6, path deformations can be carried out freely.

For details on the method of characteristic Green's functions, the reader is referred to references 10.

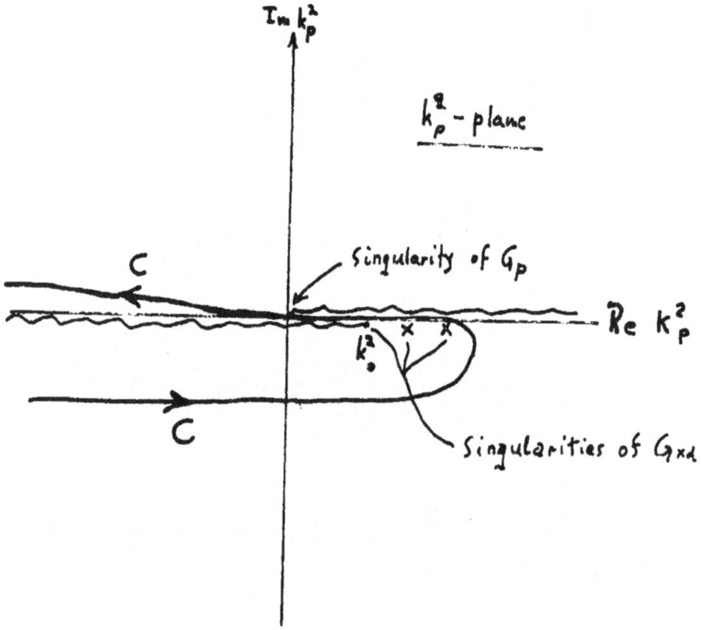

Fig.6 - Contours and singularities in complex k_ρ^2-plane

REFERENCES

1. A.Sommerfeld, " Partial Differential Equations in Physics", Academic Press, New York, 1949. Sec. 32.
2. L.B.Felsen and N.Marcuvitz, " Modal Analysis and Synthesis of Electromagnetic Fields", Microwave Research Institute, Polytechnic Institute of Brooklyn, Report R-726-59, PIB-654, June 1959. Sec.Clb.
3. cf. P.M. Morse and H.Feshbach, " Methods of Theoretical Physics", McGraw Hill Book Co., New York, 1953, p.766.
4. cf. B.Friedman, " Principles and Techniques of Applied Mathematics", J.Wiley & Sons, New York, 1956. Ch. 3 .
5. N.Marcuvitz and J.Schwinger, " On the Representation of the Electric and Magnetic Fields...", J.A.P., 22 , June 1951.
6. R.E.Collin, " Field Theory of Guided Waves" , McGraw Hill Book Co., New York, 1960.
7. Reference 2, Sec. Elb.
8. C.M.Angulo, " Diffraction of Surface Waves by a Semi-Infinite Dielectric Slab", IRE Trans. AP-5, 1957. The normalization factor for the continuous spectrum is listed incorrectly here.
9. cf. A.L.Cullen, " The Excitation of Plane Surface Waves" , Proc. I.E.E., Monograph 93, Part IV, Feb. 1954. Sec. 5.
10. For details on the method of characteristic Green's functions, see;
    a) E.C.Titchmarsh, " Eigenfunction Expansions", Oxford University Press, Vol. 1.
    b) N.Marcuvitz, Comm. Pure and App. Math., 4 (1951); p. 263
    c) B.Friedman, " Principles of Applied Mathematics", Wiley & Sons, New York (1956)
    d) N.D.Kazarinoff and R.K.Ritt, Annals of Phys., 6 (1959), p. 277
    e) reference 2.

CENTRO INTERNAZIONALE MATEMATICO ESTIVO

(C.I.M.E.)

G. GEROSA

PROPAGATION OF ELECTROMAGNETIC WAVES IN RECTANGULAR GUIDES LOADED WITH MAGNETIZED FERRITE

ROMA - Istituto Matematico dell'Università

PROPAGATION OF ELECTROMAGNETIC WAVES IN RECTANGULAR GUIDES LOADED WITH MAGNETIZED FERRITE (1)

G. GEROSA

Istituto di Elettronica
dell'Università di Roma

It is the purpose of this lecture to report on some research carried out at the Cattedra di Elettronica Applicata della Facoltà d'Ingegneria dell'Università di Roma on the propagation of e. m. waves in rectangular waveguides loaded with magnetized ferrite.

Let us consider a lossless ferrite, uniformly magnetized along the z-axis of a right hand system of rectangular coordinates x, y, z, by a d. c. magnetic field of sufficient intensity H_0 to saturate the medium.

The dielectric constant of such a medium is a scalar $\varepsilon_0 \varepsilon$ and the magnetic permeability is a tensor $\underset{\sim}{\mu}$ given by :

$$(1) \qquad \underset{\sim}{\mu} = \mu_0 \begin{vmatrix} \mu_1 & j\mu_2 & 0 \\ -j\mu_2 & \mu_1 & 0 \\ 0 & 0 & 1 \end{vmatrix}$$

where :

$$(1a) \qquad \mu_1 = 1 + \frac{\rho}{1-\tau^2} \quad ; \qquad \mu_2 = \frac{\tau\rho}{1-\tau^2} \quad ;$$

---

(1) The work reported in this document has been sponsored by the Electronics Research Directorate, AFCRL, AFRD(ARDC) through its European Office, under Contract AF 61(052)-101.

$$\text{(1b)} \qquad \rho = \frac{M_0}{\mu_0 H_0} \quad ; \qquad \tau = \frac{\omega}{\omega_0} \quad ;$$

and M_0 is the intensity of the saturation magnetization, ω and $\omega_0 = -\gamma H_0$ are the applied and the resonant circular frequencies (time dependence $e^{j\omega t}$ is assumed), γ is the gyromagnetic ratio for the electron, μ_0 and ε_0 are the permeability and the dielectric constant of the vacuum.

Let us consider a plane wave in the ferrite region:

$$\text{(2)} \qquad \begin{Bmatrix} \underset{\sim}{E}_f \\ \\ \underset{\sim}{H}_f \end{Bmatrix} = \begin{Bmatrix} \sqrt{\frac{\mu_0}{\varepsilon_0}}\ \underset{\sim}{e}_f \\ \\ \underset{\sim}{h}_f \end{Bmatrix} \cdot A\, e^{j(K_{xf}x + K_{yf}y + K_{zf}z)}$$

where $\underset{\sim}{e}_f$ and $\underset{\sim}{h}_f$ are two constant adimensional vectors and A is an amplitude factor.

By introducing (2) into Maxwell's equations written for the ferrite region :

$$\text{(3)} \qquad \begin{cases} \nabla \times \underset{\sim}{E} = -j\,\omega \underset{\sim}{\mu} \cdot \underset{\sim}{H} \\ \nabla \times \underset{\sim}{H} = j\omega \varepsilon_0 \varepsilon \underset{\sim}{E} \end{cases}$$

we obtain a system of six linear homogeneous algebraic equations. By setting equal to zero the determinant of the coefficients of such a system we obtain the following characteristic equation for a plane wave in the ferrite region :

$$\text{(4)} \qquad \mu_1 t^4 + \left[(\mu_1 + 1)K_{zf}^2 - \varepsilon(\mu_1^2 - \mu_2^2 + \mu_1)\right] t^2 + K_{zf}^4 - 2\mu_1 \varepsilon K_{zf}^2 + \varepsilon^2(\mu_1^2 - \mu_2^2) = 0$$

where

$$t^2 = K_{xf}^2 + K_{yf}^2 \,. \tag{5}$$

In (4), (5) and in what follows the propagation constants will be measured, except when otherwise specified, in the units of $\omega\sqrt{\mu_0\varepsilon_0}$. $\underset{\sim}{e}_f$ and $\underset{\sim}{h}_f$ have the following expressions :

$$\begin{aligned}
e_{xf} &= -K_{zf}\left[(\mu_1 - 1)K_{xf}K_{yf} + j\,\mu_2(\varepsilon - K_{xf}^2)\right] \\
e_{yf} &= -K_{zf}\left[K_{zf}^2 - \mu_1(\varepsilon - t^2) - (\mu_1 - 1)K_{xf}^2 - j\,\mu_2 K_{xf}K_{yf}\right] \\
e_{zf} &= j\,\mu_2 K_{xf}(\varepsilon - t^2) + K_{yf}\left[K_{zf}^2 - \mu_1(\varepsilon - t^2)\right] \\
h_{xf} &= K_{xf}^2\left[K_{zf}^2 - (\varepsilon - t^2)\right] - \varepsilon\left[K_{zf}^2 - \mu_1(\varepsilon - t^2)\right] \\
h_{yf} &= K_{xf}K_{yf}\left[K_{zf}^2 - (\varepsilon - t^2)\right] + j\,\mu_2\,\varepsilon\,(\varepsilon - t^2) \\
h_{zf} &= -K_{zf}\left[K_{xf}(\mu_1\varepsilon - K_{zf}^2 - t^2) + j\,\mu_2\,\varepsilon\,K_{yf}\right]
\end{aligned} \tag{6}$$

It should be noted that the wave (2), (6) possesses reflection symmetry along the z-direction; this however is not true for the x and y directions.

In order to obtain therefore a mode for a parallel plate guide with plates normal to the z-direction it is sufficient to impose :

$$K_{zf} = \frac{m\pi}{a} \quad ; \qquad m = 0,\ 1,\ 2, \ldots\ldots \tag{7}$$

We shall now consider a modal solution for a parallel plate guide with plates normal to the y-axis (Fig.1).

Each mode is labelled by a pair of values K_z, K_x (we delete the subscript f). For such a pair we obtain from (4), (5) four values of K_y, $\pm K_{y1}$ and $\pm K_{y2}$.

We shall therefore express our modal solution as a superposition of four waves of type (2), (6).

By imposing the boundary conditions

$$E_x(y = 0) = 0$$

$$E_z(y = 0) = 0$$

$$E_x(y = b) = 0$$

$$E_z(y = b) = 0$$

we obtain a system of four linear homogeneous equations in the four unknown amplitudes of the four waves.

By setting equal to zero the determinant of the coefficients we obtain the characteristic equation for our structure, which has the following form [(1)] :

(8) $$f(K_z^2, K_x^2) = 0$$

We shall report the principal results of the discussion of (8).

Positive real K_z^2 and K_x^2 are assumed. K_{y1} and K_{y2} are therefore real or purely imaginary . With reference to the y dependence we shall therefore classify our modes into three classes :

I) K_{y1} and K_{y2} both real

II) " " " one real and the other imaginary,

III) " " " both imaginary (surface-wave type).

These three classes correspond to three zones of the first quadrant of the K_z^2, K_x^2 plane, zone limited by the curves $K_x^2 = t_1^2$ and $K_x^2 = t_2^2$ which belong to an hyperbola.

The configuration of these three zones depends on the actual values of ε, ρ and τ. We shall fix ε and ρ and we assume τ as variable.

Following the configuration of the three zones we can divide the field of variability of τ from 0 to ∞ into four regions.

The typical configuration of the various zones for each region is depicted in Figs. 2, 3, 4, 5, in which the propagation constants are normalized with respect to $\omega\sqrt{\mu_0 \varepsilon_0 \varepsilon}$ rather than with respect to $\omega\sqrt{\mu_0\varepsilon_0}$.

In Figs. 2, 3, 4, 5 are also recorded the solution curves of equation (8).

We note that, for $K_z = 0$, (8) becomes :

$$K_x^2 = t_{1,2}^2 - \frac{n^2\pi^2}{b^2} \qquad n = 0, 1, 2, \dots \tag{9}$$

From (9) it is apparent that by decreasing b the distance between the solution curves increases.

We shall consider now a rectangular guide obtained by the parallel plate guide of Fig. 1 by adding two parallel perfectly conducting planes normal to the preceding ones.

If the planes are normal to the z-axis, in order to satisfy the new boundary conditions it is sufficient to impose (7). We can derive the following conclusions :

1) In the cases of Figs.2 and 5 we have a finite number of unattenuated propagating modes, going under cut-off for a and b sufficiently small.

2) In the cases of Figs.3 and 4 we have always an infinite number of unattenuated propagating modes.

3) The lowest modes are not in general modes with $K_z = 0$.

Consider now a rectangular guide obtained by the parallel plate guide by adding two planes normal to the x-axis. The lack of reflection symmetry of the modes of the parallel plate guide accounts for the difficulty encountered in finding a modal solution for this structure.

This is a discontinuity problem, which we have recently solved by superimposing an infinite number of modes of the parallel plate guide [2].

We shall consider now another structure : a parallel plate guide partially filled with magnetized ferrite (Fig.6).

A modal solution for this structure can be found in a similar manner as for the case of the completely filled guide.

We need now to construct also a general field for the vacuum region, associated with a pair K_z, K_x . For the vacuum we have only two values of K_y, $\pm K_{yo}$, obtained by :

$$K_x^2 + K_{yo}^2 + K_z^2 = 1 \tag{10}$$

To each triplet of values K_x, K_y, K_z there correspond however for the vacuum two independent plane waves (f.i. with reference to the x-direction a TE and a TM wave).

We can therefore express the field in the vacuum region as a superposition of four plane waves.

By imposing the boundary and continuity conditions :

$$E_{xf}(y = b_f) = 0$$

$$E_{zf}(y = b_f) = 0$$

$$E_{xo}(y = -b_o) = 0$$

$$E_{zo}(y = -b_o) = 0$$

$$E_{xf}(y = 0) = E_{xo}(y = 0)$$

$$E_{zf}(y = 0) = E_{zo}(y = 0)$$

$$H_{xf}(y = 0) = H_{xo}(y = 0)$$

$$H_{zf}(y = 0) = H_{zo}(y = 0)$$

we obtain a system of eight linear homogeneous equations.

By setting equal to zero the determinant of the coefficients we obtain a characteristic equation of the following form :[1]

(11) $$f(K_z^2, K_x) = 0$$

Since (11) contains odd powers of K_x, the structure is not reciprocal.

We have discussed equation (11)[3].

The most peculiar result of this discussion is the unidirectional character presented by the solutions for some values of the parameters.

Let us consider a rectangular waveguide obtained by adding two plates normal to the z-axis.

The solution curves (K_x versus b_f) for such a structure and for the indicated values of the parameters, are depicted in Figs.7 and 8 (the two cases differ only for the value of b, larger in the first case), from which the unidirectional character is apparent.

This character has been the starting point of the so-called thermodynamic paradox.

Another problem we have discussed is the more general one of the spectrum of modes for the parallel plate wave guide loaded with a slab of ferrite away from the side walls [(4)].

Some experiments have been carried out in order to verify the theory.

---

(1) G.Barzilai and G.Gerosa - "Modes in Rectangular Guides Filled with Magnetized Ferrite" - L'Onde Electrique, 38e Année, N° 376ter, pp.612-617, Supplément Spécial - Congrès International Circuits et Antennes Hyperfréquences, Paris 21-26 October 1957 - See also : Il Nuovo Cimento Vol.X-7, pp.685-697; March 1958.

(2) G.Barzilai and G.Gerosa - "An exact Modal Solution for a Rectangular Guide Loaded with Longitudinally Magnetized Ferrite" -

Istituto di Elettrotecnica dell'Università di Roma - Techn.Note No. 3, Contract AF 61 (052) - 101; January 2, 1961.

(3) G.Barzilai and G.Gerosa - "Modes in Rectangular Guides Partially Filled with Transversely Magnetized Ferrite" - IRE Trans. on Antennas and Propagation, vol. AP-7, Special Supplement, pp. S471-S474; December 1959 - For details : Istituto Elettrotecnico dell'Università di Roma - Techn. Note No. 1, Contract AF 61 (052) - 101; June 3, 1959.

(4) G.Barzilai and G.Gerosa - "Modes in Rectangular Guides Loaded with a Transversely Magnetized Slab of Ferrite away from the Side Walls" IRE Transactions on Microwave Theory and Techniques, vol. MTT-9, pp.403-408; September 1961.

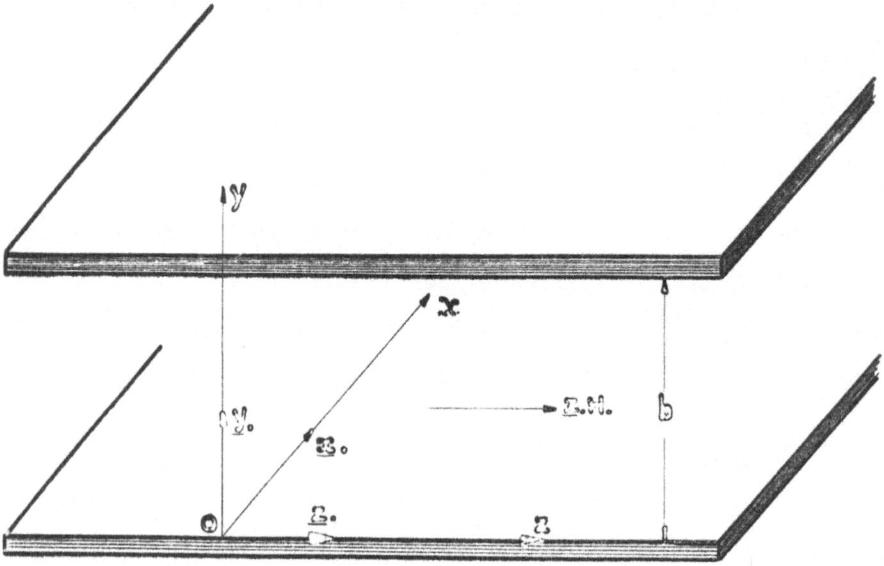

Fig. 1

Geometry of the parallel plate guide completely filled with ferrite.

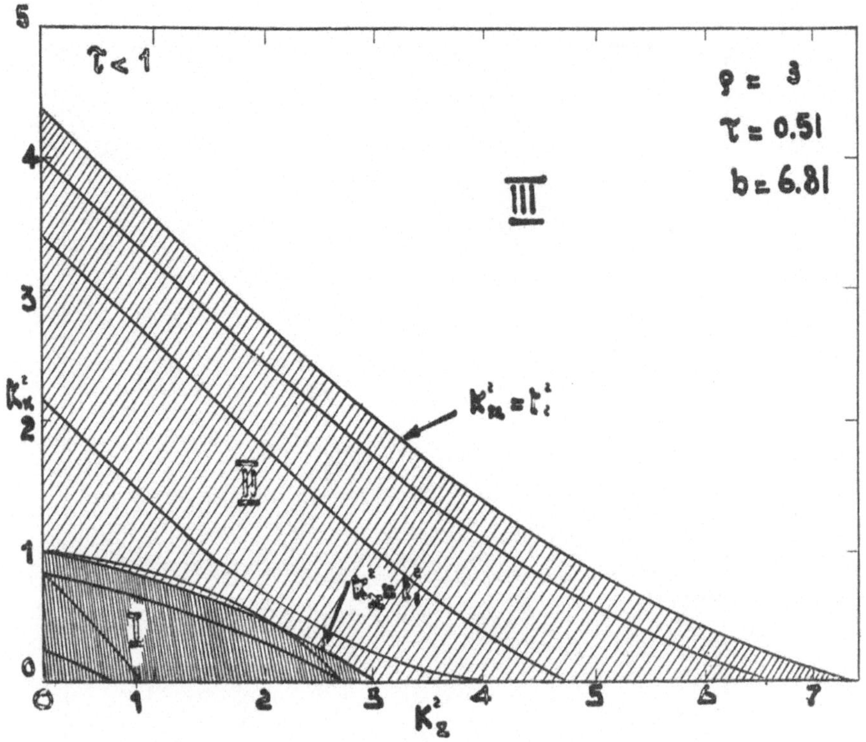

Fig. 2

The mapping of the solutions of equation (8) for $\tau < 1$. The curves have been calculated for the values of ρ, τ and b indicated. These values may be taken to correspond to $\varepsilon = 10$; $H_o = \frac{10^6}{4\pi}$ A/m ; $M_o = 0{,}3$ Wb/m^2 ; $f = \frac{\omega}{2\pi} = 1425$ Mc/s ;

$b/\omega\sqrt{\mu_o \varepsilon_o \varepsilon} = 2{,}84"$

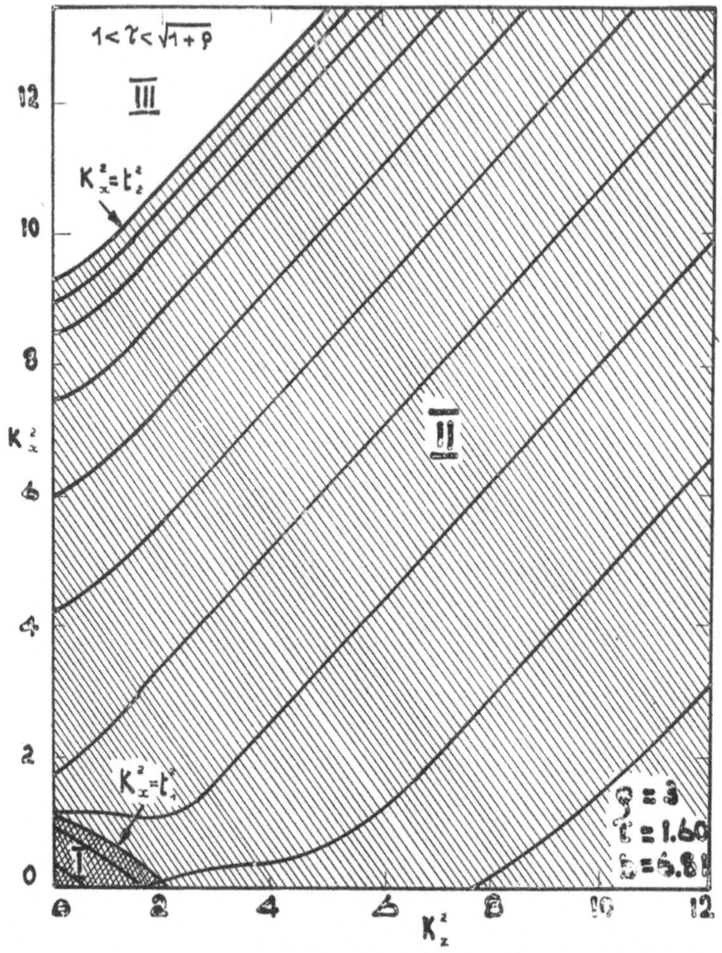

Fig. 3

The same as Fig. 2 except for $1 < \tau < \sqrt{1+\rho}$; f = 4500 Mc/s ; $b/\omega\sqrt{\mu_0 \varepsilon_0 \varepsilon}$ = 0,9"

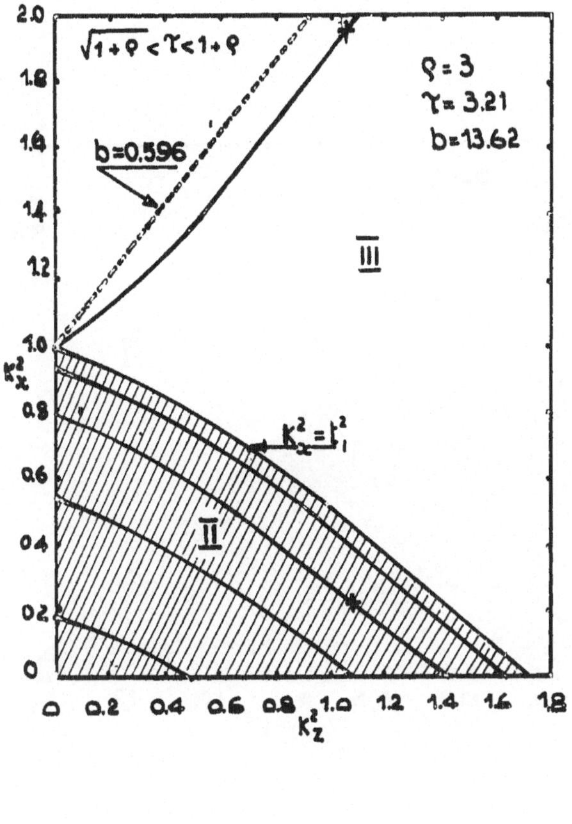

Fig. 4

The same as Fig. 2 except for $\sqrt{1+\rho} < \tau < 1+\rho$; f = 9000 Mc/s; $b/\omega\sqrt{\mu_0\varepsilon_0\varepsilon}$ = 0,9"

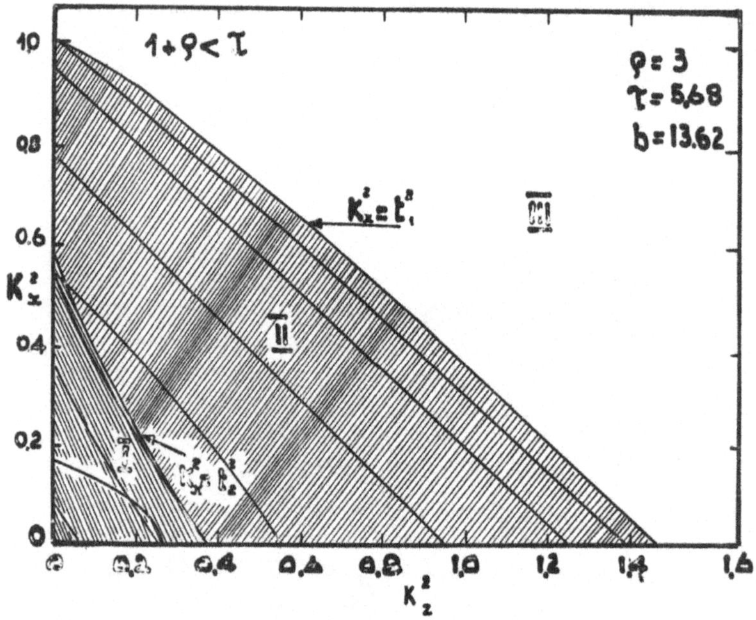

Fig. 5

The same as Fig. 2 except for $\tau > 1+\rho$; f = 15.900 Mc/s ; $b/\omega\cdot\sqrt{\mu_0\varepsilon_0\varepsilon}$ = 0,51"

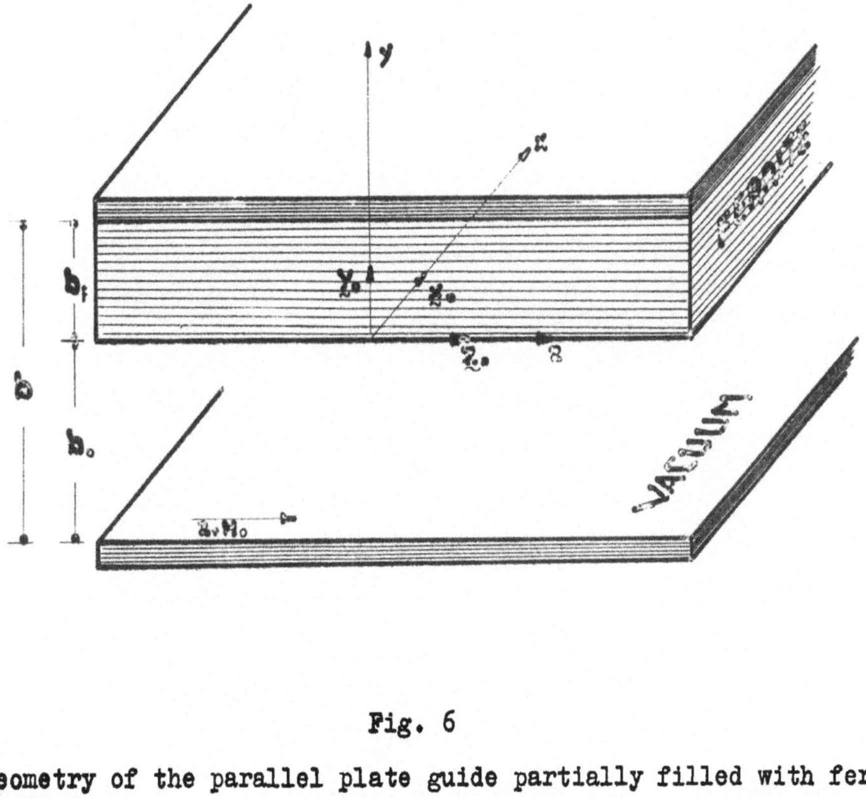

Fig. 6

Geometry of the parallel plate guide partially filled with ferrite.

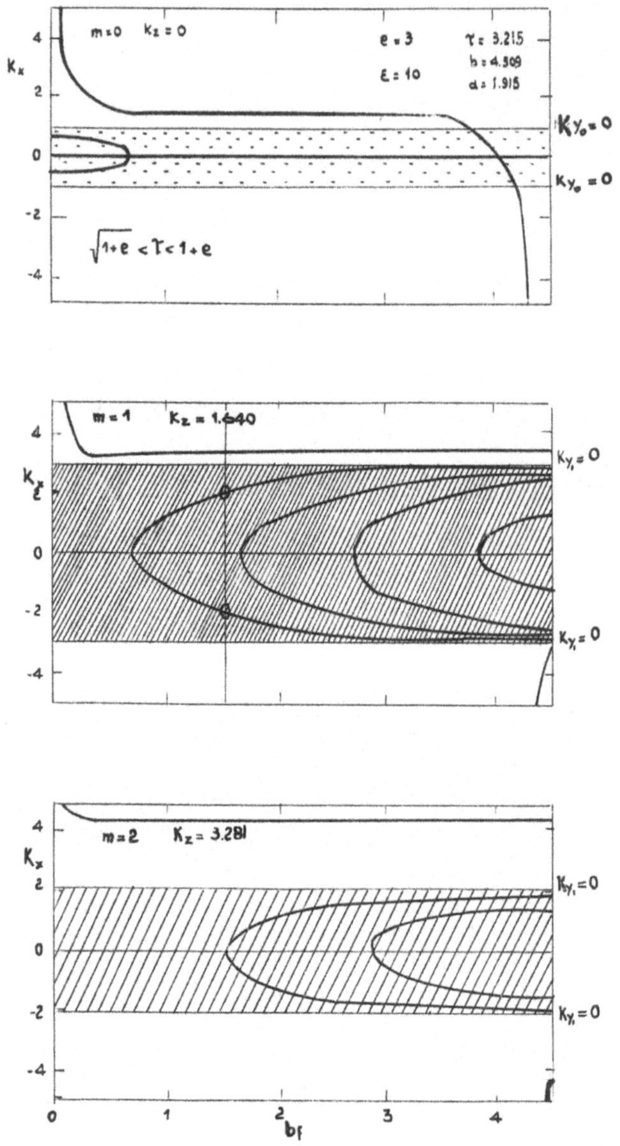

Fig. 7

The mapping of the solution of equation (11) for the indicated values of ρ , ε , τ , b and a. These values may be taken correspond to

$$H_0 = \frac{10^6}{4\pi}\ \mathrm{A/m}\ ;\ M_0 = 0{,}3\ \frac{\mathrm{Wb}}{\mathrm{m}^2}\ ;\ f = \frac{\omega}{2\pi} = 9000\ \mathrm{Mc/s}\ ;$$

$$b/\omega\sqrt{\mu_0\varepsilon_0} = 0{,}9";\quad a/\omega\sqrt{\mu_0\varepsilon_0} = 0{,}4".$$

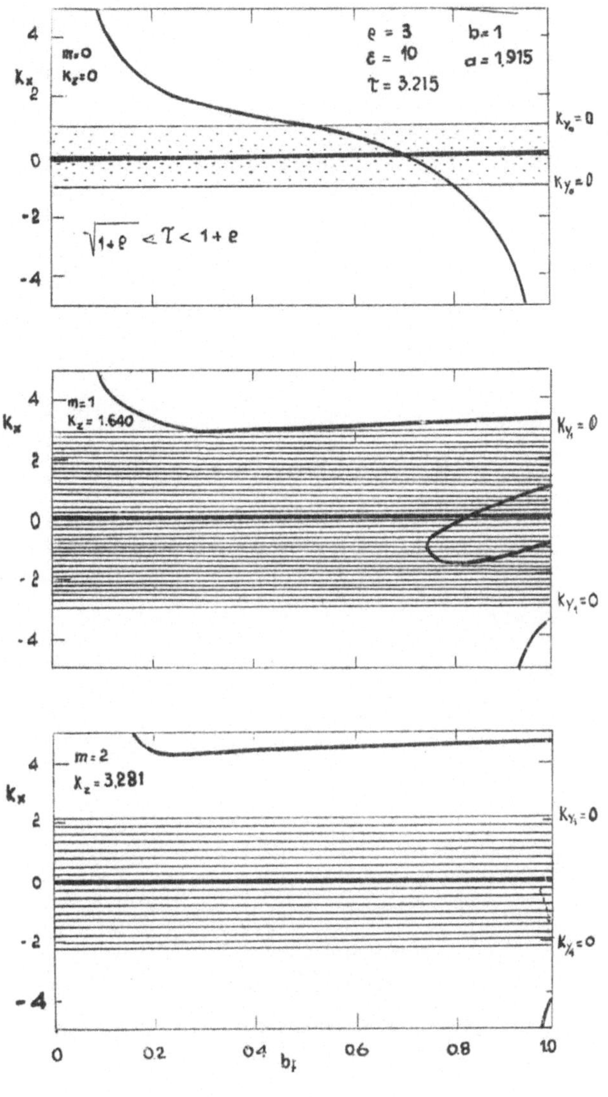

Fig. 8

The same as Fig. 7 except for $b/\omega\sqrt{\mu_0\varepsilon_0} = 0{,}209$".

CENTRO INTERNAZIONALE MATEMATICO ESTIVO

(C.I.M.E.)

DARIO GRAFFI

SULLE CONDIZIONI AL CONTORNO APPROSSIMATE NELL'ELETTROMAGNETISMO

ROMA - Istituto Matematico dell'Università

SULLE CONDIZIONI AL CONTORNO APPROSSIMATE NELL'ELETTROMAGNETISMO

di

Dario Graffi

1. Nelle lezioni tenute a questo corso del C.I.M.E. il Prof.Bremmer ha esposto la soluzione di Sommerfeld per il campo di un dipolo verticale supponendo la terra piana e omogenea. Più precisamente, ha considerato il campo di un dipolo qualora lo spazio sia occupato da due mezzi omogenei separati da un piano α, come appunto l'atmosfera e la terra (supposta piana e omogenea); ammettendo inoltre il dipolo normale ad α. Le formule ottenute dal Sommerfeld, sono, come si è visto, alquanto complicate e perciò appaiono opportune quelle approssimazioni delle formule stesse che il prof.Bremmer ci ha esposto.

Ora, lo scienziato russo Leontovitch (1) ha proposto, credo per primo, d'introdurre, fin dall'inizio, nel problema sovra citato (per brevità lo chiameremo problema di Sommerfeld) alcune approssimazioni che permettono di trattarlo in modo più agevole e inoltre conducono rapidamente alle formule semplificate a cui si è accennato. In altre parole, Leontovitch, si propone di calcolare solo il campo del dipolo nell'aria (il che ha maggior interesse dal punto di vista pratico) ed invece di considerare come fa il Sommerfeld anche il campo della terra suppone che l'ef-

(1) Cfr. G.Boudouris - Propagation tropospherique - Centre de Documentation Universitaire 5, Place de la Sorbonne, Paris, 1957. Cfr. anche l'articolo del medesimo autore inserito nel Supplemento del Nuovo Cimento 5-71-1957.

fetto di quest'ultima si possa approssimare con alcune relazioni valide alla superficie α del suolo. In tal modo, come si vedrà meglio in seguito, la determinazione del campo di un dipolo si riduce alla risoluzione dell'equazione delle onde nella regione occupata dall'aria corredata da alcune condizioni all'infinito e sul piano che limita quella regione.

Ora però condizioni al contorno approssimate si ammettono anche nella teoria delle guide d'onda. Di regola questa teoria si svolge supponendo perfetto il conduttore che limita la guida sicchè basta considerare il campo entro la guida stessa, del tutto indipendente, per le proprietà dei conduttori perfetti, dal campo esterno. Però qualora si volesse tener conto dell'imperfetta conduttività delle pareti della guida, si dovrebbe, a rigore, considerare il campo nell'interno del conduttore che limita la guida, anzi anche il campo nel suo esterno. Ora è ben noto che, per semplificare il problema, Schelkunoff (1) ha ammesso, se il conduttore che limita la guida è dotato di buona conduttività, la relazione :

$$\vec{E}_t = \sqrt{\frac{\mu}{\varepsilon_{eff}}}\ \vec{H}_t \times \vec{n} \tag{1}$$

dove $\vec{E}_t$ ed $\vec{H}_t$ sono le componenti tangenziali del campo elettrico e del campo magnetico alla superficie della guida, $\vec{n}$ è il versore normale a tale superficie, orientato verso l'interno del conduttore, ε_{eff} conforme alla notazione di Bremmer vale :

(1) Schelkunoff - Electromagnetic waves - Van Nostrand, 1943, p.320.

(2) $$\varepsilon_{eff} = \varepsilon - j\frac{\gamma}{\omega}$$

dove ω è la pulsazione del campo supposto sinoidale ε, γ, μ sono rispettivamente la costante dielettrica, la conduttività e la permeabilità del conduttore, j l'unità immaginaria [1]. E' poi ben noto che in pratica ε è trascurabile rispetto a γ/ω sicchè si ha :

(3) $$\sqrt{\frac{\mu}{\varepsilon_{eff}}} \simeq \sqrt{\frac{\mu\omega}{\gamma}j} = \sqrt{\frac{\mu\omega}{2\gamma}}(1 + j)$$

e in tal modo si scrive d'ordinario, il coefficiente che compare al secondo membro della (1).

In base alla (1) si può studiare la propagazione della guida a pareti imperfettamente conduttrici senza tener conto del campo nell'interno del conduttore che la limita. Ora è da notare che vi sono guide particolari (per esempio quella formata dall'intercapedine fra due semispazi paralleli e riempiti da buoni conduttori oppure l'altra costituita da un cilindro circolare interamente circondato da un conduttore), in cui si propagano modi che possono calcolarsi in maniera rigorosa tenendo conto solo delle condizioni di continuità del campo elettromagnetico alla loro superficie; ma i risultati che così si ottengono poco differiscono da quelli ottenuti applicando la (1) che rimane perciò confermata [2].

---

(1) Si noti che Bremmer suppone la legge di variazione del campo nel tempo secondo il fattore $\exp(-i\omega t)$ perciò le nostre formule si ottengono dalle sue scambiando ı in -j ; inoltre egli scrive σ in luogo di γ .

(2) M.De Socio "Sulle condizioni al contorno per le guide imperfettamente conduttrici" Rendiconti Lincei (8) XX (1956) - 469.

Ora per quanto esposte, di solito, in forma diversa, le condizioni di Leontovitch e quelle di Schelkunoff coincidono [1]; è però difficile stabilire la priorità dell'uno o dell'altro Autore.

2. Passiamo ora ad esporre una giustificazione ovviamente non rigorosa, della (1) che diremo relazione di Leontovitch-Schelkunoff e che per brevità scriveremo condizione L.S.; inoltre sempre per brevità di esposizione, considereremo, come avviene spesso in pratica, la condizione L.S. sulla superficie che separa l'aria da un conduttore e naturalmente sulla faccia rivolta verso l'aria. Indicheremo sempre con ε_0 la costante dielettrica dell'aria mentre la permeabilità dell'aria sarà supposta uguale a quella μ del conduttore.

Supponiamo anzitutto, come nel caso del Sommerfeld, piana la superficie (che diremo ancora α) che separa i due mezzi conduttore ed aria. Consideriamo un sistema cartesiano trirettangolo con origine in un punto O di α e con l'asse z normale ad esso e orientato verso l'interno del conduttore. Sarà perciò $\vec{k} = \vec{n}$. Allora, se nel conduttore si propaga un'onda piana parallela all'asse z è ben noto (e lo ha richiamato anche il prof. Zücker) che il campo elettrico $\vec{E}$ e il campo magnetico $\vec{H}$ sono normali all'asse z e normali fra loro; più precisamente si ha :

$$(4) \qquad \vec{E} = \sqrt{\frac{\mu}{\varepsilon_{eff}}} \; \vec{H} \times \vec{n}$$

(1) Cfr. D.Graffi "Sulle condizioni al contorno approssimate nell'elettromagnetismo" Atti Accad.Scienze Bologna (11) V (1958) - 88.

questa relazione vale anche alla superficie α sulla faccia orientata verso il conduttore; anzi si può porre nella (4) $\vec{E}_t$, $\vec{H}_t$ in luogo di $\vec{E}$ e $\vec{H}$ perchè questi vettori sono normali ad $\vec{n}$ e quindi tangenti a α . Ma $\vec{E}_t$ e $\vec{H}_t$ sono continui attraverso le superfici del conduttore; quindi sulla faccia di α rivolta verso l'aria vale la (4) con $\vec{E}$ ed $\vec{H}$ sostituite da $\vec{E}_t$ e $\vec{H}_t$. Di conseguenza la condizione L.S. risulta valida se nel conduttore si propaga un'onda piana in direzione normale alla superficie che lo limita.

Consideriamo ora un'onda piana che si propaghi nell'aria, il suo piano d'incidenza faremo coincidere con il piano x z; sia poi θ_0 l'angolo d'incidenza. E' noto che se, provvisoriamente, in luogo del conduttore poniamo un dielettrico di costante dielettrica ε_1 detto θ_1 l'angolo di rifrazione cioè l'angolo che la direzione di propagazione dell'onda forma con l'asse z si ha per il campo elettrico dell'onda trasmessa nel dielettrico l'espressione :

$$(5) \qquad \vec{E} = \vec{E}_0 \exp\left[-j\,\omega\sqrt{\varepsilon_1\mu}\,(\cos\theta_1\, z + \text{sen}\,\theta_1\, x)\right]$$

dove $\vec{E}_0$ è un vettore costante ; analoga espressione vale per il campo magnetico. Si ha poi per la legge di Cartesio :

$$(6) \qquad \frac{\text{sen}\,\theta_0}{\text{sen}\,\theta_1} = \sqrt{\frac{\varepsilon_1}{\varepsilon_0}}$$

Ora, se in luogo del dielettrico sostituiamo il conduttore, valgono le stesse relazioni (5) e (6) purchè in luogo di ε_1 si ponga ε_{eff} che in un metallo è sensibilmente uguale a $-j\,\frac{\gamma}{\omega}$.

Quindi si ha :

$$\text{(7)} \qquad \text{sen}\ \theta_1 = \text{sen}\ \theta_0 \sqrt{+j\frac{\varepsilon_0\omega}{\gamma}}$$

Ora il rapporto $\varepsilon_0\omega/\gamma$ vale il rapporto fra le correnti di spostamento e quelle di conduzione ed è sempre molto piccolo nei buoni conduttori. Per esempio in un buon conduttore γ è dell'ordine di 10^7 e perciò se $\omega = 2\pi.10^7$ si ha $\frac{\varepsilon_0\omega}{\gamma} = \frac{2\pi.10^{-9}}{36\pi} = \frac{1}{1,8}.10^{-10}$. Quindi :

$$|\text{sen}\ \theta_1| \leq |\text{sen}\ \theta_0| \frac{1}{\sqrt{1,8}}.10^{-5}$$

Cioè fino a che l'onda è ordinaria $|\text{sen}\ \theta_0| < 1$ quindi sen $\theta_1 \simeq 10^{-5}$ in altre parole l'onda nel secondo mezzo si propaga in direzione normale al piano che limita il conduttore cioè, da quanto precede si ha che per qualunque onda ordinaria vale la relazione L.S.. Anzi si può dire di più. Anche se l'onda non è omogenea o come si suol dire evanescente, cioè $|\text{sen}\ \theta_0|$ può essere maggiore di 1, risulta ancora $|\text{sen}\ \theta_1|$ trascurabile purchè $|\text{sen}\ \theta_0|$ non raggiunga valori dello stesso ordine di $\sqrt{\gamma/\varepsilon_0\omega}$.

Consideriamo ora un campo elettromagnetico qualunque. Esso può sempre ridursi, applicando per esempio il principio dell'interferenza inversa di Toraldo di Francia [1] ad una somma

---

(1) Cfr. G.Toraldo di Francia "La diffrazione della luce", Einaudi, Torino, 1958, pag.76. Intenderemo superficie di riferimento la superficie su cui si propone assegnato il campo elettromagnetico che poi si decompone in onde piane.

d'infinite onde piane ordinarie ed evanescenti. Se è possibile assumere come piano di riferimento un piano nell'aria parallelo ma abbastanza lontano dalla superficie del conduttore in modo che, su quest'ultimo, si possano trascurare le onde evanescenti o comunque sia trascurabile il contributo di queste onde con $|\text{sen}\,\theta_0|$ paragonabile o più grande a $\sqrt{\gamma/\varepsilon_0\omega}$, la condizione L.S. risulta valida per ogni onda componente, quindi anche per la loro somma cioè per il generico campo elettromagnetico.

E' da notare però che il piano di riferimento dovrebbe essere compreso fra la sorgente del campo elettromagnetico e la superficie del conduttore, altrimenti non sarebbe possibile applicare il principio dell'interferenza inversa. Perciò se la sorgente è molto prossima alla superficie del conduttore la relazione L.S. può non risultare valida; s'intende nelle vicinanze della sorgente, perchè nei punti di a lontani dalla sorgente stessa il piano di riferimento potrebbe sostituirsi con altra superficie opportuna o perchè, come osserva Bremmer (1), nel caso della propagazione lungo la terra, il campo elettromagnetico consiste principalmente nell'onda ordinaria che si propaga parallelamente al suolo. Risulta perciò giustificata (salvo condizioni eccezionali) la validità della condizione L.S.. Prendiamo infine in considerazione un conduttore limitato da superfici curve. Com'è noto entro tale conduttore il campo elettromagnetico è sensibilmente

(1) Bremmer "Propagation of electromagnetic waves" Handbuch der Physik, vol.XVI Springer, Berlin, p.532.

diverso dallo zero in uno straterello aderente alla superficie del conduttore e di spessore δ (profondità di penetrazione) dell'ordine di $1/2\sqrt{\mu\omega\gamma}$. Quindi nell'intorno di un punto P della superficie in cui i raggi principali di curvatura sono grandi rispetto a δ, si può, agli effetti del calcolo del campo elettromagnetico sostituire alla superficie il piano tangente in P. Si possono perciò ripetere le considerazioni precedenti, in tal modo resta estesa anche alle superfici curve la relazione L.S.

E' bene notare che una dimostrazione molto più rigorosa è stata data da Panic (1) il quale ha ottenuto formule più approssimate tenendo anche conto della curvatura della superficie che limita il metallo.

3. Ammesse valide le condizioni L.S. stabiliremo un teorema di unicità per il campo elettromagnetico. Ci riferiremo al problema di Sommerfeld esaminato in principio, ma le considerazioni che seguono si estendono facilmente a casi più generali. Siano $\vec{E}_i$ ed $\vec{H}_i$ il campo del dipolo qualora il mezzo fosse omogeneo, $\vec{E}_r$ ed $\vec{H}_r$ sia il campo (nell'atmosfera) dovuto all'azione del suolo, il campo totale sarà $\vec{E} = \vec{E}_i + \vec{E}_r$, $\vec{H} = \vec{H}_i + \vec{H}_r$. Questo campo soddisferà all'equazione di Maxwell e sul piano α, che rappresenta il suolo, fra $\vec{E}$ ed $\vec{H}$ passerà la relazione L.S.. Ammetteremo inoltre che $\vec{E}$ ed $\vec{H}$ soddisfino all'infinito alle cosidette condizioni di radiazione. Più precisamente, posto per comodità

(1) Doklady Akd. Nauk USSR 70 (1950), 589.

l'origine degli assi nel punto O del suolo in cui si proietta il dipolo e detta R la distanza da O del punto generico P, in cui si calcola $\vec{E}$ ed $\vec{H}$, ed $\vec{r}$ il vettore unitario parallelo e dello stesso verso di P - O (ossia $\vec{r} = \frac{P-O}{R}$) ammetteremo che [1]

$$(8) \qquad \lim_{R\to\infty} R\vec{H} = \vec{u} \qquad\qquad (9) \qquad \lim_{R\to\infty} R(\vec{E} - \sqrt{\mu/\varepsilon_0}\,\vec{H}\times\vec{r}) = 0$$

dove $\vec{u}$ è un vettore finito. Queste condizioni determinano in modo univoco $\vec{E}_r$ ed $\vec{H}_r$ (e quindi $\vec{E}$ ed $\vec{H}$ perchè $\vec{E}_i$ ed $\vec{H}_i$ sono noti) purchè si suppongono regolari in tutto il semispazio occupato dall'aria.

Per dimostrare il teorema enunciato supponiamo per assurdo che esistano due valori diversi del campo dovuto all'azione del suolo $\vec{E}_r$, $\vec{H}_r$, $\vec{E}_r + \vec{e}$, $\vec{H}_r + \vec{h}$. Allora, per la linearità delle equazioni di Maxwell, delle condizioni L.S. e di radiazione, si ha che $\vec{e}$ ed $\vec{h}$ oltre ad essere regolari soddisfano le stesse condizioni di $\vec{E}$ e $\vec{H}$.

Consideriamo ora la regione del semispazio occupato dall'aria compresa fra la semisfera Σ di centro O e raggio R e la parte σ del piano α interna a Σ. Per il teorema di Poynting applicato a $\vec{e}$ e $\vec{h}$ si ha (Re indica parte reale, $\vec{h}^*$ è il coniugato di $\vec{h}$):

$$(10) \qquad \mathrm{Re}\left[\int_\Sigma \vec{e}\times\vec{h}^*\,.\,\vec{n}\,d\Sigma + \int_\sigma \vec{e}\times\vec{h}^*\,.\,\vec{n}\,d\sigma\right] = 0$$

---

(1) Cfr. Silver "Microwave antenna theory and design" Mc Graw, 1947, p.85. Si potrebbe dimostrare che le (8), (9) non dipendono dalla posizione di O. I limiti che compaiono in (8) e (9) s'intendono uniformi rispetto a tutte le direzioni in cui P può tendere all'infinito.

e per la condizione L.S. e per la condizione di radiazione detto $\vec{\xi}$ un vettore che tende allo zero uniformemente al tendere all'infinito di R :

$$(11) \quad \mathrm{Re}\left[\int_{\Sigma}\sqrt{\frac{\mu}{\varepsilon_0}}\,(\vec{h}\times\vec{r})\cdot(\vec{h}^*\times\vec{r})\,d\Sigma + \int_{\Sigma}\frac{\vec{\xi}\times\vec{h}^*}{R}\cdot\vec{r}\,d\Sigma\right] +$$
$$+\ \mathrm{Re}\int_{\sigma}\sqrt{\frac{\mu}{\varepsilon_{ff}}}\,(\vec{h}\times\vec{n})\cdot(\vec{h}^*\times\vec{n})\,d\sigma = 0$$

Se R tende all'infinito per le ipotesi (8) e (9) il secondo integrale esteso a Σ tende allo zero [1]. Ora il primo termine dell'equazione che così si ottiene è positivo ed è pure positivo il secondo perchè $\sqrt{\mu/\varepsilon_{ff}}$ ha parte reale positiva. Si conclude perciò, in particolare, $(\vec{h}\times\vec{n})\cdot(\vec{h}^*\times\vec{n}) = 0$ ossia $\vec{h}_t = 0$ su tutto il piano α che definisce il suolo; di conseguenza per la relazione L.S. è nullo su questo piano anche $\vec{e}_t$. Allora, per un noto teorema [2], $\vec{e}$ e $\vec{h}$ sono nulle in tutto lo spazio e il teorema di unicità è così provato.[3]

4. Esprimiamo ora la condizione L.S. nella forma più adatta per risolvere il problema di Sommerfeld e gli altri indicati dal prof.Bremmer.

---

(1) Infatti detta Ω una sfera di raggio uno l'integrale in discorso può scriversi $\int_{\Omega}\vec{\xi}\times R\vec{h}^*\cdot\vec{n}\,d\Omega$ che tende allo zero per la (8) e per la (9) se R tende all'infinito.

(2) M.De Socio "Alcuni teoremi di unicità per le equazioni di Maxwell" Boll.UMI (3) VIII, 1953, p.196. - Claus Muller "Grundprobleme der mathematischen Theorie elektromagnetischen Schiwingungen", Springer Verla, 1957, p.135.

(3) Per un analogo teorema di unicità cfr. I. Ferrari "Un teorema ./.

Nel caso di un dipolo verticale dalle formule del prof.Bremmer (pag.2) abbiamo, ricordando $\vec{\Pi} = \Pi\,\vec{k}$ (ora $\vec{k}$, come l'asse z, è verticale e orientato verso l'alto quindi ora $\vec{n} = -\vec{k}$) e che in luogo di ε_{eff} porremo ε_o e $-j$ in luogo di i abbiamo :

$$\vec{E} = -j\,\omega\varepsilon_o\mu\Pi\vec{k} - \frac{j}{\omega}\,\mathrm{grad}\,\frac{\partial\Pi}{\partial z}\,,\quad \vec{H} = \varepsilon_o\,\mathrm{rot}\,(\Pi\vec{k})$$

Quindi (1)

$$\vec{E}_t = \frac{-j}{\omega}\,\mathrm{grad}_\sigma\frac{\partial\Pi}{\partial z}\qquad \vec{H}_t = -\,\varepsilon_o\,\mathrm{grad}\,\Pi\times\vec{n}$$

e per la (1)

$$-\frac{j}{\omega}\,\mathrm{grad}_\sigma\,\frac{\partial\Pi}{\partial z} = -\,\varepsilon_o\sqrt{\frac{\mu}{\varepsilon_{eff}}}\;(\mathrm{grad}\,\Pi\times\vec{n})\times\vec{n} =$$

$$= +\,\varepsilon_o\sqrt{\frac{\mu}{\varepsilon_{eff}}}\;\left[\mathrm{grad}\,\Pi - (\mathrm{grad}\,\Pi\,.\,\vec{n})\,\vec{n}\right] = \varepsilon_o\sqrt{\frac{\mu}{\varepsilon_{eff}}}\;\mathrm{grad}_\sigma\Pi$$

relazione certamente soddisfatta se è :

$$(12)\qquad \frac{\partial\Pi}{\partial z} = j\,k_o\sqrt{\frac{\varepsilon_o}{\varepsilon_{eff}}}\;\Pi$$

con $k_o = \sqrt{\varepsilon_o\mu\omega^2}$.

Perciò la determinazione del campo di un dipolo nell'atmosfera si riduce al cercare una soluzione dell'equazione :

$$\Delta\Pi + k_o^2\,\Pi = 0$$

soddisfacente alle condizioni al contorno (12) sul piano che rappresenta il suolo, a opportune condizioni all'infinito e tale da comportarsi come il vettore di Hertz di un dipolo in un mezzo o-

---

di unicità per un filo percorso da corrente alternata in prossimità di un mezzo conduttore". Atti Accad. Torino XCIV, 1959-60,p.77.

(1) Si ha : $\mathrm{grad}_\sigma\Pi = \frac{\partial\Pi}{\partial x}\,\vec{i} + \frac{\partial\Pi}{\partial y}\,\vec{j}$

mogeneo, dove si trova il dipolo che genera il campo. La soluzione completa del problema in discorso si trova nel citato libro del Boudouris pag.132 e segg.[1].

Passiamo ora al caso della terra sferica (il cui centro indichieremo con O) pure considerato dal prof.Bremmer.

Converrà porre sulla sfera un sistema di coordinate geografiche θ , φ (l'asse polare contenga il dipolo) e siano $\vec{\theta}$ e $\vec{\varphi}$ i vettori unitari tangenti rispettivamente ai meridiani e paralleli. Sia ancora $\vec{r}$ un vettore unitario normale alla sfera e orientato verso il suo esterno. $\vec{\theta}$ e $\vec{\varphi}$ siano poi orientati in modo che $\vec{r}$, $\vec{\theta}$ e $\vec{\varphi}$ formino un triedro destro. Ora dalle formule di pag.43 del testo di Bremmer si ha (si tenga presente lo scambio di i in -j, di ε_{eff} in ε_0, che la $\vec{r}$ di Bremmer vale il nostro $\vec{r}$ moltiplicato per R e infine che R sostituisce r) :

$$\vec{E} = -j\,\omega\,\mu\varepsilon_0\,\Pi\,R\,\vec{r} - \frac{j}{\omega}\,\mathrm{grad}\,\frac{\partial}{\partial R}(R\,\Pi)$$

$$\vec{H} = \varepsilon_0\,\mathrm{rot}\,(R\,\Pi\,\vec{r})$$

E quindi poichè Π non dipende da φ , ma solo da R e θ,

$$\vec{E}_t = \frac{-j}{\omega}\,\frac{1}{R}\,\frac{\partial^2}{\partial R\,\partial\theta}(R\,\Pi)\,\vec{\theta}\,, \quad \vec{H}_t = \varepsilon_0\,\mathrm{grad}(R\,\Pi)\times\vec{r} =$$

$$= -\,\varepsilon_0\,\mathrm{grad}(R\,\Pi)\times\vec{n}$$

---

(1) Oppure l'articolo dello stesso Autore inserito nel Supplemento del Nuovo Cimento e già citato.

Si ha poi

$$\vec{H}_t \times \vec{n} = -\varepsilon_0(\text{grad}\, R\Pi \times \vec{n}) \times \vec{n} = \varepsilon_0\left[\text{grad}(R\Pi) - \{\text{grad}(R\Pi).\vec{n}\}\vec{n}\right]$$

$$= \varepsilon_0\left[\text{grad}(R\Pi) - \{\text{grad}(R\Pi).\vec{r}\}\,\vec{r}\right] = {}^{(1)}$$

$$= \varepsilon_0 \frac{1}{R} \frac{\partial(R\Pi)}{\partial\theta}\vec{\theta}$$

quindi sostituendo nella (1) questa equazione risulta soddisfatta se :

$$\frac{\partial(R\Pi)}{\partial R} = j\sqrt{\frac{\varepsilon_0}{\varepsilon_{eff}}}\, k_0 (R\Pi) \tag{13}$$

formula che coincide con la (3) di pag.45 del testo di Bremmer purchè si trascuri come è lecito k_1^2 rispetto a k_2^2 e si tenga conto del solito combiamento di notazioni.

5. Terminerò osservando che, come nella teoria delle guide d'onda, si può ancora approssimare la (1) sostituendo in luogo di $\vec{H}_t$ il valore di questa grandezza che si otterrebbe qualora il conduttore fosse perfetto (2). Il valore di E_t che così si ottiene permette di calcolare, con relativa facilità, l'influenza sul campo elettromagnetico dell'imperfetta conducibilità dei conduttori.

---

(1) Ovviamente $\text{grad}(R\Pi) - [\text{grad}(R\Pi).\vec{r}]\,\vec{r}$ non è altro che la componente tangenziale di grad.$R\Pi$, cioè l'espressione del testo.

(2) Cfr. per esempio G.Toraldo di Francia "Onde elettromagnetiche" Zanichelli, Bologna, 1953, p.232.

Applicando queste considerazioni il prof.Marziani [1] ha calcolato l'energia irradiata da un dipolo verticale in presenza della terra piana ritrovando per altra via una formula approssimata di Sommerfeld. E' probabile che con questo metodo si possono ottenere risultati sul problema alquanto complesso della propagazione sulla terra non omogenea.

---

(1) M.Marziani "Sull'energia irradiata da un dipolo" Boll.UMI (3) XVI, 1961, pag.117.

CENTRO INTERNAZIONALE MATEMATICO ESTIVO

(C.I.M.E.)

M. A. MILLER and V. I. TALANOV

THE USE OF THE SURFACE IMPEDANCE CONCEPT IN THE THEORY OF ELECTROMAGNETIC SURFACE WAVES

ROMA - Istituto Matematico dell'Università

THE USE OF THE SURFACE IMPEDANCE CONCEPT IN THE THEORY OF ELECTROMAGNETIC SURFACE WAVES ( )

(A Review)

M.A. Miller and V.I. Talanov.

Certain general problems in the theory of electromagnetic surface waves related to the impedance description of the guiding properties of interfaces are reviewed here. It is assumed that, in general, the surface impedance may have spatial dispersion, i.e., that it may depend on the structure of the field which it determines. The value of such a description is demonstrated both for the study of free waves and for the solution of the problem of surface field excitation by means of various sources (including diffraction). Only those studies are discussed in this review that deal directly with the use of the surface impedance concept. The appended bibliography is more complete : it contains references to nearly all the main articles on electromagnetic surface waves which have been published during the last few years.

( ) This paper was submitted by the Author as the lecture notes for the C.I.M.E. course on Surface Waves in Varenna, Sept. 1961. It was first published (in Russian) in Izvestia VUZ, Radiofisika, vol.IV, No.5, 1961, pages 795-830.

Translated jointly by Dr. D.Dubrovsky for the Centro Internazionale Matematico Estivo, and by Emmanuel College under contract with Air Force Cambridge Research Laboratories, Office of Aerospace Research, U.S.A.

INTRODUCTION

Scope of the review. The abundance of papers dealing with microwave slow-wave systems, and in particular with surface waveguides, makes it difficult to write a review moderate in length and yet throwing light on all directions taken by these researches, and on their results. Therefore we have considered it advisable to concentrate on a certain group of problems, those related to the impedance description of the guiding properties of interfaces between different media *.

The existence of a number of other reviews (Zucker [19-21], Harvey [18], Neumann [11], Miller and Talanov [9], and others [7, 15-17]), justifies the self-imposed restriction in subject matter since some topics omitted by us are quite fully dealt with there. Finally, it was our aim to preserve a logical unity in the review and to demonstrate at the same time that through suitable generalization of the surface impedance concept it is possible to approach a survey of diverse systems from a single point of view.

Classification of problems. If we consider the practical applications of different systems, several groups of problems may be distinguished in which surface waves are guided by interfaces between media: 1) problems pertaining to the propagation of waves under natural conditions, as, for instance, along the surface of the earth, or in a region with rather sharp variations in the refractive index, either in the atmosphere or in the ionosphere, etc; 2) problems pertaining to the transmission of

---

* In order not to overload our bibliography, we shall use the bibliography given in Harvey's review [18], employing a double numbering method for corresponding cross references (for example, Elliot's article [18, 77]).

energy over long distances, in particular with so-called single-wire lines; 3) problems pertaining to the transmission of energy or transformation of the field on a laboratory scale, i.e., over small regions of open (unshielded) or closed (shielded) systems; 4) problems pertaining to the interaction between various surface waves, as well as between surface waves and beams of moving charged particles, or beams of a moving quasi-neutral medium; 5) problems pertaining to the study of the radiation of electromagnetic waves and the creation of required pattern distributions by means of so-called surface wave antennas.

Of course, different characteristics of surface electromagnetic fields move to the fore according to the type of application. However, also from the point of view of these characteristics, it is possible to group the problems under a few more or less distinctive topics: 1) propagation of "free" surface waves; 2) excitation by prescribed sources; 3) transformation of surface waves into space waves and of space waves into surface waves, i.e., diffraction and radiation of electromagnetic waves in systems that admit the existence of surface fields. This in fact is the classification which we shall follow below.

1. FREE SURFACE WAVES [22 - 125, 18]

Surface impedance. It is well known that the effect of a closed boundary surface on the formation of an electromagnetic field inside a volume can in the final analysis be reduced to a homogeneous equation connecting the tangential vector components of the field on this surface:

$$E_i = \sum_{k=1}^{2} Z_{ik} \left[nH\right]_k \,, \tag{1.1}$$

where n is the normal to the surface (pointing into the volume), and i k are indices characterizing the orthogonal coordinate directions on the surface. For istance, for Cartesian coordinates, we obtain by replacing $1 \to x$, $2 \to y$, $n \to z$

$$\begin{aligned} E_x &= -Z_{xx} H_y + Z_{xy} H_x \,; \\ E_y &= -Z_{yx} H_y + Z_{yy} H_x \,. \end{aligned} \tag{1.2}$$

The tensor $Z_{ik} = R_{ik} + jX_{ik}$ has the dimension of impedance in the rationalized MKS system (used in this review); it is called the surface impedance tensor, while the surface on which equations (1.1) or (1.2) are satisfied is called impedance surface.

If equations (1.1) or (1.2) are regarded as boundary conditions for the field, formulated on a closed surface which contains all sources, then we must further demand that the energy flowing into the region from the outside be zero. This is known to be satisfied (i.e., satisfied at every point of the surface and not only for the total flux) for pure reactance surfaces:

$$Z_{ik} = -Z_{ki}^{*} \qquad (1.3)$$

and for absorbing surfaces :

$$R_{11} > 0, \quad R_{22} > 0, \quad R_{11} R_{22} - (R_{12} + R_{21})^2/4 > 0 . \qquad (1.4)$$

Equations (1.3), (1.4) are identical with the conditions imposed on the impedances of passive linear quadrupoles [13] . This analogy follows naturally from the boundary conditions (1.1) or (1.2), and therefore the impedance boundary may be treated as a linear transformer of the field components.* However, not all surfaces have impedances that satisfy equations (1.3) or (1.4): the presence of negative (active) resistances (energy flows into the volume) is quite possible in some areas, and this fact, when solving the original problem of finding a field satisfying such conditions, can lead to ambiguity.

Finally, let us make one more remark concerning methods of satisfying equations (1.1) and (1.2) on boundary surfaces. It is possible to assume with a certain degree of accuracy that the tensor Z_{ik} is, in some systems, independent of the field to be determined, and we shall call such surfaces impedance surfaces; but as a rule this, of course, is not so and equations (1.1), (1.2) are not boundary conditions in the usual sense. This situation is similar to the one that occurs in media with spatial dispersion, where the material constants

* Hence the use of equivalent circuits for the impedance boundary in computations and in the interpretation of results is as appropriate as for any linear transformation.

themselves depend on the field that they determine [6]. A description based on the introduction of surface impedance with spatial dispersion possesses a useful universality and, at the same time, does not lead to substantial complications, at least not in those cases in which it is possible to determine beforehand a certain complete system of "elementary" fields, and to relate these to the terms in the surface tensors Z_{ik}. Thus, when studying surface waves guided by a plane boundary, it is expedient to take a plane inhomogeneous wave as an "elementary" field, while for cylindrical surfaces a cylindrical inhomogeneous wave should be taken. Many studies dealing with surface waves are dedicated to the investigation of the guiding properties of surfaces of precisely this type; in this manner it becomes possible to explain almost all basic results.

Waves guided by impedance surface. By using the impedance description it is possible to solve problems in two stages: first, determine the characteristics of waves guided by surfaces with a fixed impedance, i.e., by surfaces that satisfy equation (1.1); second, study the various means that assure this condition. In the case of plane-guided systems the first part of the problem was discussed in [8, 43, 44]. Below we shall reproduce some results, generalized for systems with tensors that are not reducible to a diagonal form, (for instance, with complex rotating principal axes, as in gyrotropic media).

Let us introduce the Cartesian coordinates x, y, z letting the guiding boundary coincide with the coordinate surface $z = 0$, and assuming that conditions (1.2) are specified on it. The field in the isotropic half-space $z > 0$, satisfying (1.2), will be written as a superposition of transverse-magnetic (TM) and transverse-electric (TE) fields (transverse with respect to the z-direction). The expressions for the components are written as follows

$$
\text{(TM)} \qquad
\begin{aligned}
E_z &= A k_\perp^2 \varphi^{TM} e^{\mp j k_z z}; \\
E_\perp &= \mp j A k_z \nabla \varphi^{TM} e^{\mp j k_z z}; \\
H_\perp &= - j A \frac{k}{Z_0} \left[z_0 \nabla \varphi^{TM} \right] e^{\mp j k_z z};
\end{aligned}
\qquad (1.5)
$$

$$
\text{(TE)} \qquad
\begin{aligned}
H_z &= B k_\perp^2 \varphi^{TE} e^{\mp j k_z z}; \\
H_\perp &= \mp j B k_z \nabla \varphi^{TE} e^{\mp j k_z z}; \\
E_\perp &= j B k Z_0 \left[z_0 \nabla \varphi^{TE} \right] e^{\mp j k_z z}.
\end{aligned}
\qquad (1.6)
$$

Here A and B are amplitudes, ε, μ are the permittivities of the upper $(z > 0)$ half-space, $Z_0 = \sqrt{\mu/\varepsilon}$, $k = \omega\sqrt{\varepsilon\mu}$, z_0 is the unit vector in the z-direction, the index $\perp$ refers to the transverse components along x, y.

The functions φ^{TM} and φ^{TE} satisfy Helmholtz's two-dimensional equation

$$
\Delta_\perp \varphi^{\binom{TM}{TE}} + k_\perp^2 \varphi^{\binom{TM}{TE}} = 0 , \qquad (1.7)
$$

the solution of which we shall write as a superposition of plane waves of the type

$$
e^{\pm(j k_x x \pm j k_y y)} \qquad (1.8)
$$

so that

$$k^2 = k_\perp^2 + k_z^2 = k_x^2 + k_y^2 + k_z^2 \; . \tag{1.9}$$

The substitution of (1.5) and (1.6) in (1.2) leads to an equation connecting the wave numbers k_x, k_y, k_z with the components of the surface impedance Z_{ik} :

$$\begin{gathered}(\tilde{\gamma}\alpha - \alpha Q_{xx} - \beta Q_{xy})(\alpha - \beta\tilde{\gamma}Q_{yx} + \tilde{\gamma}\alpha Q_{yy}) + \\ +(\beta + \beta\tilde{\gamma}Q_{xx} - \alpha\tilde{\gamma}Q_{xy})(\tilde{\gamma}\beta - \alpha Q_{yx} - \beta Q_{yy}) = 0 \; ,\end{gathered} \tag{1.10}$$

which may also be given the following expression :

$$\begin{gathered}(1+\tilde{\gamma}^2)\left\{Q_{22}\left[\tilde{\gamma}^2 - 2M\tilde{\gamma} - N - \beta^2(1-N)\right] + \right. \\ \left. + \tilde{\gamma}\, Q_{12} Q_{21} - \alpha\beta(Q_{12} + Q_{21})\right\} = 0 \; .\end{gathered} \tag{1.11}$$

Here the following symbols have been introduced :

$$\begin{gathered}Z_{ik} = jQ_{ik} Z_o \; ; \qquad i, k \to 1, 2 \quad \text{or} \quad xy \; ; \\ 2M = Q_{11} - 1/Q_{22} \; ; \qquad N = Q_{11}/Q_{22} \; ; \\ k_x = \alpha k \; ; \qquad k_y = \beta k \; ; \qquad k_z = \gamma k \; ; \\ \tilde{k}_z = jk_z = \tilde{\gamma} k \; .\end{gathered} \tag{1.12}$$

<u>Basic properties of the surface field.</u> As is directly clear from (1.5) and (1.6), the field becomes localized in the vicinity of the surface $z = 0$ only on condition that

$\mathrm{Re}\,\tilde{\gamma}k > 0$; therefore if the upper half-space is considered lossless ($\mathrm{Im}\,\varepsilon = \mathrm{Im}\,\mu = 0$), then the condition for the existence of a surface field is reduced to the inequality*

$$\mathrm{Re}\,\tilde{\gamma} = -\mathrm{Im}\,\gamma > 0\ . \qquad (1.13)$$

Hence, there immediately follows one of the most important properties of surface waves with real propagation constants ($\mathrm{Im}\,\alpha = \mathrm{Im}\,\beta = 0$). Such waves are always slow, in the sense that their phase velocities v_ϕ do not exceed the propagation velocities of waves of corresponding structure in the upper half-space (for two-dimensional waves simply the velocities of light $c = 1/\sqrt{\varepsilon\mu}$). Actually, with $\tilde{\gamma}^2 > 0$

$$v_\phi = \omega/k\sqrt{\alpha^2+\beta^2} = 1/\sqrt{\varepsilon\mu}\sqrt{1+\tilde{\gamma}^2} < 1/\sqrt{\varepsilon\mu}\ . \qquad (1.14)$$

We shall study the other properties of the solution of systems (1.10), (1.11) using as example a two-dimensional surface wave, assuming $\beta = 0$ in (1.10) and (1.11), which, of course, does not limit the generality of the discussion, as the axes of the tensor Z_{ik} remain arbitrarily oriented in relation to the direction of the wave propagation. By disregarding the solution $\tilde{\gamma}^2 = -1$ from (1.11) as inappropriate

---

* The structure of a surface wave in a lossy medium ($\mathrm{Im}\,\varepsilon \neq 0$ or $\mathrm{Im}\,\mu \neq 0$) does not offer any essential interest since the localization of the field is always, in fact, assured at the expense of absorption. For instance, let $\varepsilon = \varepsilon_D - j\varepsilon_M$ ($\varepsilon_M \ll \varepsilon_D$); then even when the roots of (1.11) are purely imaginary, $\tilde{\gamma} = j|\gamma|$, the field will prove to be confined to the surface $z = 0$:

$$\mathrm{Re}\,\tilde{\gamma}k = \frac{1}{2}\,\omega\sqrt{\varepsilon_D\mu}\,|\gamma|\,\varepsilon_M/\varepsilon_D\ .$$

when studying localized fields, we obtain :

$$\tilde{\gamma}_{\pm} = \tilde{M} \pm \sqrt{\tilde{M}^2 + N} \ , \qquad (1.15)$$

where

$$\tilde{M} = M - Q_{12}Q_{21}/2Q_{22} = Q_{22}^{-1}(Q_{11}Q_{22} - 1 - Q_{12}Q_{21})/2. \qquad (1.16)$$

Let parameters $\tilde{M}$ and N be purely real which, by the way, is always the case for reactance boundaries (1.3). In this case the classification of all the possibilities for the formation of surface fields is carried out with relative ease. The corresponding location of the plane of these $\tilde{M}$ and N parameters in five different regions is shown in Fig.1.

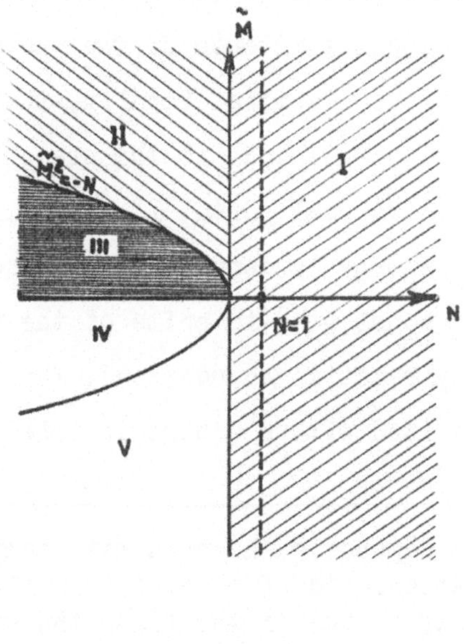

Figure 1.

The condition (1.13) is satisfied in the first three regions (hatched in the figure). In region I($N>0$) there is only one positive root ($\tilde{\gamma}_+ > 0$, $\tilde{\gamma}_- < 0$) and, therefore, a surface wave of only one type can exist here. The case of identical diagonal reactances ($N=1$, $Q_{11}=Q_{22}=Q$) is marked by a dotted line. In the absence of anisotropy the tensor Z_{ik} is reduced to a diagonal form: $Z_{ik} = jQ\,\delta_{ik} Z_o$. For such an isotropic boundary, equation (1.15) yields :

$$\tilde{\gamma}_+ = Q \ , \qquad \tilde{\gamma}_- = -1/Q \ . \qquad (1.17)$$

These well known solutions belong to separately existing TM and TE waves, so that the first are guided by boundaries with inductive surface reactance ($Q>0$), and the second by boundaries with a capacitive reactance ($Q<0$). In region II ($\tilde{M}^2 > -N$, $N<0$, $\tilde{M}>0$,) simultaneous existence of two propagating waves ($\tilde{\gamma}_+ > 0$, $\tilde{\gamma}_- > 0$) is permissible. Finally, the third region ($\tilde{M}^2 < -N$, $N<0$, $\tilde{M}>0$) corresponds to the presence of two complex roots with positive real parts. It is easily verified that no transfer of energy is connected with such waves having complex propagation constants (in spite of the fact that they are guided by purely reactive boundaries). This is a peculiar surface field, localized within a certain impedance angle ($x=\text{const}$, $z=0$)*.

*As noticed by A.V.Gaponov, such waves with complex propagation constants are characteristic of coupled transmission lines (see, for instance, [33]), if in one of the lines the unperturbed wave is in the forward direction (the phase and group velocities coincide in direction) but in the other, is backward (the phase and group velocities are opposite). Under certain conditions the energy flux is cancelled out and the normal wave in coupled lines becomes non-propagating.

The fourth and fifth regions correspond to spatial, not localized fields, furthermore in region IV ($\tilde{M}^2 < -N$, $N < 0$, $\tilde{M} < 0$) both roots of equation (1.15) are complex, while in region V ($\tilde{M}^2 > -N$, $N < 0$, $\tilde{M} < 0$) they are real and negative. In spite of the fact that these waves (as well as the waves determined by the boundary values $\mathrm{Re}\,\tilde{\gamma} = 0$, $\mathrm{Im}\,\tilde{\gamma} \neq 0$)* do not satisfy the conditions (1.13) and, strictly speaking, are not realizable in infinite systems, nevertheless, when analyzing fields in finite regions of space, they can be used to approximate the correct field distribution. The simplest example is an impedance plane with a dielectric coating, inside of which there are always solutions propagating along z. In systems limited with respect to x or y these may even be fast waves, partly emitting electromagnetic energy into the surrounding space. Therefore such waves are called " waves with leakage" or " leaky " waves [204, 206] .

The more complicated cases of the formation of surface fields near boundaries with complex impedances are analyzed analogously. For instance, for a two-dimensional TM wave over an isotropic plane with a vector impedance $Z_{xx} = Z_{yy} = jZ_0 Q_D (1 - jQ_M/Q_D)$, possessing a small active component $(Q_M/Q_D \ll 1)$, we find from (1.17) and (1.11) that

$$\tilde{\gamma} = Q_D - jQ_M \;;$$
$$\alpha = \alpha_D (1 - jQ_D Q_M \alpha_D^{-2}) , \tag{1.18}$$

where $\alpha_D = \sqrt{1 + Q_D^2}$. Assuming that $Q_D > 0$, $Q_M > 0$, we shall get along the direction of propagation an exponentially

* The solution $\tilde{\gamma}^2 + 1 = 0$, omitted above, relates, incidentally, to them too.

damped wave $e^{-j\alpha_D x - \alpha_M x}$, $\alpha_M = \mathrm{Im}\,\alpha = Q_D Q_M (1 + Q_D^2)^{-1/2}$. If instead of absorption, we admit the presence of a finite energy flux flowing from beneath the surface into the upper half-space, i.e., if we assume that $Q_M < 0$, then the solution increases exponentially so that in both cases, the corresponding surface waves may be both fast and slow*.

Let us note, finally, that although strictly speaking the foregoing discussion pertained to systems with a impedance boundary, actually they remain valid also when Z_{ik} has spatial dispersion. However, the analysis of dispersion equations (1.10) (1.11) and (1.15) is in general complicated because the quantities Q_{ik}, M, N entering into them are themselves functions of the wave vector components k_x, k_y, k_z**. As a result, a family of curves $\tilde{\zeta}$ = const becomes distinguishable in the plane of parameters $\tilde{M}$ and N, whose distribution in regions I to V permits us to judge the possibilities of surface field formation and to find out the critical values of parameters corresponding to transitions from one region to another.

The surface impedances of plane boundaries. Below we shall mention expressions for surface impedances of certain practical guiding systems, referring the reader to the original papers whenever it becomes necessary to become aquainted with specific dispersion characteristics.

---

* This refers to phase velocity of propagation in the directions x and y.

** It may happen also that the reduction of tensor Z_{ik} to a diagonal form, even in non-gyrotropic systems, is inexpedient, because of the dependence of the direction of the principal axis on the field structure.

We will actually deal with the "input impedance" of a plane-parallel structure with permeabilities depending on the z coordinate. Here, of course, there is a certain arbitrariness in the choice of the surface considered as guiding. However, if starting with some $z = 0$, the medium becomes homogeneous, it is natural to combine the plane $z=0$ with the interface between homogeneous $(z > 0)$ and inhomogeneous $(z < 0)$ half-spaces. Then expressions (1.5), (1.6) written for a homogeneous medium bounded by an impedance plane, can be used without any changes or recalculations.

Let us note beforehand that it is profitable, for isotropic media, to calculate first and separately the impedances for TM and TE waves (since they prove to be scalar), and only then to formulate the generalization for waves of an arbitrary structure. Considering that the tangential components of the vector field are distributed on the $z = 0$ plane according to (1.8), it is not difficult to express the components of the Z_{ik} tensor by using the scalar impedances Z^{TM} and Z^{TE}(*):

$$Z_{xx} = \frac{k_x k_y}{k_\perp^2} \left(\frac{k_x}{k_y} Z^{TM} + \frac{k_y}{k_x} Z^{TE} \right) ;$$

$$Z_{yy} = \frac{k_x k_y}{k_\perp^2} \left(\frac{k_y}{k_x} Z^{TM} + \frac{k_x}{k_y} Z^{TE} \right) ; \qquad (1.19)$$

$$Z_{xy} = Z_{yx} = \frac{k_x k_y}{k_\perp^2} \left(Z^{TM} - Z^{TE} \right) .$$

(*) In deriving (1.19), account was taken of the fact that from equation $\mathrm{div}\, H=0$ for TM waves it follows that $k_x H_x + k_y H_y = 0$ (and correspondingly for TE waves $k_x E_x + k_y E_y = 0$).

Hence, among other things, it follows that for isotropic boundaries $(Z^{TM}Z^{TE} = Z_0^2,\ k_z^2 Z^{TE} = k^2 Z^{TM})$ the equality

$$Z_{xx} - \frac{k_x}{k_y} Z_{xy} = Z_{yy} - \frac{k_y}{k_x} Z_{yx} , \qquad (1.20)$$

must be observed.

Certain versions of plane systems are shown in Fig.2 that allow (for certain values of parameters) the localization of electromagnetic fields near the $z=0$ boundary. The simplest system is the one illustrated in Fig.2a, showing two homogeneous half-spaces with different permeabilities. Let us suppose a field in the lower half-space in the form of the following combination of waves

$$e^{+jk_z^{(-)}z \pm (jk_y^{(-)}y \pm jk_x^{(-)}x)} . \qquad (1.21)$$

Here, unlike (1.8), all the values are related to the lower half-space. But if we foresee, at once, the need to provide for continuous matching of tangential field components at the $z=0$ boundary, then we must assume that*

$$k_x^{(-)} = k_x^{(+)} = \alpha k^{(+)} , \qquad k_y^{(-)} = k_y^{(+)} = \beta k^{(+)} . \qquad (1.22)$$

---

* Where necessary, the values relating to the upper half-space will be given index «+» ; those of the lower half-space will be given index « - » .

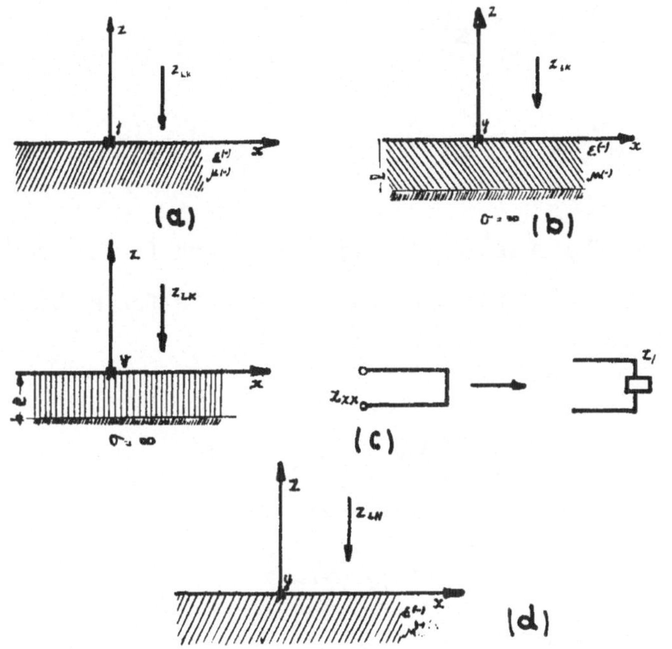

Fig. 2.

At the same time the z-component of the wave number (we shall also use the notation $\tilde{k}_z^{(-)} = jk_z^{(-)}$ must be limited by the radiation conditions

$$\operatorname{Re} \tilde{k}_z^{(-)} = - \operatorname{Im} k_z^{(-)} > 0 \ . \qquad (1.23)$$

Then the input impedance of the lower half-space for waves of the form of (1.5) or (1.6) will be expressed by the following formulas *

* We seek the impedance on the side of the lower half-space; therefore, in formulas (1.2), we must consider $n = -z_0$.

$$Z^{TM} = Z_0^{(-)} k_z^{(-)} / k^{(-)} = k_z^{(-)} / \omega \varepsilon^{(-)} ; \qquad (1.24)$$

$$Z^{TE} = Z_0^{(-)} k^{(-)} / k_z^{(-)} = \omega\mu^{(-)} / k_z^{(-)} . \qquad (1.25)$$

The substitution of (1.24) and (1.25) in (1.19) permits us to determine the Z_{ik} components for a field of arbitrary type :

$$Z_{xx} = Z_0^{(-)} \frac{k_x^{(-)2} - \tilde{k}_z^{(-)2}}{-jk^{(-)}\tilde{k}_z^{(-)}} ;$$

$$Z_{yy} = Z_0^{(-)} \frac{k_y^{(-)2} - \tilde{k}_z^{(-)2}}{-jk^{(-)}\tilde{k}_z^{(-)}} ; \qquad (1.26)$$

$$Z_{xy} = Z_{yx} = -Z_0 \frac{k_x^{(-)} k_y^{(-)}}{-jk^{(-)}\tilde{k}_z^{(-)}} .$$

In the absence of absorption ($\mathrm{Im}\, \varepsilon^{(-)} = 0$) and when (1.23) is satisfied, the Z^{TM} impedance, as is clear from (1.24), has purely inductive character, provided

$$\varepsilon^{(-)} < 0 . \qquad (1.27)$$

If moreover the inequality $|\varepsilon^{(-)}| > \varepsilon^{(+)}$ is also satisfied,*

*It is not difficult to verify that the propagation constant of a TM wave with $\mu^{(+)} = \mu^{(-)}$ is $k_x = \omega\sqrt{\mu^{(+)}\varepsilon^{(+)}|\varepsilon^{(-)}|}\,(|\varepsilon^{(-)}| - \varepsilon^{(+)})^{-\frac{1}{2}}$, whence follows the requirement that $|\varepsilon^{(-)}| > \varepsilon^{(+)}$.

then the condition (1.13) for localization of a field above the surface can likewise be satisfied. Thus, a surface TM wave can propagate along the boundary of two lossless media whose dielectric permeabilities have opposite signs [38]. The situation is analogous with TE waves, but here the magnetic permeability of one of the media must be negative. These assertions are physically simple: electromagnetic waves are damped exponentially in media with negative permeabilities ($\varepsilon < 0$ or $\mu < 0$, but not $\varepsilon < 0$ and $\mu < 0$) and, at the same time, these waves do not radiate electromagnetic energy in the upper half-space, since their phase velocities there do not exceed the velocity of light.

In as much as the exponential decrease of amplitudes takes place also in media with complex permittivities, it is not difficult to construct a surface field on the interface between an absorbing and a lossless medium. This is the so-called Zenneck wave [18, 291], with which, strictly speaking, the study of surface waves in electromagnetic systems began. In the limiting case of very large values of conductance ($\varepsilon_M = \omega/\sigma \gg \varepsilon_D$) and of fields that change smoothly in comparison with $1/|k^{(-)}|$, ($\sqrt{\alpha^2+\beta^2} \ll |k^{(-)}|/k^{(+)}$), the surface impedance ceases to depend on the field structure:

$$Z_{xy} = Z_{yx} = 0 \;; \qquad (1.28)$$

$$Z_{xx} = Z_{yy} = Z^{TM} = Z^{TE} \simeq \sqrt{\frac{\mu^{(-)}}{\varepsilon_M^{(-)}}}\,\frac{1+j}{\sqrt{2}}\,,$$

and condition (1.1) becomes an impedance boundary condition (Leontovitch's condition).

In an analogous way, other surface wave systems are examined.

For an ideally conducting sheet, covered with a dielectric layer (Fig.2b),

$$Z^{TM} = jZ_o^{(-)} \frac{k_z^{(-)}}{k^{(-)}} \operatorname{tg}(k_z^{(-)} \ell) ;$$

$$Z^{TE} = jZ_o^{(-)} \frac{k^{(-)}}{k_z^{(-)}} \operatorname{tg}(k_z^{(-)} \ell) . \qquad (1.29)$$

The components of the Z_{ik} tensor are obtained by substituting in (1.26)

$$Z_o^{(-)} \rightarrow jZ_o^{(-)} \operatorname{tg}(k_z^{(-)} \ell) . \qquad (1.30)$$

It is obvious that such systems, even when the permeabilities are real and positive, guide surface waves of different types [18, 32] . In the limiting case of large permeabilities $k^{(-)} = \omega^{(-)} \sqrt{\varepsilon^{(-)} \mu^{(-)}} \gg k_{\perp}^{(-)}$, and for optical thickness of dielectric coating $k_z^{(-)} \ell \lesssim 1$, the impedances (1.29) cease to have spatial dispersion ($k_z^{(-)} \simeq k^{(-)}$), which gives the prescription for the principal realization of any impedance boundary condition.*

*In connection with this it is of interest to note that there is no need to generalize any of the electrodynamic theorems for systems with impedance boundaries, since the latter can always be realized by suitable choice of ε and μ.

For a perfectly conducting corrugated surface (Fig.2c), with the idealization of fixed field structure between the plates (waves of the TE_{on} type), we have

$$Z_{xx} = j Z_o^{(-)} \frac{k^{(-)}}{k_z^{(-)}} \operatorname{tg} (k_z^{(-)} \ell) ;$$
$$Z_{xy} = Z_{yx} = Z_{yy} = 0 . \tag{1.31}$$

This is a limiting case (the distance between plates is small in comparison with wavelengh) of a periodic comb of finite dimensions, whose strict calculation is included, for instance, in Weinstein's work [31].

A magneto-anisotropic half-space ($\varepsilon^{(-)} \mu_{ik}^{(-)}$) (Fig.2d) has the following input impedance:

$$Z_{xx} = Z^{TM}(1 + k_y^{(-)} T_y / k_z^{(-)} T_z) ;$$
$$Z_{yy} = Z^{TM}(1 + k_x^{(-)} T_x / k_z^{(-)} T_z) ; \tag{1.32}$$
$$Z_{xy} = -Z^{TM} k_y^{(-)} T_x / k_z^{(-)} T_z \quad ; \quad Z_{yx} = -Z^{TM} k_x^{(-)} T_y / k_z^{(-)} T_z ,$$

where

$$Z^{TM} = k_z^{(-)} / \omega \varepsilon^{(-)} \quad ; \quad T_k = \sum_i k_i^{(-)} \mu_{ik}^{(-)} . \tag{1.33}$$

On the basis of these expressions we can, in particular, show the possibility of a field localized on both sides of the boundary between vacuum and the gyrotropic medium.

The case of two-dimensional waves ($\beta = 0$), propagating in a direction perpendicular to the magnetizing field ($H_0 \parallel \gamma$), has been studied in [34, 104]. There exist regions of parameter values where such surface waves are uni-directional, but this of course does not lead to a thermodynamic paradox, because, side by side with a surface field in these regions, spatial fields are admissible, which carry the energy in the opposite direction [72, 95].

It is self-evident that all results related to media μ_{ik} can be transferred, by substitution according to the principle of duality, to media ε_{ik} and can be used, for instance, when studying waves guided by gyrotropic plasma [58].

A remark on impedance boundary conditions. Let us note once more that what we call impedance boundary conditions are those that are identical for any field structure; i.e., that do not have spatial dispersion. A few examples were mentioned above. A common prescription for the constructions of such boundaries should be based upon the independence of a field's local structure in one of the media from its local structure in another, i.e., on the independence from the distribution of the sources of excitation. Of course, this can be carried out only with a certain approximation, particularly in two contrasting limiting cases: in the geometrical optics approximation when locally the field has the structure of plane homogeneous waves, and in the quasi-static approximation, when the local structure of a field is determined by the electromagnetic geometry of the system. Leontovich's condition (1.28) is an example of an impedance condition of the first kind: when the conductivity goes to infinity, the

Leontovich condition degenerates into the ordinary condition on a perfectly conducting boundary $E_T = 0$, which thus also belongs to the category of $(Z_{ik} = 0)$ impedance conditions. Conditions in the form of (1.29) are an example of the second kind. In general, every time that the field changes smoothly over the length of the characteristic dimensions of the boundary and the latter can be broken up into sections of quasi--static structure (these sections can be loaded on the other side with arbitrary impedances, see Fig.2c), the approximate introduction of impedance boundary conditions proves admissibile. It is interesting to note that if the system that forms the boundary impedance becomes resonant, then, even over sections of the order of magnitude of the wavelength, the field structure is essentially determined by the resonance oscillation; however, this occur only in a narrow frequency range near the resonant values of impedance $Z = 0$ or $Z = \infty$.

Let us consider now one other question, connected with transformations of impedance boundary conditions from one surface into another. Here we will not work out these transformations in a general form but will explain the essence of the matter by a very simple recalculation of the condition $E_T = 0$ [22, 135]. Let this condition be fulfilled on a certain curvilinear surface $s_3 = -\ell$. By introducing a curvilinear orthogonal system of coordinates ξ_1, ξ_2, ξ_3 and corresponding metric coefficients h_1, h_2, h_3 ($ds_i = h_i\, d\xi_i$), we shall find the relations between the components on the $s_3 = 0$ surface, which lies close to the initial surface $s_3 = -\ell$. Expanding in powers of ℓ, and retaining terms of the first order, we obtain :

$$E_1(-\ell) = E_1(0) - \ell \frac{\partial E_1}{\partial s_3}\Big|_{s_3=0} = E_1(0) - k\ell \times$$

$$\times \left(-jZ_oH_2 + \frac{1}{h_3k}\frac{\partial(h_3E_3)}{\partial s_1} - E_1\frac{1}{h_1k}\frac{\partial h_1}{\partial s_3}\right) ; \qquad (1.34)$$

$$E_2(-\ell) = E_2(0) - k\ell\left(jZ_oH_1 + \frac{1}{h_3k}\frac{\partial(h_3E_3)}{\partial s_2} - E_2\frac{1}{h_2k}\frac{\partial h_2}{\partial s_3}\right).$$

Setting $E_T(-\ell)$ equal to zero, we find (when $k\ell \ll 1$; $(\ell/h_i)\partial h_i/\partial s_j \ll 1$; $\ell\partial E_j/\partial s_i \ll E_i$) the approximate boundary conditions on surface $s_3 = 0$, which can also be considered as impedance conditions, suitable (within the limits of accepted approximation) for a field of any structure. Expansion (1.34) can be continued further; therefore boundary conditions with space dispersion in principle always admit a notation in the form of impedance conditions, connecting on the boundary both the values of field vectors themselves and also their derivatives of high orders. Usually, it is true, we have to restrict ourselves to the approximation (1.34), but, by using it, we succeed in analysing some interesting problems, for instance, about waves in multi-rod systems (Armand [22])*.

---

* Analogous calculations for multiconductor lines are cited in [24] .

Systems with cylindrical boundaries. Despite a somewhat different formalism used to solve problems on surface waves guided by cylindrical boundaries, qualitatively the results are similar to those obtained above for plane boundaries. This too can be done in two steps: first we examine the waves guided by a given impedance, and then we determine the values of surface impedances for specific arrangements. We shall limit ourselves to quoting only the general dispersion equation. Let us suppose on cylinder $r = a$ the boundary conditions of form (1.2), expressed in components of cylindrical coordinates r, θ, x $(z \to r,\ x \to \theta,\ y \to x)$:

$$E_\theta = -Z_{\theta\theta} H_x + Z_{\theta x} H_\theta \ ; \qquad E_x = -Z_{x\theta} H_x + Z_{xx} H_\theta \ . \tag{1.35}$$

Let us assume a field where TM and TE waves are superposed in relation to the axial x-coordinate, and let us express the x-components of these fields in the form of:

$$E_x = A K_n(\tilde{k}_r r)\, e^{-jn\theta}\, e^{\mp j k_x x} \qquad H_x = B K_n(\tilde{k}_r r)\, e^{-jn\theta}\, e^{\mp j k_x x} \tag{1.36}$$

Here K_n is Macdonald's function of order n, but $\tilde{k}_r$ is the radial wave number, connected with the propagation constant k_x by the relation

$$k_x^2 = k^2 + \tilde{k}_r^2 = k^2(1 + \tilde{\delta}^2) = \alpha^2 k^2 \, . \tag{1.37}$$

Determining other components from (1.36) and substituting them in (1.35) we obtain an equation with respect to $\tilde{k}_r = \tilde{\delta} k$ $(x \to 2\ ,\ \theta \to 1)$:

$$Q_{22}\left[N\,\tilde{\delta}^4 F_n^2 + 2M\,\tilde{\delta}^3 F_n + \tilde{\delta}^2\left(\frac{n^2}{p^2}F_n^2 - 1\right) + \frac{n^2}{p^2}F_n^2\right] + \\ + \tilde{\delta}^2\left[\frac{n\alpha}{p}F_n^2(Q_{21}+Q_{12}) + Q_{12}Q_{21}\,\tilde{\delta}\,F_n\right] = 0\ , \qquad (1.38)$$

where

$$Z_{ik} = j\,Q_{ik}\,Z_o\ ;$$

$$F_n = K_n(\tilde{\delta} p)\,/\,K_n'(\tilde{\delta} p)\ ; \qquad (1.39)$$

$$p = ka\ ; \quad 2M = Q_{11} - 1\,/\,Q_{22}\ ; \quad N = Q_{11}/Q_{22}$$

For large values of argument $(\tilde{\delta} kr \gg n)$, Macdonald's functions behave in the following manner :

$$K_n(\tilde{\delta} kr) \sim \sqrt{\pi/2\,\tilde{\delta} kr}\; e^{-\tilde{\delta} kr}\ .$$

Therefore the condition of localization reduces to the inequality

$$\mathrm{Re}\,(\tilde{\delta} k) > 0\ . \qquad (1.40)$$

By virtue of (1.37) this condition is more stringent than (1.13); it follows that surface waves guided by a circular cylinder (or by a closed cylindrical surface of any other form), have a phase velocity less than the velocity of light independently of the structure of their field.

The general examination of the characteristics of equation (1.38) is very complicated and, apparently, up to the present time has not been carried out. But there exist many practically important particular cases, for which (1.38) is analysed without any special difficulty. This, for instance, is the case of an isotropic boundary $Q_{11} = Q_{22}$, $Q_{12} = Q_{21} = 0$, or of an anisotropic boundary with principal axes along the θ- and x-directions ($Q_{12} = Q_{21} = 0$), or the case, among others, where the cylinder has a large radius.

Surface waves along inhomogeneous boundaries. The problem concerning surface waves guided by curvilinear boundaries with inhomogeneous impedance is a natural generalization of the previous description. It is possible to approach the solution of such problems in various ways. First, a start may be made from some already formulated surface field, searching in it for the values of the components in the tensor impedance on particular surfaces. It is obvious that, when such an impedance distribution is satisfied on this surface, a field of the original form can be synthesized. Secondly, it is possible, without predefining the field structure from the very beginning, to find it by choosing simultaneously both the surface form and the corresponding impedance distribution functions. Finally, in the most general case it is necessary, apparently, to solve the problem directly by resorting to direct rigorous or approximate methods.

Constructive approach. Let us cite examples illustrating the possibilities indicated above. Let a symmetric surface wave guided by an isotropic impedance cylinder be the starting point for the construction of a surface field above the impedance plane (Fig. 3).

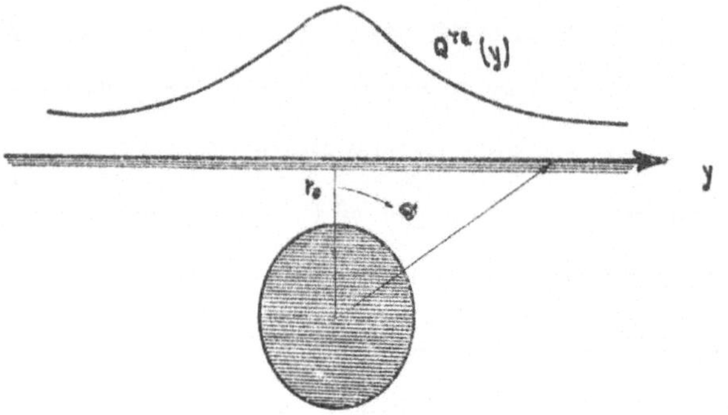

Fig. 3.

Let us find the distribution of surface impedance along the plane $z = 0$ situated at a distance $r = r_0$ from the axis of the cylinder. After simple calculations, taking into account (1.36) and (1.39), we obtain for the case of a symmetric initial TM_0-field :

$$Q_{yy} = \infty \; ; \quad Q_{xy} = Q_{yx} = 0 \; ;$$
$$Q_{xx} = -\tilde{\delta} \sqrt{1+y^2/r_0^2} \; F_0(\tilde{\delta} k r_0 \sqrt{1+y^2/r_0^2}) \tag{1.41}$$

and for an initial TE-field :

$$Q_{xx} = 0 \; ; \quad Q_{xy} = Q_{yx} = 0 \; ;$$
$$Q_{yy} = \left(\tilde{\delta} \sqrt{1+y^2/r_0^2}\right)^{-1} F_0^{-1}(\tilde{\delta} k r_0 \sqrt{1+y^2/r_0^2}) \; . \tag{1.42}$$

One of the characteristic dependences of impedance distribution $Q^{TE} = Q_{yy}$ upon y is shown in Fig.3. If we take into consideration that (1.41) or (1.42) are impedance boundary conditions, then it is obvious that the corresponding impedance boundary will support the localized field of the given structure. Of course, the number of similar constructions can be enlarged arbitrarily.

Systems with separable variables. [8, 12, 56, 80]. An example illustrating the second possibility mentioned at the beginning of the preceding paragraph is the application of the method of separation of variables to the calculation of fields in systems with curved boundaries simultaneously possessing inhomogeneous surface impedances, and in particular to the finding in such systems of waves analogous to surface waves in cylindrical systems with homogeneous impedance.

For the sake of simplicity let a field in a certain region V, bounded by the surface S, be described by a single scalar function $u(\xi_1, \xi_2, \xi_3)$, that satisfies the wave equation and boundary conditions of the type

$$\frac{\partial u}{\partial n} + kqu = 0 \Big|_S \qquad (1.43)$$

on the boundary of the region S. Here $\partial/\partial n$ is the normal derivative with respect to S. The condition (1.43), under certain assumptions regarding the geometry of the system, the dependence of surface impedance upon coordinates, and the character of the field, is equivalent to the impedance boundary condition. The parameter q, determined by the surface impedance, is the function of coordinates on the surface S .

Let ξ_1, ξ_2, ξ_3 be a system of orthogonal curvilinear coordinates in which the variables in the wave equation are separated, and let the boundary S coincide with the coordinate surfaces $\xi_i = \bar{\xi}_i$. Then, assuming that $u = U_1(\xi_1)\, U_2(\xi_2)\, U_3(\xi_3)$ and considering that the function q in (1.43) is connected with Lame's parameter h_i corresponding to the coordinate ξ_i, by the relation

$$kq = q_i/h_i = \mathrm{const}/h_i \Big|_{\xi_i = \bar{\xi}_i} \quad . \tag{1.44}$$

we can reduce the boundary condition (1.43) to the form:

$$dU_i(\xi_i)/d\xi_i + q_i U_i(\xi_i) = 0 \,\Big|\, \xi_i = \bar{\xi}_i \ . \tag{1.45}$$

The relation (1.44) determines the dependence of impedance upon the coordinates in different orthogonal curvilinear systems which permit the separation of variables. For instance, in the case of the propagation of TM waves inside an impedance wedge, it is necessary to assume, according to (1.44), that the impedance of the faces vary in inverse proportion to the distance from the vertex. It is interesting to note that here we can obtain solutions of the type [56, 80]

$$H_z = H_{iq}^{(1.2)}(kr)\, e^{-q\varphi} \tag{1.46}$$

(r, φ, z are cylindrical coordinates, $H_{iq}^{(1.2)}$ is the Hankel function), describing fields that decrease exponentially with respect to φ. When the decrease is sufficiently fast, it is possible to consider these fields localized near the impedance faces. In analogy with plane surface waves it is possible to call waves of type (1.46) azimuthal localized cylindrical surface waves. For both types of waves an exponential amplitude variation along the wave front is characteristic.*

Usually the problem of wave propagation between two surfaces with inhomogeneous impedance, defined by relation (1.44), is reduced to the following problem of the Sturn--Liouville type :

$$M\left[U_i, \xi_i, \beta\right] = 0 \quad ; \tag{1.47}$$

$$\left.\frac{\partial U_i}{\partial \xi_i} + q_1 U_i = 0\right|_{\xi=\bar{\xi}_i}, \quad \left.\frac{\partial U_i}{\partial \xi_i} - q_2 U_i = 0\right|_{\xi=\bar{\bar{\xi}}_i}, \tag{1.48}$$

where $M\left[U_i, \xi_i, \beta\right]$ is the linear differential operator of the second order for the function $U(\xi_i)$, describing the field distribution along coordinate ξ_i in the space between surfaces $\xi_i = \bar{\xi}_i$ and $\xi_i = \bar{\bar{\xi}}_i$, β is eigenvalue.

The non-propagating solution of equation (1.47) should correspond to waves of type (1.46). When the parameters

---

* The system examined here has interest from the point of view of certain applications [203]. Thus, for instance, the use of systems with impedances changing according to the law r^{-1} in plane surface wave antennas gives not only the possibility to improve the antenna parameters but, what is no less important, to design surface wave antennas by a method analogous to the one used in the design of horn antennas.

q_1 and q_2 have definite values, such solutions, generally speaking, exist, at least in a definite interval of variable ξ_i. If the conditions

$$\left|U(\bar{\xi}_i)\right| \gg \left|U(\bar{\bar{\xi}}_i)\right| \quad \text{or} \quad \left|U(\bar{\xi}_i)\right| \ll \left|U(\bar{\bar{\xi}}_i)\right| , \qquad (1.49)$$

are also satisfied, then the fields described by the function $U(\xi_i)$ will be localized on the corresponding impedance surfaces.* The field localization will be maintained to some extent also in the case of a single impedance surface. As a rule, such a field, however, will not be a purely surface field, as the $U(\xi_i)$ function may become propagating in points sufficiently removed from the surface, a fact which corresponds to the presence of an energy flux component in the ξ_i direction. Surface wave propagation of this kind will take place also when there is distortion of the guiding surface, even if the surface impedance does not depend on coordinates. The unstable azimuth waves analyzed in articles [45, 59, 18, 77] may serve as an example.

A boundary with sinusoidally varying impedance. In an arbitrary case, when both the form of the guiding surface and the law of impedance variation are specified, it is necessary to use direct methods in determining the fields. The problem of waves guided by a cylindrical boundary with periodically changing surface impedance is one of the interesting illustrations. Here the method of space harmonics, based upon the application of Floquet's theorem [10], proves to be greatly

---

* As a rule, conditions (1.49) are fulfilled only for sufficiently slowly changing impedance of suitable sign and magnitude.

effective. Let us explain this by an example of two-dimensional TM wave propagation over a surface with sinusoidally changing impedance.

$$Q^{TM}(y) = Q_o + Q_1 \sin(\frac{2\pi}{D} y) = Q_o + Q_1 \sin(bky) . \qquad (1.50)$$

Such waves take the form of:

$$H(y, z) = \sum_{-\infty}^{\infty} a_n e^{-jk_{yn} y - jk_{zn} z} , \qquad (1.51)$$

where $k_{yn} = k_{yo} + kbn$, $k_{zn} = \sqrt{k^2 - k_{yn}^2}$. The substitution of (1.51) in boundary conditions with an impedance of (1.50) results in the following characteristic equation with respect to k_{yo} , containing continued fractions :

$$\sigma_o = \frac{Q_1^2/4}{\sigma_1 -} \frac{Q_1^2/4}{\sigma_2 - \dots} + \frac{Q_1^2/4}{\sigma_{-1} -} \frac{Q_1^2/4}{\sigma_{-2} - \dots} , \qquad (1.52)$$

where

$$\sigma_n = Q_o - i\gamma_n , \quad \gamma_n = \sqrt{1 - \beta_n^2} , \quad \beta_n = k_{yn}/k .$$

A detailed analysis of this equation is carried out in papers [103, 188, 190] . Let us note here only one important peculiarity : equation (1.52) may have purely real as well as complex roots. In the first case (1.51) represents an undamped surface wave localized in a surface layer; all the space-harmonics of such waves are slow ($|\mathrm{Re}\,k_{yn}| > k$). To the complex roots of the characteristic equation there correspond fields containing also fast ($|\mathrm{Re}\,k_{yn}| < k$) space-harmonics. From the view point of antenna application, systems in which the first harmonic is the only fast one, ($|\mathrm{Re}\,k_{y-1}| < k$) are of basic interest.

For clarity's sake, let us limit ourselves to the examination of just this case. Strictly speaking, we have to distinguish here two possibilities: $\mathrm{Re}\,k_{y-1} > 0$ and $\mathrm{Re}\,k_{y-1} < 0$. It is not difficult to be convinced that, when $\mathrm{Re}\,k_{y-1} < 0$ (and $\mathrm{Im}\,k_{yo} < 0$ is a wave damped in the y-direction), the imaginary parts of all are negative, and we have a purely surface wave with a complex propagation constant; about the pecurialities of such wave we spoke above. As for the case where $\mathrm{Re}\,k_{y-1} > 0$ (all other conditions being equal), then $\mathrm{Im}\,k_{z-1} > 0$, so that the first space harmonic grows exponentially away from the plane. Such a field is of the " leaky wave " type [204] and can be named quasi-surface in the sense that it has the structure of a surface wave only in the layer which adjoins the guiding surface.

A peculiar one is the case where $k_{y-1} = 0$ and it occurs at the frequency of the so-called " 2π " oscillation [190]. Here fields (1.51), corresponding to the two roots $k_{yo} = \pm kb$ of the characteristic equation, prove to be linearly dependent and give only one solution, which has the character of a purely periodic standing surface wave, where $a_{-1} = 0$, so that surface radiation is completely absent. Let us designate this solution by $p(\underline{y}, \underline{z})$. A second linearly independent solution in this case is a function of the form

$$H_2 = yp(y,z) + \psi(y,z) \ , \tag{1.53}$$

where $\psi(y,z)$ is a periodic function of y (with period D). The coefficients of this function in the Fourier-series expansion can be determined easily, after substituting (1.53) in the wave equation for H, and after a solution of the resultant inhomogeneous wave equation for $\psi(y,z)$ with the same boundary conditions which are satisfied by the complete field H_2. A peculiarity of the solution of (1.53) is the fact that the function $\psi(y,z)$ contains a non-vanishing first space harmonic, which has the character of a plane wave radiating normally to the surface ($k_{y-1} = 0$, $k_{z-1} = k$). It proves to be necessary to use both solutions (both the first, periodic, and the second, aperiodic one) when explaining radiation of finite dimension structures at frequencies of type "2π"*.

* A periodic law of the (1.53) type may also be obtained at the frequencies of π -oscillations described in [190].

2. THE EXCITATION OF SURFACE WAVES BY SOURCES

Beginning with the well-known Sommerfield paper [153], a whole series of publications, a considerable part of which relates to recent years, has been devoted to the problem of the excitation of surface waves by sources situated near the interfaces of media.

The determination of the surface field itself is, in principle, a very simple problem, which does not differ in any way from the analogous problem of determining a field of a certain type within a shielded transmission line. In this case, well-developed method in the theory of waveguides, for instance those based on a generalization of Lorentz's lemma or on the reciprocity theorem [5, 18, 96], can be utilized.

However, in practice any real sources (with the exception of sources with specially selected distribution) will, besides exciting a surface wave field, also excite a whole complex of other fields forming the so-called space wave, which is only weakly connected with the guiding boundary (more precisely, not localized near it). If we continue the comparison with the shielded waveguide, then the latter must be considered as multimode and absorbing, i.e., characterized by a continuous spectrum of wave numbers. This, of course, complicates the problem, but not essentially, for it is usually only required to determine the far field*, in particular,

* When calculating the far field of a space wave one may also use the reciprocity theorem, for which it is sufficient to know only the field of the free space-waves, i.e., waves created in the system by remote sources [146]. Such a method is also suitable for systems with gyrotropic media, but then the auxiliary free wave must be taken not in the real, but in some fictious transposed ($\varepsilon_{ik} \to \varepsilon_{ki}$, $\mu_{ik} \to \mu_{ki}$) medium.

the ratios of energy flux, carried off by surface and space waves.

In this section we shall examine only fields of sources situated near impedance boundaries. Problems connected with diffraction excitation, in which there are induced sources so that it is necessary to carry out a continuous matching of fields on the aperture of the exciter, are referred to in the next section of this review.

Radiation of currents near regular guiding surfaces. Problems of this kind are very simple and are reduced to the search for a solution of the inhomogeneous system of Maxwell's equations, satisfying boundary conditions on a guiding surface, and the radiation conditions. In general, their solution can be obtained by a Fourier transformation. A space wave is obtained, as a rule, by the saddle point method, but for the calculation of surface wave amplitudes the residue theorem is used.

Let us illustrate this method by an example of finding the field excited by an arbitrary distribution of electric and magnetic sources j^e, j^m over a plane, the surface impedance of which has, in general, spatial dispersion, so that boundary condition (1.2) can be stipulated. It is not difficult to see that in this formulation the problem under consideration involves practically all the problems related to the radiation of prescribed sources above a half-space filled with a stratified, plane-parallel medium $\varepsilon(z)$, $\mu(z)$.

Let us write the "initial field" (corresponding to the source field in an unlimited space)

$$A_1^e = \frac{\mu}{4\pi} \int_V j^e \frac{e^{-jkR}}{R} dV , \qquad A_1^m = \frac{\varepsilon}{4\pi} \int_V j^m \frac{e^{-jkR}}{R} dV, \tag{2.1}$$

in the form of a plane wave expansion, similar to what is done in the well-known Weyl method in the solution of the problem (dipole above a plane interface) in the theory of radio wave propagation above the earth [1] :

$$A_1^e = \frac{\mu}{8\pi^2 j} \int_{-\infty}^{\infty}\int_{-\infty}^{\infty} \frac{\int_V j^e(x',y',z') e^{-jk_x(x-x')-jk_y(y-y')-jk_z|z-z'|} dV'}{k_z} dk_x dk_y$$

$$A_1^m = \frac{\varepsilon}{8\pi^2 j} \int_{-\infty}^{\infty}\int_{-\infty}^{\infty} \frac{\int_V j^m(x',y',z') e^{-jk_x(x-x')-jk_y(y-y')-jk_z|z-z'|} dV'}{k_z} dk_x dk_y . \tag{2.2}$$

Here $R = \sqrt{(x-x')^2+(y-y')^2+(z-z')^2}, \qquad k_x^2 + k_y^2 + k_z^2 = k^2 .$

The secondary field, which is a solution of the homogeneous wave equations, we shall seek in a form analogous to (2.2) :

$$A_2^e = \frac{\mu}{8\pi^2 j} \int_{-\infty}^{\infty}\int_{-\infty}^{\infty} f^e(k_x,k_y) \frac{e^{-jk_+ r}}{k_z} dk_x dk_y ; \tag{2.3}$$

$$A_2^m = \frac{\varepsilon}{8\pi^2 j} \int_{-\infty}^{\infty}\int_{-\infty}^{\infty} f^m(k_x, k_y) \frac{e^{-jk_+ r}}{k_z} dk_x \, dk_y , \qquad (2.4)$$

where k_+ is a vector with components (k_x, k_y, k_z), and r has components (x, y, z).

Let us note that, for a complete description of the reflected field, it is sufficient to employ only two components of the vector-potential. For these we may choose, for instance, A_{2z}^e and A_{2z}^m, by assuming correspondingly in (2.3) and (2.4) that

$$f^e = f^e z_0 , \quad f^m = f^m z_0 . \qquad (2.5)$$

Expressing the fields E and H by the potentials $A_1^e + A_2^e$ and $A_1^m + A_2^m$, substituting them in the boundary condition (1.2), and solving the equations thus obtained with respect to f^e and f^m, we find that

$$f^e = \frac{P_x D - P_y B}{\Delta} ; \qquad (2.6)$$

$$f^m = \frac{P_y A - P_x C}{\Delta} , \qquad (2.7)$$

where P_x, P_y are the components of the vector $P = T + j\hat{Q}[z_0 S]$;

$$T = \left[\frac{k_-}{k}\mathcal{D}^e\right] + \frac{1}{Z_0}\mathcal{D}^m ; \quad S = \mathcal{D}^e - \frac{1}{Z_0}\left[\frac{k_-}{k}\mathcal{D}^m\right] ;$$

$$\mathcal{D}^{e,m} = \left[k_- F^{e,m}\right] ; \quad F^{e,m}(k_x, k_y) = \int_V j^{e,m}(r')e^{jk_- r'}dV',$$

k_- is a vector with components $(k_x, k_y, -k_z)$, $\hat{Q}$ is a tensor of dimensionless impedance (1.12);

$$\Delta = AD - BC ;$$

$$A = -\frac{k_x k_z}{k} - jk_x Q_{11} - jk_y Q_{12} ; \quad B = -\frac{1}{Z_0}(k_y + j\frac{k_y k_z}{k} Q_{11} - jQ_{12}\frac{k_x k_z}{k}); \quad (2.8)$$

$$C = -\frac{k_y k_z}{k} - jk_x Q_{21} - jk_y Q_{22} ; \quad D = -\frac{1}{Z_0}(-k_x + \frac{jk_y k_z}{k} Q_{21} - j\frac{k_x k_z}{k} Q_{22}). \quad (2.9)$$

The relations obtained completely solve the probelm of an arbitrary impedance plane. Substituting in these relations various values of the impedance components, it is possible to find the field distribution in different specific systems.

The analysis of solution (2.3) and (2.4) is complicated in its general form even when the distribution of sources and the surface impedance components are specified. However, the relations do permit us to find both the surface wave fields as well as the space wave field at great distances from the source, which in practice suffices to determine the degree of efficiency of the source from the point of view of the ratio of energy fluxes carried off by surface and space waves.

The surface wave can be found by determining the residue of the poles of the integrand in the expression for the secondary field. Let us examine, for instance, the expression for A^e_{2z} :

$$A^e_{2z} = \frac{\mu}{8\pi^2 j} \int_{-\infty}^{\infty}\int_{-\infty}^{\infty} \frac{N}{\Delta} \frac{e^{-jk_+ r}}{k_z} dk_x dk_y , \qquad (2.10)$$

where

$$N = P_x D - P_y B . \qquad (2.11)$$

The roots of the equation

$$\Delta = 0 , \qquad (2.12)$$

will be the poles of the integrand in (2.10) which, as is evident, coincides with the characteristic equation (1.10) for free surface waves above the impedance plane.

Equation (2.12) permits us to find the values of $k_y^{(s)}(k_x)$ at a fixed k_x or, vice versa, of $k_x^{(s)}(k_y)$ at a fixed k_y. Let us fix k_x, for instance, and let us examine the integral in (2.10) with respect to k_y. By applying the residue theorem, we obtain, for the surface field, in the region outside the sources the following expression :

$$A_{2z}^{e(s)} = \frac{\mu}{4\pi j} \int\limits_{c} \frac{N\left[k_x, k_y^{(s)}(k_x)\right]}{\Delta'_{k_y}\left[k_x, k_y^{(s)}(k_x)\right]\tilde{k}_z^{(s)}} e^{-\tilde{k}_z^{(s)} z - jk_x x - jk_y^{(s)} y} dk_x, \qquad (2.13)$$

where $\tilde{k}_z^{(s)} = \sqrt{k_x^2 + k_y^{(s)2} - k^2}$. Let us stress that the integral in (2.13) is taken only over the values of k_x for which $\tilde{k}_z^{(s)} > 0$. In case equation (2.12) has not one, but several roots, corresponding to different surface waves, then instead of (2.13) we shall get, obviously, the sum of the analogous expressions corresponding to these waves.

Publications devoted to the excitation of surface waves, as a rule, restrict themselves to the examination of those cases only where the analysis is simplified on account of a symmetry in the distribution of sources, permitting us to reduce the matter to the calculation of only a finite number of surface waves on a fixed structure. However, expression (2.13) permits us to find, in a general form, the surface wave field at great distance from the sources and, in particular, to obtain a radiation pattern of the surface field in the x, y, plane by using the saddle-point method :

$$A^{(s)}(\rho,\varphi) = \frac{\mu}{4\pi j}\sqrt{\frac{2\pi}{jk_y^{''(s)}(k_{xc})\sin\varphi}}\, \frac{N\, k_{xc}, k_y^{(s)}(k_{xc})}{\Delta'_{k_y}\left[k_{xc}, k_y^{(s)}(k_{xc})\right]\tilde{k}_z^{(s)}}\, \times$$

$$\times \frac{e^{-\tilde{k}_z^{(s)}(k_{xc})z - j\rho\left[k_{xc}\cos\varphi + k_y^{(s)}(k_{xc})\sin\varphi\right]}}{\sqrt{\rho}}. \qquad (2.14)$$

Here ρ, φ are the cylindrical coordinates of the point of observation $(x=\rho\cos\varphi,\ y=\rho\sin\varphi)$, k_{xc} is the stationary point, which is the root of the equation

$$dk_y^{(s)} / dk_x = -\mathrm{ctg}\,\varphi . \tag{2.15}$$

Formula (2.14) describes a wave with a non-cylindrical phase front.

In the case of isotropic impedance, when

$$k_x^2 + k_y^{(s)2} = \Gamma^{(s)2} = k^2 + \tilde{k}_z^{(s)2} = \mathrm{const} \neq \mathrm{const}(k_x)$$

we have

$$k_{xc} = \Gamma^{(s)} \cos\varphi, \qquad k_{yc}^{(s)} = \Gamma^{(s)} \sin\varphi$$

and

$$A^{(s)}(\rho,\varphi) = -\frac{\mu}{2\sqrt{2\pi}} \frac{N(\Gamma^{(s)}\cos\varphi, \Gamma^{(s)}\sin\varphi)\sin\varphi\, \Gamma^{(s)}}{\Delta'_{k_y}(\Gamma^{(s)}\cos\varphi, \Gamma^{(s)}\sin\varphi)\, \tilde{k}_z^{(s)}} \times \tag{2.16}$$

$$\times \frac{e^{-j(\Gamma^{(s)}\rho - \pi/4)}}{\sqrt{\Gamma^{(s)}\rho}} ,$$

i.e., the surface wave, at considerable distances from the sources, has the character of a diverging cylindrical wave with the propagation coefficient $\Gamma^{(s)}$.

One may calculate analogously the contribution to

a surface wave from the z-component of the magnetic vector-potential A_2^m.

At long distances we shall get a surface wave if we apply the saddle-point method for the asymptotic estimate of the double integrals (2.3) and (2.4):

$$A_{2r}^e = \frac{\mu}{4\pi} \frac{N(k_{xc}, k_{yc})}{\Delta(k_{xc}, k_{yc})} \frac{e^{-jkR}}{R} \qquad (2.17)$$

An analogous expression is also obtained for A_{2r}^m. In (2.17) the stationary point is determined by the relations

$$k_{xc} = k \cos\varphi \sin\theta, \quad k_{yc} = k \sin\varphi \sin\theta, \quad k_{zc} = k \cos\theta \qquad (2.18)$$

(R, φ, ϑ are spherical coordinates of the observation point).

Various particular cases of the problem of surface wave excitation above a plane interface between media are examined in papers [129, 132, 133, 135-137, 149, 153-155]. It is shown in [18.61, 18.87, 18.168, 18.273], in particular, that a high efficiency in surface wave excitation can be had even by means of a point source, if the latter is situated in a suitable way above the boundary guiding the surface wave. The problem is solved similarly also in the case of cylindrical configuration of the guiding boundary. The general solution of such a problem when $Z_{\theta x} = Z_{x\theta} = 0$, $Z_{\theta\theta}$ and Z_{xx} are independent of k_x and n, is cited in [192].

Besides the foregoing general method of finding fields in surface wave systems, based on Fourier transformation, other methods also are possible in principle, whose domain of application is limited to various specific systems. For instance, to determine the field of sources situated above an impedance plane with no spatial dispersion, a very useful approach is that by Malyuzhinets [129], and Khaskind [135, 136], consisting in the introduction of a certain auxiliary function, which satisfies the simplest boundary condition on the guiding surface (for instance, vanishes). But, of course, this happens at the expense of some change on the right side of the wave equation, which the new function obeys in the half-space above the plane. Thus we deal with the transformation of the initial problem, formulated for a region with a unknown Green's function, into a problem for a region with a known Green's function.

The influence of weakly curved guiding surfaces. In many respects this problem is similar to the problems examined in the diffraction theory of waves on conducting bodies of large dimensions (Fock [14]), and, consequently, it can be solved by an analogous method. In particular, it is possible here to reduce the weakly converging series to corresponding integrals, and subsequently calculate the space (saddle-point method) and surface (residue method) fields.

Papers [131, 158] are devoted to the question of excitation of surface waves on curved surfaces. A surface wave excited on a spherical reactance surface by a radial electric dipole is examined in paper [158], and its basic characteristics (phase velocity, attenuation, field distribution in the radial direction) are analysed in the case of sufficiently large radii of curvature. The geometric theory of propagation and excitation of surface waves, which is an extension of the methods of ray optics to the case of waves propagating at complex angles, is developed in paper [148]. This theory permits us to get an asymptotic solution for a whole series of problems (surface waves on a wedge with inhomogeneous impedance faces, waves on impedance cylinder).

3. DIFFRACTION OF SURFACE WAVES

Although problems involving fields of prescribed sources situated near the guiding boundaries are very useful for the solution of many important questions, they are of course not always adequate for a real problem. For instance, in the diffraction excitation of surface waves by the open end of a waveguide, it is necessary to specify not the outside sources, but the structure and amplitude of the incident wave field inside the waveguide. Moreover, the whole range of problems connected with perturbations of the regularity of structures that guide surface waves cannot be reduced to problems with prescribed sources*. As in the study of free surface waves, the use of the impedance approach turns out to be very effective also in the investigation of problems of this kind, since it permits us to explain, by using several simplified models, the main features of practically useful systems.

* The need to consider a certain aggregate of specified sources can arise at individual stages in the solution of such problems.

Surface wave excitation by the open end of a waveguide. Individual variants of such an excitation method can be studied by means of integral equations of the Wiener-Hopf type: surface wave excitation above an impedance plane or a dielectric slab [168, 169, 175], excitation of symmetric surface waves guided by an impedance cylinder, [159] etc..

On making the transition to the Fourier transforms of currents or fields $F(k_x)$, the problems, cited above, reduce to a pair of integral equations of the same type as in the problem concerning the diffraction of electromagnetic waves by the open end of a semi-infinite waveguide (Weinstein [4])*:

$$\int_{-\infty}^{+\infty} F(k_x)\, e^{-jk_x x}\, dk_x = A\, e^{-jk_{x_0} x} \qquad (x>0) ;$$
$$\int_{-\infty}^{+\infty} F(k_x)\, L(k_x)\, e^{-jk_x x}\, dk_x = 0 \qquad (x<0) . \qquad (3.1)$$

The solution of equation (3.1) depends on the practical possibility of factorizating the kernel $L(k_x)$. The form of the function $L(k_x)$ depends on the character of the boundary conditions on the impedance surface (usually having spatial dispersion) for $x<0$ and $x>0$, and the equation

$$L(k_x) = 0 \qquad (3,2)$$

* The question is that of diffraction of a normal mode in a shielded system.

is the characteristic equation of the system for $x < 0$, while

$$\left[L(k_x)\right]^{-1} = 0 \tag{3.3}$$

holds for $x > 0$.

For several two-dimensional systems, we here give examples of characteristic functions in terms of which the kernel $L(k_x)$ can be expressed, and the roots of which are the eigenvalues for these systems. It is assumed that the direction of the propagation of the wave coincides with one of the principal axes of the surface impedance tensor.

1) Impedance plane:

$$\varphi(k_x, f) = 1 + jkf / k_z , \tag{3.4}$$

where $k_z = \sqrt{k^2 - k_x^2}$, and the parameter f is determined by one of the components of the dimensionless surface impedance:

$$f = \begin{cases} Q_{TM} = jZ_{TM} / Z_0 & \text{for TM waves} \\ -1/Q_{TE} = -j Z_0 / Z_{TE} & \text{for TE waves} \end{cases}$$

2) Waveguide formed by an impedance plane and a perfectly conducting plane:

a) TM waves:

$$\psi_{TM}(k_x, Q_{TM}) = 1 + jkQ_{TM}/k_z - (1 - jkQ_{TM}/k_z)e^{-2jk_z d} ; \tag{3.5}$$

b) TE waves:

$$\psi_{TE}(k_x, Q_{TE}) = 1 - jk/k_z Q_{TE} + (1 + jk/k_z Q_{TE})e^{-2jk_z d} . \tag{3.6}$$

3) Plane waveguide with perfectly conducting walls:

a) TM waves:

$$\tilde{\psi}_{TM} = \psi_{TM}(k_x, Q_{TM} = 0) = 1 - e^{-2jk_z d} ; \qquad (3.7)$$

b) TE waves:

$$\tilde{\psi}_{TE} = \psi_{TE}(k_x, Q_{TE} = \infty) = 1 + e^{-2jk_z d} . \qquad (3.8)$$

When the impedances are spatially dispersionless, these functions are easily factorized by the usual methods [4, 162, 168, 12].

Let us cite the results of field calculation in some systems.

Surface wave excitation on an impedance plane by the open end of a plane, semi-infinite waveguide (impedance plane is continued inside the waveguide).

a) TM waves [168]:

$$L(k_x) = k_z \, \psi_{TM}(k_x, Q_{TM}) / \varphi(k_x, Q_{TM}) . \qquad (3.9)$$

The power transmission coefficient of an incident wave of the fundamental mode into the surface wave on the impedance plane (only one mode is propagating in the waveguide) is:

$$|T_{os}|^2 = 2 \operatorname{sh} \tilde{p}_0 \operatorname{th} \tilde{p}_0 \, e^{-q} \frac{2\Gamma_0}{\Gamma_0 + \Gamma_s} \frac{\Gamma_0 + D}{\Gamma_s + D} e^{-(\Gamma_0 - \Gamma_s)} , \qquad (3.10)$$

where $\tilde{p}_0$ is the root of the equation $\tilde{p}_0 \operatorname{tg} \tilde{p}_0 = q$,

$$D = kd, \quad q = kQ_{TM}d, \quad \Gamma_o = k_{xo}d, \quad \Gamma_s = k_{xs}d ,$$

d is the height of the waveguide aperture, k_{xo} and k_{xs} are, respectively, the propagation constants of the fundamental mode in the waveguide and of the surface wave on the impedance plane.

The power reflection coefficient of the fundamental mode is:

$$|R_{oo}|^2 = \left(\frac{\Gamma_o + D}{\Gamma_s + D}\right)^2 \frac{1}{ch^2 \tilde{p}_o} e^{-2\Gamma_o} \qquad (3.11)$$

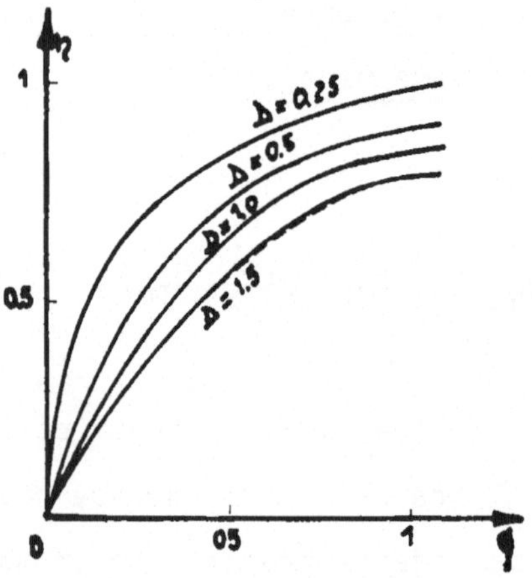

Fig. 4.

Fig.4 shows the dependence of surface wave excitation efficiency

$$\eta = \frac{|T_{os}|^2}{1-|R_{oo}|^2} \qquad (3.12)$$

on q for different values of aperture height D. With increasing D, the efficiency tends to its limiting value

$$\eta_{lim} = 1-e^{-2q}$$

(dashed curve in Fig.4).

b) TE waves [12] :

$$L(k_s) = \frac{1}{k_z} \frac{\psi_{TE}(k_x, Q_{TE})}{\varphi(k_x, -1/Q_{TE})} \quad . \qquad (3.13)$$

The transmission coefficient (with the same assumptions as in a)) is :

$$|T_{os}|^2 = 4\mathrm{ch}\,\tilde{p}_o \frac{\Gamma_o}{\Gamma_o+\Gamma_s} e^{-(\Gamma_o-\Gamma_s)} \quad . \qquad (3.14)$$

The reflection coefficient of the fundamental wave is:

$$|R_{oo}|^2 = \frac{\Gamma_s-\Gamma_o}{\Gamma_s+\Gamma_o} e^{-2\Gamma_o} \quad . \qquad (3.15)$$

Here p_0 is the root of the equation

$$\tilde{p}_0 \operatorname{cth} \tilde{p}_0 = -kd/Q_{TE} \; .$$

In the above problems, and also in problems with an impedance cylinder [159], the simplicity of the factorization of the integral equation kernel was due to the absence of spatial dispersion of the impedance. The Wiener-Hopf method may be applied to the solution of even more complicated problems, when the calculation of impedance spatial dispersion of the slow-wave system becomes, in principle, a necessity. Here we touch upon problems of dielectric waveguide excitation [169, 175, 18.7] .

Diffraction by an impedance step on a plane [171, 180, 183 . The kernel of the integral equations (3.1) here is the function

$$L(k_x) = \varphi(k_x, f_1) / \varphi(k_x, f_2) \; , \tag{3.16}$$

where indices 1, 2 are related to the regions along both sides of the impedance jump. The reflection and transmission coefficients of the surface wave at the step are described as follows:

$$|R|^2 = k^4 f_1^2 (f_1 - f_2)^2 / k_{x_1}^2 (k_{x_1} + k_{x_2})^2 \; ; \tag{3.17}$$

$$|T|^2 = 4 f_1 f_2 / (f_1 + f_2)^2 \; . \tag{3.18}$$

<u>Diffraction by an impedance half-plane</u> [12, 160, 172, 179]. Here we must distinguish between cases in which the surface impedance has equal signs ($Z^{(1)} = Z^{(2)}$) or opposite signs ($Z^{(1)} = -Z^{(2)}$) on the two sides of the half-plane.*

---

* It is interesting to note that the solution of the problem of diffraction on an impedance half-plane with identical surface impedances can be obtained by the image method, based on the well-known solution for wave diffraction on an impedance step [12]. An idea of the method is given in Fig.5, using the example of a TM wave. In order to apply the results obtained for the case of diffraction by an impedance step to the impedance half-plane, it is sufficient to: 1) examine the problem of plane wave diffraction on an impedance step $Z_{TM}^{(1)}$, $Z_{TM}^{(2)} = 0$ and, using the image method, pass over to the impedance half-plane, letting the magnetic field in the half-space $z \leqslant 0$ have the same sign as for $z \geqslant 0$ (Fig. 5a);

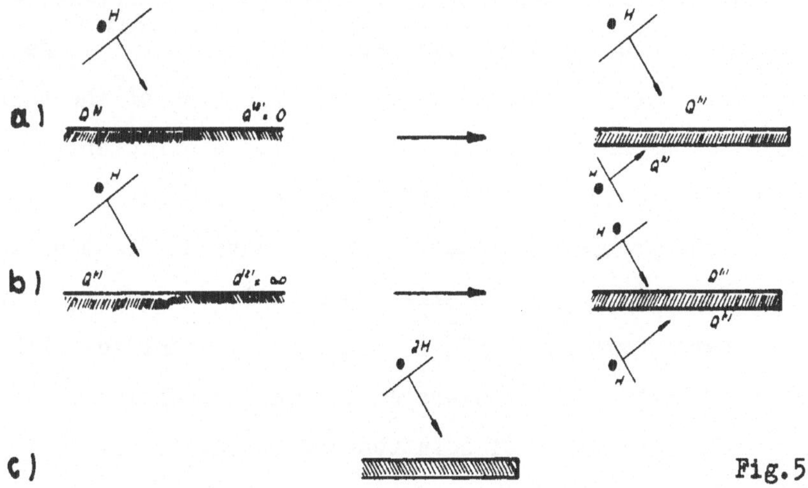

Fig.5.

2) examine the problem of diffraction of the same wave at a step $Z_{TM}^{(1)} \neq 0$, $Z_{TM}^{(2)} = \infty$ and, using the image method, pass over to the half-plane by changing the sign of the magnetic field in the half-space $z \leqslant 0$ into its opposite (Fig.5b);
3) add the results of the solution of the first two problems (Fig.5c).

For the modulus of the reflection coefficient the following simple relations are obtained:

$$
|R| = \begin{cases} \sqrt{\dfrac{k_x/k-1}{2k_x/k}} & \text{when } Z^{(1)} = Z^{(2)} ; \\[2ex] \sqrt{\dfrac{k_x/k-1}{k_x/k+1}} & \text{when } Z^{(1)} = -Z^{(2)} , \end{cases} \tag{3.19}
$$

where k_x/k is the relative wavenumber of the incident wave. When the incident wave phase velocities are equaly retarded on a half-plane with identical side-impedances, and on a half-plane with impedances of opposite signs, the normalized antenna radiation patterns coincide. When $Z^{(1)} = Z^{(2)}$, the surface wave also penetrates to the reverse side of the half--plane, and the modulus of the transmission coefficient is equal to the modulus of the reflection coefficient.

The problem of electromagnetic wave diffraction by a semi-infinite array of parallel, perfectly conducting, thin wires or narrow ribbons [160] is analogously examined, with the help of Moizhes' [48] averaged boundary conditions.

In its method of solution and character of the results, the problem of the diffraction of a wave at a step in surface impedance on one of the walls of a plane parallel waveguide, comes very close to the one discussed above. But, of course, in this case the field reradiated by the step is distributed over the discrete set of waves propagating in the waveguide, and only in the limiting case of a remote wall does it turn into a radiation field with a continuous spectrum of wave numbers; thus the problem consists in the calculation of the reflection and transmission coefficients for an arbitrary type wave incident on the step [170]. In as much as in this problem the kernel of the integral equation is expressed in terms of the same functions ψ_{TM} and ψ_{TE}, as in the problem of the surface wave excitation by the open end of a waveguide, no additional difficulties arise in the factorization. The reflection coefficient of the fundamental mode, in the simple case when higher order modes cannot propagate, is given by the expression

$$\left|R_{oo}\right| = \frac{\left|k_{x_o}^{(1)} - k_{x_o}^{(2)}\right|}{k_{x_o}^{(1)} + k_{x_o}^{(2)}} , \qquad (3.20)$$

and the transmission coefficient by

$$\left|T_{oo}\right| = \frac{2\sqrt{k_{x_o}^{(1)}\, k_{x_o}^{(2)}}}{k_{x_o}^{(1)} + k_{x_o}^{(2)}} . \qquad (3.21)$$

It is interesting to note that these power coefficients depends only on the wavenumbers of propagating modes. This holds also in the case where several modes can propagate in the waveguide simultaneously. This result is analogous, in particular, to the result obtained in the problem of the junction of a coaxial cable with a round waveguide (Weinstein [162]), and also in the problem of the junction of plane waveguides partially filled with dielectric [169] .

Diffraction on an impedance wedge. The examples given above concerning diffraction by an impedance step on a plane and on an impedance half-plane, can be considered as special variants of a certain general problem concerning diffraction on a wedge, on which, in addition to the impedance step, there is also a break in the surface. By presenting the solution in the form of a Sommerfeld integral (Malyuzhinets [165]), one can calculate the space wave field at long distances from the vertex of the wedge, and also the amplitude of the surface waves that propagate along the faces of the wedge, when the impedance values admit of the existence of these waves.

Diffraction at the open end of a cylindrical impedance line. With suitable idealizations, it is possible to handle successfully the solution of problems with boundaries of a cylindrical geometry. Thus, for the case of surface wave diffraction at the end of a hollow impedance cylinder, it is possible to obtain a complete and rigorous solution of the problem by using the Wiener-Hopf technique, and for an impedance rod with a flat end the problem can be reduced to the solution (by successive approximations) of an infinite system of algebraic equations (Guan - Ding - Hua [164]).

The Wiener-Hopf technique may be used with good results also for the solution of the problem of wave diffraction on a discontinuity in a helix (Mikazan [166]), if we assume an averaged boundary condition for the longitudinal components of the electric field on a ribbon wound in a helix. In this case, by taking into account the phenomenon of so-called "spatial resonance" (Weinstein [5]), it is possible, by retaining in the infinite series only the "resonance" terms, to arrive at comparatively simple expressions for the reflection and transmission coefficients of surface waves of various types, and also for the radiation in the far field. The above mentioned method is applicable also when studying wave diffraction on the junction between helical waveguides, and also on the junction between a helically conducting waveguide and an ordinary isotropically conducting waveguide. All these questions are interesting from the point of view of finding (or of making more precise) the characteristics of cylindrical radiators, as also of the matching elements of waveguide lines with retardation systems.

Propagation of waves in lines with random variations in wall impedance. The variation in the parameters of the surface wave guiding structure was considered above to be a prescribed function of the coordinates. However, as has already been mentioned, these variations can have random character, if they are caused by inaccurancy in the manufacture of the structure. It is worthwhile to investigate by statistical methods the influence of such random perturbations on the propagation of surface waves. As before, a waveguide with impedance walls can be chosen as the simplest model, assuming that the impedance is subject to small variations in space. Then, using the perturbation method, it is possible (Bespalov [161 a]) to obtain general formulas for the reflection coefficients of the initial incident wave and for the coefficients of its transformation into waves of other types. In the case of open lines, this transformation is equivalent to energy losses to radiation, and, consequently, if such lines are used as traveling wave antennas, this leads to the distortion of the antenna radiation pattern. Moreover the scattered field has on an average the same distribution in space as the radiation field of the individual discontinuities, whose dimensions are determined by the correlation radius of the random perturbations.

4. REMARKS ON SURFACE WAVE ANTENNAS

One of the most promising directions in the development of surface wave systems turned out to be their application to antenna technique, permitting the extension of the possibilities of creating radiating devices which meet certain structural, electrodynamic and tactical demands. Here too the application of the impedance approach facilitates to a considerable extent the investigation by simplifying the calculation of equivalent source distributions on the aperture of such antennas. Papers [18, 76] can be used as an example, where the antenna radiation pattern of the impedance structure, excited by the open end of a waveguide, is calculated by Kirchhoff's approximation. The influence of individual factors on the pattern (such as the direct radiation of the feed, the termination of the impedance structure at the end of the antenna, the introduction of reradiating elements, and so on) can be ascertained in detail by using the results pertaining to the diffraction problems discussed in the preceding section.

The impedance approach is used very fruitfully when studying surface wave antennas with periodically changing parameters of the retarding system [188, 190].

For instance, let there be a radiator with sinusoidally modulated impedance on the guiding surface, for which the fast first harmonic is the radiating one (see section 1). The distribution of the amplitude $A(x)$ of this harmonic along the antenna with variable depth of impedance modulation is described by the function

$$A(x) = \sqrt{\xi}\ \exp\left[-kG\int_0^x \xi(x')dx' - k\delta x\right], \qquad (4.1)$$

where $\xi = Q_1^2(x)/4$, $Q_1(x)$ is the amplitude of the dimensionless impedance modulation (for definiteness, the case of TM waves is being considered), $k\delta$ is the coefficient of wave attenuation on the antenna due to ohmic losses, kG is the coefficient determining the dependence of attenuation (due to losses through radiation) on the average values of the impedance structure parameters. Expression (4.1) permits us to find the variation of $\xi(x)$ as a function of the required amplitude distribution $A(x)$:

$$\xi(x) = \frac{A^2(x)\exp(2k\delta x)}{1-2kG\int_0^x A^2(x')\exp(2k\delta x')\,dx'}, \qquad (4.2)$$

and also permits us to calculate the efficiency and the gain of radiators of this type [190].

Thus, the problem of the synthesis of surface impedance distribution for periodically modulated impedance structures can be solved comparatively simply, at least for systems with sufficiently smoothly changing average parameters.

In principle this is also possible in the more common case of impedance antennas with arbitrary laws of surface impedance change. But a practical approach to the solution of such a problem is possible only for impedance surfaces of the most simple configuration (impedance cylinder [192], impedance plane [191]), when the distribution of the impedance over the surface can be readily related, for specified sources, to the complex radiation pattern. However, in most cases a specified pattern corresponds to non-realizable (or realizable only with difficulty) impedance distributions: the impedance turns out to be complex or, on some portions, even active. Therefore it is necessary to formulate beforehand the conditions for realizable impedance distribution, and thereby limit the class of prescribed patterns. The search for the pattern most accurately approximating the specified one in this restricted class is, as a rule, a very complicated problem. This leads to the necessity of resorting to an approximate examination of separate particular cases only, making definite assumptions about the arrangement of sources and the form of the radiation pattern.

Here we shall limit ourselves to these few remarks concerning surface wave antennas, and refer the reader for more detailed information to the extensive literature available at the present time on this question [18.20, 21, 185-227].

The problem which we set ourselves when writing this review consisted not only in the unification of information on surface electromagnetic waves, but also in the demonstration of the fruitfulness of the use of the surface impedance concept. As a matter of fact, impedance relations between field components on closed surfaces are the simplest from among those that ensure uniqueness of the solution inside the region. Their use also permits a unique characterization of interface guiding properties. As a result it becomes possible to obtain results that are easily interpreted even for complicated electromagnetic systems, and consequently to analyze these systems from general points of view.

From this point of view, the methods and results described above should have a wider range of application. We are thinking not only about analogous effects, let us say, in shielded electromagnetic waveguide systems with partially localized fields near the interfaces, but also about the translation of these effects into adjacent fields of investigation, particularly into acoustics [29, 36].

B i b l i o g r a p h y

M.A.Miller and
V.I.Talanov

GENERAL QUESTIONS

1. Ya. L. Alpert, V.L. Ginzburg, E.L. Feinberg, The propagation of radio waves, Gov. ed. of theor. techn. literature, Moscow, 1953.

2. L.M. Brekhovskikh, Waves in layered media, by Academic Press, New York, 1960

3. L. Brillouin, M. Parodi, Wave-Propagation in periodic structures, electric Filters and crystal lattices, Dover Publishing Co. New York, 1953.

4. L.A. Weinstein, Diffraction of electromagnetic and sonic waves on the open end of a waveguide, published by Soviet Radio, Moscow, 1953.

5. L. A. Weinstein, Electromagnetic waves, published by Soviet Radio, Moscow, 1957.

6. V. L. Ginzburg, Propagation of electromagnetic waves in plasma, Phys. Math. State Publications, Moscow, 1960.

7. The Channeling of electromagnetic energy by means of surface waves (a survey), Problems of Radar Engineering, No. 1 (7), 14 (1952).

8. M. A. Miller, Application of uniform boundary conditions in the solution of the problem of the propagation of surface electromagnetic waves, and in the investigation of the oscillations of thin antennas, Thesis, Gorky University, 1953.

9. M.A. Miller, V.I. Talanov, Surface electromagnetic waves (a survey of Soviet work), a report for the XIIIth General Assembly of URSI, London, 1960

10. F.M. Morse, G. Feshbach, Methods of theoretical physics, I, Mc Graw-Hill Book Company, New York, 1953.

11. M.S. Neumann, The problem of surface electromagnetic waves, Bulletin of Higher Education Institutions, Radio Engineering, 1-2, No. 1, 7 (1958).

12. V.I. Talanov, Problems of the diffraction and excitation of electromagnetic waves in slow-wave systems, Thesis, Gorky University, 1959.

13. Transmission line theory for super-high frequencies, 1, published by Soviet Radio, Moscow, 1951.

14. V.A. Fock, Diffraction of radio waves around the earth's surface, published by the USSR Academy of Sciences, M.-L., 1946.

15. H.M. Barlow, A.L. Cullen, Surface waves, Proceeding of the Institute of Electrical Engineers, 100, III, 329 (1953).

16. M.F. Bracey, Surface-wave research in Sheffield, Transactions of the Institute of Radio Engineers, AP-7, XII, Special Suppl., 219 (1959).

17. H.V. Cottony, R.S. Elliott, E.C. Jordon, V.H. Rumsey, K.M. Siegel, J.R. Wait, O.C. Woodyard, U.S.A. National Committee Report to URSI Sub-commission 6. 3. Antennas and waveguides, and annotated bibliography.

18. A.F. Harvey, Periodic and guiding structures at microwave frequencies (a survey), IRE, MTT-8, 30 (1960).

19. F.J. Zucker, The Guiding and radiation of surface waves, 1954 Proc. Symp. on Modern Advances in Microwave Tech., page 403. Polytechnic Inst. of Brooklyn, 1955.

20. F.J. Zucker, Surface and leaky-wave antennas, Chapter 16 in the Handbook of Antenna Engineering, H. Jasik, Mc Graw-Hill Book Co., 1960.

21. F.J. Zucker, Progress in surface and leaky-wave antennas during the last three years, J. Res. NBS, 64 D, 6 (1960).

FREE SURFACE WAVES

22. N.A. Armand, The propagation of electromagnetic surface waves along a multiwire system, Journal of Engineering Physics, 29, 107 (1959).

23. F.G. Bass, The boundary conditions for an average electromagnetic field on a surface with random irregularities and with fluctuations of impedance, Bulletin of Higher Education Institutions, Radio Physics, 3, 72 (1960).

24. A.M. Belyantsev, Toward a theory of multiwire systems with surface waves, Bulletin of Higher Education Institutions, Radio Physics, 1, No. 5-6, 112 (1958).

25. V.I. Bespalov, M.A. Miller, Electromagnetic waves in rectangular grooves with a dielectric coating of the bottom, Scientific notes of Gorky Geophysics Institute, Physics Series, 30, 61 (1959).

26. L.S. Benenson, Anisotropic properties of corrugated retarding systems, Radio Engineering and Electronics, 4, 517 (1959).

27. M.S. Bobrovnikov, E.A. Babin, The maximum rated power transmittible along a single cylindrical wire, Bulletin of Higher Education Institutions, Physics, No. 1, 175 (1957).

28. D.M. Bravo-Zhivotovski, Surface waves with circular polarization in corrugated systems, Bulletin of Higher Education Institutions, Radio Physics, 2, 829 (1959).

29. L.M. Brekhovskikh, Surface waves in acoustics, Journal of Acoustics, 5, 4 (1959).

30. B.M. Bulgakov, V.P. Shestopalov, L.A. Shishkin and I.P. Yakimenko, Symmetrical surface waves in a spiral waveguide, located in a ferrite medium, Radio Engineering and Electronics, 5, 1818 (1960).

31. L.A. Weinstein, Surface electromagnetic waves over a comb structure, Journal of Engineering Physics, 26, 385 (1956).

32. T.A. Vereshchakova, V.V. Tyazhelov, Experimental investigation of space-beats in a two-wire line with decimetric waves, Bulletin of Higher Education Institutions, Radio Engineering, 2, 217 (1959).

33. B.N. Gershman, Ordinary waves in a uniform plasma with a magnetic field, Collection in memory of A.A.Andronova, Moscow, 1955, page 599.

34. M.A. Ginzburg, Surface waves on the boundary of a gyrotropic medium, Journal of Experimental and Theoretical Physics, 34, 1635 (1958); Transactions of the Graduate School, Radio Engineering and Electronics, No. 3, 38 (1958).

35. A.A. Denisov, Determination of the propagation constant of a surface TM (transverse magnetic) wave, propagating along a cylindrical conductor with a ring-type corrugated structure, Transactions of the Leningrad Polytechnic Institute, Radic Physics, No. 181, 68 (1955).

36. K.M. Ivanov-Schitz, F.V. Rozhin, Toward an investigation of surface waves in the air, Journal of Acoustics, 5, 495 (1959).

37. B.Z. Katzenelenbaum, Asymmetrical oscillations of an infinite dielectric cylinder, Journal of Engineering Physics, 19, 1182 (1949).

38. L.D. Landau, E.M. Lifschitz, Electrodynamics of solid media, Gov. ed. of theor. techn. literature, Moscow , 1957, page 364.

39. N.N. Malov, A transverse electric wave in a metallic trough, Radio Engineering and Electronics, 2, 1289 (1957).

40. L.I. Mandelstam, The propagation of waves along a surface and the guiding action of conductors, A complete collection of transaction, 3, published by the USSR Academy of Sciences, Moscow, 1950, page 365.

41. A.L. Mikaelyan, The application of ferrites in high-frequency systems engineering, Thesis, Radio Electronics Publications of the USSR Academy of Sciences, 1955.

42. A.L. Mikaelyan, A.K. Stolyarov, Surface waves in ferrite waveguides, Radio Engineering and Electronics, 4, 1079 (1959).

43. M.A. Miller, The propagation of electromagnetic waves over a flat surface with anisotropic homogeneous boundary conditions, Transactions of the USSR Academy of Sciences, 87, 571 (1952).

44. M.A. Miller, Surface electromagnetic waves in rectangular grooves, Journal of Engineering Physics, 25, 1972 (1955).

45. M.A. Miller, V.I. Talanov, Surface electromagnetic waves guided by a boundary with slight curvature, Journal of Engineering Physics, 26, 2755 (1956).

46. D.I. Mirovitsky and G.G. Valieev, Surface wave directional couplers, Radio Engineering and Electronics, 5 , 1078 (1960).

47. D.I. Mirovitsky and G.G. Valieev, Hybrid connections on surface wave lines, Radio Engineering and Electronics, 5, 1179 (1960).

48. B. Ya. Moizhes, Averaged electromagnetic boundary conditions for metallic grids, Journal of Engineering Physics, 25, 1 (1955).

49. V.I. Molotkov, Investigation of the propagation of surface waves along cylindrical conductors, Transactions of the Leningrad Polytechnic Institute, Radio Physics, No. 181, 60 (1955).

50. M.S. Neumann, The nature of surface electromagnetic waves and methods for their calculation, Transactions of Moscow Institute of Aeronautics, issue 50, 93 (1955).

51. B.A. Poperechenko, Surface electromagnetic waves on a flat layer, Transactions of the Graduate School, Radio Engineering and Electronics, No. 2, 36 (1958).

52. B.A. Poperechenko, Surface electromagnetic waves on a coated cylinder, Transactions of the Graduate School , Radio Engineering and Electronics, No. 2, 42 (1958).

53. N.A. Semienov, Attenuation in a dielectric waveguide, Transactions of the Graduate School, Radio Engineering and Electronics, No. 1, 83 (1959).

54. N. A. Semienov, Wave types of a dielectric waveguide, Transaction of the Graduate School, Radio Engineering and Electronics, No. 4, 60 (1958).

55. V. Ya. Smorgonsky, Calculation of the phase and group velocities of surface waves, Radio Engineering, 10, No. 5, 25 (1955).

56. V. I. Talanov, Surface electromagnetic waves in systems with non-uniform impedance, Bulletin of Higher Education Institutions, Radio Physics, 2, 132 (1959).

57. V.V. Tyazhelov, An experimental investigation of the interaction of single-wire transmission lines, Radio Engineering and Electronics, 4, 592 (1959).

58. Ya. B. Feinberg, M. F. Gorbatenko, Electromagnetic waves in plasma with a magnetic field, Journal of Engineering Physics, 29, 549 (1959).

59. P. R. Cheriep, Bends in surface waveguides, Kiev, 1957.

60. V. P. Shestopalov, K. P. Yatsuk, The use of slow surface waves for measurement of the specific inductive capacitance of material at super-high frequencies, I, Journal of Engineering Physics, 29, 819 (1959).

61. V. P. Shestopalov, K. P. Yatsuk, The use of slow surface waves for measurement of the specific inductive capacitance of material at super-high frequencies, II, Journal of Engineering Physics, 29, 1090 (1959).

62. V. P. Shestopalov, K. P. Yatsuk, I. P. Yakimenko, The use of slow surface waves for measurement of the specific inductive capacitance of material at super-high frequencies, III, Journal of Engineering Physics, 29, 1130 (1959).

63. A. I. Shtirov, On the question of channeling uniform retarding systems, Radio Engineering and Electronics, 2, 244 (1957).

64. A. I. Shtirov, On the propagation of cylindric surface waves in periodic structures, Radio Engineering and Electronics, 4, 903 (1959).

65. K. P. Yatsuk, G.N. Bichkova, Application of resonant retarding systems for the measurement of specific inductive capacitance of materials at super-high frequencies, Journal of Engineering Physics, 30, 165 (1959).

66. H. M. Barlow, Surface waves, Proceedings of the Institute of Radio Engineers, 46, 1413 (1958).

67. H. M. Barlow, Surface waves: A proposed definition, Proceedings of the Institute of Electrical Engineers, 107 B, 240 (1960).

68. H. M. Barlow, A. E. Karbowiak, An investigation of the characteristics of cylindrical surface waves, Proceedings of the Institute of Electrical Engineers, 100, III, 321, 341 (1953).

69. H. M. Barlow, The power radiated by a surface wave, circulating around a cylindrical surface, Proceedings of the Institute of Electrical Engineers, B-106, 180 (1959).

70a. T. Berceli, Propagation of the surface wave along an insulated conductor, Acta tech. Acad. sci. hung., 17, 219 (1957).

70b. T. Berceli, Examination of surface wave lines, Acta tech. Acad. sci. hung., 25, 257 (1959).

71. F. Bertein, W. Chahid, On obtaining retarded electromagnetic waves by means of cylindric current systems, C. R. Acad. sci., 242, 2918 (1956).

72. A. D. Bresler, On the TE_{no} mode of a ferrite slab loaded rectangular waveguide and the associated thermodynamic paradox, Transactions of the Institute of Radio Engineers, MTT-8, 81 (1960).

73. Butterfield, Dielectric sheet radiators, Transactions of the Institute of Radio Engineers, AP-3, 152 (1954).

74. B. Chiron, Influence of surrounding conditions on the propagation of a surface wave. An example of the construction of an antenna feeder system consisting of a rectangular waveguide and of a line with a surface wave, Cable and transm., 11, 237 (1957).

75. Cu Fu - nian, Electromagnetic waves propagating along a spiral, Acta phys. sinica, 15, 637 (1959).

76. J. Dain, The propagation of slow waves, Electronic Engr., 30, 388 (1958).

77. R. S. Elliott, Spherical surface wave antennas, Transactions of the Institute of Radio Engineers, AP-4, 422 (1956).

78. B. Epstein, G. Mourier, Determination, measurement and characteristics of phase velocities in systems with periodic structure, Ann. radioelectr., 10, 39, 64 (1955).

79. P. S. Epstein, On the possibility of electromagnetic surface waves, Proc. Nat. Acad. Sci. USA, 40, 1158 (1954).

80. L.B. Felsen, Electromagnetic properties of wedge and cone with a linearly varying surface impedance, Transactions of the Institute of Radio Engineers, AP-7, Special Suppl., 231 (1959).

81. B. Friedman, Surface waves over a lossy conductor, Transactions of the Institute of Radio Engineers, AP-7, Special Suppl., 227 (1959).

82. L. O. Goldstone, A. A. Oliner, A note on surface waves along corrugated structure, Transactions of the Institute of Radio Engineers, AP-7, 274 (1959).

83. G. Goubau, Design of surface wave transmission lines, Electronics, 28, 6A, 60 (1955).

84. G. Goubau, C. E. Sharp, Investigations with a model surface wave transmission line, Transactions of the Institute of Radio Engineers, AP-5, 242 (1957).

85. G. Goubau, Some characteristics of surface wave transmission lines for long-distance transmission, Proceedings of the Institute of Electrical Engineers, B 106, Suppl. 13, 166 (1959).

86. R. C. Hansen, Single slab arbitrary polarization surface wave structure, Transactions of the Institute of Radio Engineers, MTT-5, 115 (1957).

87. F. R. Huber, H. Rudat, Operating characteristics and uses of the Goubau line, Broadcasting technical Informations, 3, XII, 277 (1959).

88. C. Jauquet, Transverse magnetic surface waves on an infinite conducting cylinder, Bull. cl. Sci. Acad. roy. Belgique, 42, 1178 (1956).

89. A. E. Karbowiak, The concept of heterogeneous surface wave impedance and its application to the cylindrical cavity resonators, Proceedings of the Institute of Electrical Engineers, C 105, 1 (1958).

90. A. E. Karbowiak, The elliptic surface wave, Brit. J. Appl. Phys., 5, 328 (1954).

91. A. E. Karbowiak, Surface E H-wave, Wireless Engr., 31, 71 (1954).

92. A. E. Karbowiak, Theory of composite guides: Stratified guides for surface waves, Proceedings of the Institute of Electrical Engineers, III, 101, 72, 238 (1954).

93. B. Kockel, Sommerfeld's surface waves, Ann. Physik, 1, 145 (1958).

94. Kikuchi Hiroshi, Yamashite Eikichi, Hybrid waves propagating on a line with Goubau's surface wave, J. Inst. Electr. Commun. Engr. Japan, 43, 39 (1960).

95. B. Lax, K. Button, Theory of ferrites in rectangular waveguide, Transactions of the Institute of Radio Engineers, AP-4, 531 (1956).

96. G. G. Mac Farlane, Surface impedance of an infinite parallel-wire grid at oblique angles of incidence, Journal of the Institute of Electrical Engineers, III A, 93, 1523 (1946).

97. D. Markuse, Examination of power exchange and field distribution in parallel surface waveguides, Arch. electr. Ubertrag, 10, 117 (1956).

98. Matsuo Iukito, Edisutani Keisuko, Cho Ioscio, Phase velocity in Karp's retarded system, Mem. Inst. Scient. and Industr. Rec. Asaka Univ., 15, 9 (1958).

99. A. G. Mungall, D. Morris, Surface wave propagation over a sand-covered conducting plane, Canad. J. Phys., 37, 1349 (1959).

100. A. G. Mungall, D. Morris, The group velocity of plane surface waves, Canad. J. Phys., 38, 779 (1960).

101. D. Morris, A. G. Mungall, TE surface waves guided by a dielectric-covered metal plane, Canad. J. Phys., 38, 1553 (1960).

102. V. Oehrl, Propagation of slow electromagnetic waves along a heterogeneous layer of plasma, Journal of Applied Physics, 9, 164 (1957).

103. A. A. Oliner, A. Hessel, Guided waves on sinusoidally modulated reactance surfaces, Transactions of the Institute of Radio Engineers, AP-7, Special Suppl., 201 (1959).

104. R.L. Pease, On the propagation of surface waves over an infinite grounded ferrite layer, Transactions of the Institute of Radio Engineers, AP-6, 13 (1958).

105. T. E. Roberts, An experimental investigation of the single-wire transmission line, Transactions of the Institute of Radio Engineers, AP-2, 46 (1954).

106. Shimmel, Latest publications on the propagation of surface waves, Communication Engineering, 4, 279 (1954).

107. G. Schulten, Novel method for measuring impedance on surface wave transmission lines. Proceedings of the Institute of Radio Engineers, 47, 76 (1959).

108. S. A. Schelkunoff, Anatomy of " surface waves ", Transactions of the Institute of Radio Engineers, AP-7, XII, Special Suppl., 133 (1959).

109. K. P. Sharma, The estimation of the reactance resistence of a loss-free surface supporting surface wave, Proceedingsof the Institute of Electrical Engineers, B 106, 427 (1959).

110. M. Sugi, T. Nakahara, O-guide and X-guide: An advanced surface wave transmission concept, Transactions of the Institute of Radio Engineers, MTT-7, 366 (1959).

111. P. Szulkin, A theory of Goubau's surface waves, Arch. electrotechniki, 8, 313 (1959).

112. Tashio Hosono, Surface resistence of corrugated conductors, Proceeding of the Institute of Radio Engineers, 48, 247 (1960).

113. Uchida Hidenari, Hichida Shigeo, Surface and space waves on a transmission line with a surface wave, Sci. Repts. Res. Insts. Tohoku Unn. Ser. B, Electr. Commun., 6, No. 3-4, 217 (1955).

114. J. Von Bladed, O. Jr. Von Rohr, Semicircular ridges in rectangular waveguides, Transactions of the Institute of Radio Engineers, MTT-5, 103 (1957).

115. J. R. Wait, Guiding of electromagnetic waves directed by uniformly rough surfaces, Transactions of the Institute of Radio Engineers, AP-7, Special Suppl., 154 (1959).

116. J. R. Wait, Propagation of electromagnetic waves along a thin sheet of plasma, Canad. J. Phys., 38, 1586 (1960).

117. E. Weissberg, Experimental determination of wavelengths in dielectric-filled periodic structures, Transactions of the Institute of Radio Engineers, MTT-7, 480 (1959).

118. J. C. Walling, Interdigital slow-wave structures, Onde electr., 37, 136 (1957).

119. J. C. Walling, Interdigital and other slow wave structures, Electron and Control, 3, 239 (1957).

120. Sato Risaburo, Khariu Tokio, Tiba Dziro, Calculation of surface wave transmission line and matching devices with spiral elements, Techn. J. Japan Broadcast. Corp. 11, 24 (1959).

121. Sato Risaburo, Determination of transmission losses in Goubau lines and spiral lines in UHF and SHF bands, Television, 13, 248 (1959).

122. Sugi, Nakahara, Left-handed (counterclockwise) polarized waves from dielectrics of different cross sections, J. Inst. Electr. Commun. Engrs. Japan, 42, 731 (1959).

123. Iosida, Shielded transmission line with a surface wave, Toshiba Rev., 11, 840 (1956).

124. Utida, Nishida, Nagasavo, Uda, Shunting reactive elements in surface wave transmission lines, J. Inst. Electr. Commun. Engrs. Japan, 33, 353 (1955).

125. Yakhagi, Forked transmission lines with a surface wave, Bull. Electrotechn. Lab. , 23, 269 (1958).

EXCITATION OF SURFACE WAVES

126. M. S. Bobrovnikov, Lumped excitation of cylindric conductors covered with a dielectric, Transactions of the Siberian Institute of Engineering Physics, Tomsk University, 36, 37 (1958).

127. S. M. Verevkin, Excitation of an infinite cylinder with non-uniform boundary conditions by Leontovich's magnetic current loop, Transactions (Scientific Reports) of Higher School, Radio Engineering and Electronics, No.3, 54 (1958).

128. V. V. Vladimirsky, Propagation of electromagnetic waves on a single-wire line, Bulletin of the USSR Academy of Sciences, Physics Series, 8, 139 (1944).

129. G. D. Malyuzhinets, Generalization of Veilya's formula for waveguide field over an absorbing surface, Transactions of the USSR Academy of Sciences, 60, 367 (1948).

130. B. A. Poperechenko, Excitation of a cylinder with layer, Transactions of Higher School, Radio Engineering and Electronics, No. 4, 46 (1958).

131. B. A. Poperechenko, Excitation of a large diameter cylinder with layer, Transactions of Higher School, Radio Engineering and Electronics, No. 1, 62 (1959).

132. L. S. Tartakovsky, Radiation of dipole over a flat homogeneous ground, Radio Engineering, 13, No.4, 36 (1958); 14, No.8, 8 (1959).

133. O. N. Tereshin, Application of a fictitious magnetic current to the solution of the problem, due to Leontovich, of radiation from an aerial above a plane with non-homogeneous boundary conditions, Radip Engineering, 12, No. 4, 24 (1957).

134. V. A. Filonenko, Excitation of a double-layer dielectric cylinder of infinite length by electric or magnetic dipole, Transactions of the Siberian Institute of Engineering Physics at Tomsk University, 36, 364 (1958).

135. M. D. Khaskind, Excitation of surface electromagnetic waves on flat dielectric coatings, Radio Engineering and Electronics, 5, 188 (1960).

136. M. D. Khaskind, Propagation of sonic and electromagnetic waves in a half space, Journal of Acoustics, 5, 464 (1959).

137. D. V. Shannikov, Relationship of power transferred by waves and excited by a slot on a plane covered by a layer of dielectric, Radio Engineering, 15, No.2, 27 (1960).

138. H. Bremmer, The surface-wave concept in connection with propagation trajectories associated with the Sommerfeld problem, Transactions of the Institute of Radio Engineers, AP-7, Special Suppl., 175 (1959).

139. J. Brown, Some theoretical results for surface wave launchers, Transactions of the Institute of Radio Engineers, AP-7, Special Suppl., 169 (1959).

140. R. H. Clarke, A method of estimating the power radiated directly at the feed of a dielectric-rod aerial, Proceedings of the Institute of Electrical Engineers, B 104, 511 (1957).

141. A. L. Cullen, Surface wave resonance effect in a reactive cylindrical structure excited by an axial line source, J. Research NBS, 64 D, 13 (1960).

142. J. W. Duncan, The efficiency of surface wave excitation on a dielectric cylinder, Transactions of the Institute of Radio Engineers, MTT-7, 257 (1959).

143. A. D. Frost, P. R. McGeoch, C. R. Mungins, The excitation of surface waveguides and radiating slots by strip-circuit transmission lines, Transactions of the Institute of Radio Engineers, MTT-4, 218 (1956).

144. K. Furutsu, On the electro-magnetic radiation from a vertical dipole over the surface of arbitrary surface impedance, J. Radio Res. Labs., 6, 25, 269 (1959).

145. K. Furutsu, Wave propagation over an irregular terrain, Transactions of the Institute of Radio Engineers, AP-7, Special Suppl., 209 (1959).

146. G. Goubau, Relationship between surface and space waves, Onde electr., 37, 482 (1957).

147a. C. Jauquet, Excitation in a round dielectric rod of an electric-type surface wave, Ann. telecommuns., 12, No. 6, 217 (1957).

147b. C. Jauquet, Excitation of a transverse magnetic surface-wave propagated on a dielectric cylinder, Bull. cl. sci. Acad. roy. Belgique, 42, No. 7, 802 (1956).

148. J. B. Keller, F. C. Karal, Surface wave excitation and propagation, J. Appl. Phys., 31, 1039 (1960).

149. G. J. Rich, R. H. Du Hamel, Discussion on "the launching of a plane surface wave", Proceedings of the Institute of Electrical Engineers, B 103, 787 (1956).

150. J. H. Richmond, Flat surface waves and theorems of reciprocity. I. Dispersion by dielectric and metallic objects, II. Surface waves on flat and laminated dielectric strips, III. Excitation of surface waves on flat dielectric strips. Bull. Engng. Experim., Stat. Coll. Engng. Ohio State Univ., 28, 176, VIII (1959).

151. J. Robieux, General laws of linkage between wave radiators. Application to surface waves and propagation, part I, Ann. radioelectr., 14, 187 (1959).

152. J. Robieux, General theorems on the transmission coefficient from a transmitting to a receiving system , Transactions of the Institute of Radio Engineers, AP-7, Special Suppl., 118 (1959).

153. A. Sommerfeld, On propagation of electromagnetic waves in radio telegraphy, Ann. Phys., 28, 665 (1909).

154. C.T. Tai, The effect of a grounded slab on the radiation from a line source, J. Appl. Phys., 22, 405 (1951).

155. B. Van Der Pol, On discontinuous electromagnetic waves and the occurrence of a surface wave, Transactions of the Institute of Radio Engineers, AP-4, 288 (1956).

156. J. R. Wait, A note on the distribution of a non--stationary surface wave, Canad. J. Phys., 35, 1146 (1957).

157. J. R. Wait, A. M. Conda, The resonance excitation of a corrugated-cylinder antenna, Proceedings of the Institute of Electrical Engineers, C-107, 234 (1960).

158. J. R. Wait, On the excitation of electromagnetic surface waves on a curved surface, Transactions of the Institute of Radio Engineers, AP-8, 445 (1960).

DIFFRACTION OF SURFACE WAVES

159. N. A. Armand, Excitation of electromagnetic surface waves by an open-end coaxial line, Radio Engineering and Electronics, 4, 1609 (1959).

160. A. E. Bezmenov, Diffraction of electromagnetic waves on a semi-infinite array, Bulletin of Higher Education Institutions, Radio Engineering, 1, 271 (1958).

161a. V. I. Bespalov, Propagation of waves in transmission lines with heterogeneous surface impedance, Bulletin of Higher Education Institutions, Radio Physics, 1, No. 3, 54 (1958).

161b. V. I. Bespalov, E. Ya. Daume, Propagation of electromagnetic waves in a special line with small discontinuities, Bulletin of Higher Education Institutions, Radio Physics, 2, 213 (1959).

162. L. A. Weinstein, On the excitation of E_0 wave in a round waveguide by means of a coaxial line, Transactions of the USSR Academy of Sciences, 59, 1421 (1948).

163. G. A. Grinberg, V. A. Fock, Toward a theory of coastal electromagnetic wave refraction, Examinations on radio wave propagation, collection 2, edited by B.A.Vvedensky, published by the USSR Academy of Sciences, M. - L., 1948.

164. Guan-Din-Khia, Sonic surface wave diffraction on semi-infinite tube and rod impedances, Transactions of the USSR Academy of Sciences, 124, 559 (1959).

165. G. D. Malyuzhnets, Excitation, reflection and radiation of surface waves on a taper with predictable boundary impedances, Transactions of the USSR Academy of Sciences, 121, 436 (1958).

166a. P. S. Mikazan, Electromagnetic wave diffraction on an open-end spiral waveguide, Transactions of the USSR Academy of Sciences, 129, 502 (1959); Radio Engineering and Electronics, 5, 403 (1960).

166b. P. S. Mikazan, Electromagnetic wave diffraction on the butt joint of a spiral continuous waveguide, Radio Engineering and Electronics, 5, 597 (1960).

167. A. N. Sivov, Incidence of a flat electromagnetic wave on a planar array (when vector H is parallel to the conductors), Radio Engineering and Electronics, 6, 58 (1961).

168. V. I. Talanov, Surface wave excitation by the open-end of a flat waveguide, Journal of Engineering Physics, 28, 1275 (1958).

169. V. I. Talanov, Excitation of dielectric waveguides, Bulletin of Higher Education Institutions, Radio Physics, 2, 902 (1959).

170. V. I. Talanov, Electromagnetic wave diffraction by a step in waveguide surface impedance, Bulletin of Higher Education Institutions, Radio Physics, 2, 132 (1959).

171. N. G. Trenev, Electromagnetic surface wave diffraction on an impedance step, Radio Engineering and Electronics, 3, 27 (1958).

172. N. G. Trenev, Electromagnetic surface wave diffraction on a semi-infinite impedance plane, Radio Engineering and Electronics, 3, 163 (1958).

173. V. V. Tyazhelov, Approximate calculation of the influence of discontinuities on single-wire transmission lines, Bulletin of Higher Education Institutions, Radio Physics, 3, 89 (1960).

174. C. M. Angulo, W. S. Chang, A variational expression for the terminal admittance of a semi-infinite dielectric rod, Transactions of the Institute of Radio Engineers, AP-7, 207 (1959).

175. C. M. Angulo, W. S. Chang, The launching of surface waves by a parallel plate waveguide, Transactions of the Institute of Radio Engineers, AP-7, 359 (1959).

176. C. M. Angulo, Diffraction of surface waves by a semi--infinite dielectric slab, Transactions of the Institute of Radio Engineers, AP-5, 100 (1957).

177. A. E. Heins, Green's function for periodic structures in the diffraction theory: Application to parallel plate, part 1, J. Math. and Mech., 6, 401 (1957).

178. A.E. Heins, Green's function for periodic structures in the diffraction theory: Application to parallel plate structure region, part 2, J. Math. and Mech., 6, 629 (1957).

179. J. Kane, The efficiency of launching surface waves on a reactive half plane by an arbitrary antenna, Transactions of the Institute of Radio Engineers, AP-8, 500 (1960).

180. A. F. Kay, Scattering of a surface wave by a discontinuity in reactance, Transactions of the Institute of Radio Engineers, AP-7, 22 (1959).

181. S. N. Karp, F. C. Karal, Launching of surface waves on both surfaces of a rectangular taper, Communs. Pure and Appl. Mathem., 12, 435 (1959).

182. D. K. Ralph, Theory of diffraction on compound cylinder, J. Res. NBS 65 D, No. 1, 19 (1961).

183. G. Weill, Study of a diffraction problem of electromagnetic surface waves: Application to the theory of the dielectric aerial, Ann. radioelectr., 10, 228 (1955).

184. Morivaki Kavamura, Surface wave radiator. The matching of electromagnetic waves propagating on the conductor surface, Seisan kenkyu, 8, 11, 1-4 (1958), Japan.

SURFACE WAVE ANTENNAS

185. N. N. Govorun, Integral equations for antenna body of revolution with impedance surface, Transactions of the USSR Academy of Sciences, 126, 49 (1959).

186. K. I. Grineva, Surface wave antenna with a swinging beam, Radio Engineering, 14, 10, 15 (1959).

187. K. I. Grineva, Implementation of surface wave antennas, Bulletin of Higher Education Institutions, Radio Engineering, 2, 109 (1959).

188. G. A. Evstropov, Ground waves over a ribbed surface with periodic change of impedance, Problems of Radioelectronics, Series of General Technology, Edit. 13, 13 (1960).

189. M. A. Miller, The matching of uniform boundary conditions in the theory of thin aerials, Journal of Engineering Physics, 24, 1483 (1954).

190. V. I. Talanov, Antenna radiation with periodically changing surface impedance, Bulletin of Higher Education Institutions Radio Physics, 3, 802 (1960).

191a. O. N. Tereshin, An inverse electrodynamic problem applicable to an unlimited flat impedance surface, Transactions of Higher School, Radio Engineering and Electronics, No. 4, 32 (1958).

191b. O.N. Tereshin, A.S. Belov, Slot antenna decoupling by means of an impedance structure in the slot plane, Bulletin of Higher Education Institutions, Radio Engineering, 3, 359 (1960).

192. O.N. Tereshin, A.F. Chaplin, Inverse electrodynamic problem regarding a symmetrically excited impedance cylinder, Transactions of Higher School, Radio Engineering and Electronics, No. 2, 51 (1958).

193. H. W. Cooper, M. Hoffman, S. Isaacson, Image line surface wave antenna, Nat. Conv. Rec. of the Institute of Radio Engineers, 6, 230 (1958).

194. J. W. Duncan, R. H. Du Hamel, A technique for controling the radiation from dielectric rod waveguides, Transactions of the Institute of Radio Engineers, AP-5, 284 (1957).

195. M. J. Ehrlich and others, Corrugated surface antennas, Nat. Conv. Rec. of the Institute of Radio Engineers, p. 2, 18 (1953).

196. H. Ehrenspeck, W. Gerbes, F.J.Zucker, Surface wave antennas, Nat. Conv. Rec. of the Institute of Radio Engineers, 1, 25 (1954).

197. H. W. Ehrenspeck, The backfire antenna, a new type of directional line source, Proceedings of the Institute of Radio Engineers, 48, 109 (1960).

198. H. W. Ehrenspeck, W. Kearns, Two-dimensional endfire array with increased gain and side lobe reduction, Wescon Convention Record of the Institute of Radio Engineers, p. 1, 217 (1957).

199. H. W. Ehrenspeck, H. Poehler, A new method for obtaining maximum gain from Yagi antennas, Transactions of the Institute of Radio Engineers, AP-7, 379 (1959).

200. R. S. Elliott, Serrated waveguide. I. Theory, Transactions of the Institute of Radio Engineers, AP-5, 270 (1957).

201. R. S. Elliott, Spherical surface wave antennas, Transactions of the Institute of Radio Engineers, AP-4, 422 (1953).

202. R. S. Elliott, E. N. Rodda, Parasitic arrays excited by surface wave, Transactions of the Institute of Radio Engineers, AP-3, 140 (1955).

203. L. B. Felsen, Radiation from a tapered surface wave antenna, Transactions of the Institute of Radio Engineers, AP-8, 577 (1960).

204. L. O. Goldstone, A. A. Oliner, Leaky-wave antennas. I. Rectangular waveguides, Transactions of the Institute of Radio Engineers, AP-7, 307 (1959).

205. W. Hersch, The surface-wave aerial, Proceedings of the Institute of Electrical Engineers, C 107, 12, 202 (1960).

206. R. C. Honey, A flush-mounted leaky-wave antenna with predictable patterns, Transactions of the Institute of Radio Engineers, AP-7, 320 (1959).

207. R. W. Hougardy, R. C. Hansen, Scanning surface wave antennas - oblique surface waves over a corrugated conductor, Transactions of the Institute of Radio Engineers, AP-6, 370 (1958).

208. R. F. Hyneman, R. W. Hougardy, Waveguide loaded surface wave antenna, Nat. Conv. Rec. of the Institute of Radio Engineers, p. 1, 6, 225 (1958).

209. R. Jähn, Investigation of technical application of cylindrical surface wave aerials as radar aerials, Science of communications, 9, 418 (1959).

210. E. M. T. Jones, An annular corrugated-surface antenna, Proceedings fo the Institute of Radio Engineers, 40, 721 (1952).

211. E. M. T. Jones, R. A. Folsom, A note on the circular dielectric-disk antenna, Proceedings of the Institute of Radio Engineers, 41, 798 (1953).

212. K. C. Kelly, R. S. Elliott, Serrated waveguide, II. Experiment, Transactions of the Institute of Radio Engineers, AP-5, 276 (1957).

213. R. G. Malech, S. J. Blank, Experiments and calculations on surface wave antennas, Nat. Conv. Rec. of the Institute of Radio Engineers, 7, p. 1, 74 (1959).

214. L. W. Mickey, G. G. Chadmick, Closely spaced high dielectric constant polyrod arrays, Nat. Conv. Rec. of the Institute of Radio Engineers, p. 1, 213 (1958).

215. K. E. Niebuhr, E. H. Scheibe, Surface-wave dielectric--disk antenna, Proceedings of the Nat. Electronics Conf., 14, 135 (1958).

216. R.E. Plummer, Surface wave beacon antennas, Transactions of the Institute of Radio Engineers, AP-6, 105 (1958).

217. R. L. Pease, Radiation from modulated surface wave structures, Nat. Conv. Rec. of the Institute of Radio Engineers, 5, p. 1, 161 (1957).

218. F. Reggia, E. G. Spencer, R. D. Hatcher, J. E. Tompkins, Ferrod radiator system, Proceedings of the Institute of Radio Engineers, 45, 344 (1957).

219. W. Rotman, A. A. Oliner, Asymmetrical trough waveguide antennas, Transactions of the Institute of Radio Engineers, AP-7, 153 (1959).

220. D. L. Sengupta, On the phase velocity of wave propagation along an infinite Yagi structure, Transactions of the Institute of Radio Engineers, AP-6, 234 (1958).

221. D. L. Sengupta, The radiation characteristics of a zig-zag antenna, Transactions of the Institute of Radio Engineers, AP-6, 191 (1958).

222. F. Serrachioli, C. A. Levis, The calculated phase velocity of long end-fire uniform dipole arrays, Transactions of the Institute of Radio Engineers, AP-7, Special Suppl., 424 (1959).

223. J. S. Simon, G. Weill, A new type of end-fire aerial, Ann. radioelectr., 8, 33, 183 (1953).

224. J. O. Spector, An investigation of periodic rod structure for Yagi aerials, Proceedings of the Institute of Electrical Engineers, B 105, 38 (1958).

225. J. R. Wait, A. M. Conda, The radiation patterns and conductances of slots cut on rectangular metal plates, Proceedings of the Int. Congress on UHF (ultrahigh frequency) Circuits and Antennas, October, 1957.

226. C. H. Walter, Surface-wave Luneberg lens antennas, Transactions of the Institute of Radio Engineers, AP-8, 508 (1960).

227. W. L. Weeks, Coupled waveguide excitation of traveling wave antenna, Wescon Convent. Rec. of the Institute of Radio Engineers, 1, VIII, p. 1, 236 (1957).

Scientific Experimental Institute of Radio Physics, Gorky University.

Entered at the editorial office on June 27, 1961.

CENTRO INTERNAZIONALE MATEMATICO ESTIVO
(C.I.M.E.)

E R I C H   S P I T Z

NOTE ON CONTINUOUS COUPLING OF SURFACE WAVES

ROMA - Istituto Matematico dell'Università

NOTE ON CONTINUOUS COUPLING OF SURFACE WAVES

ERICH SPITZ

Dept. de Physique Appliquée, Comp.Gén. de TSF, Paris

Beside the possibility of launching a surface wave by a device at one end of the surface wave structure, one can launch it by a continuous coupling to another transmission line. This transmission line can be also a surface wave structure but it can be also an open line as a two-wire line etc. This can be very interesting especially for surface wave antennas because the transversal dimension of the launching device completely disappears.

Let us suppose two parallel open transmission lines (for ex. surface wave structures or an open two wire line etc.).

Let each of these lines propagate energy along the z direction in only one mode. If both lines are near together there will be of course interaction between them.

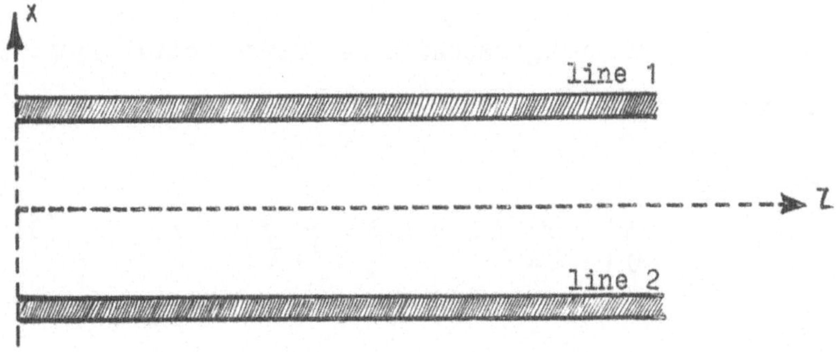

If E_1 and E_2 are one of the components of the field of line 1 and 2, then we can write the following equation

$$\frac{dE_1}{dz} = -\gamma_1 E_1 + kE_2$$
$$\frac{dE_2}{dz} = -\gamma_2 E_2 + kE_1 \qquad (1)$$

where γ_1 and γ_2 are the complex propagation constants of line 1 and 2 in presence of both lines.

$$\gamma_1 = \alpha_1 + j\beta_1$$
$$\gamma_2 = \alpha_2 + j\beta_2 \qquad (2)$$

If we suppose that there is no radiation along the whole structure

$$|E_1|^2 + |E_2|^2 = \text{const.}$$

and from that condition k must be imaginary

$$k = jc$$

From (1) the following second order differential equation can be written

$$\frac{d^2E_1}{dz^2} + \frac{dE_1}{dz}(\gamma_1 + \gamma_2) + E_1(c^2 + \gamma_1\gamma_2) = 0 \qquad (3)$$

and the same for E_2.

If we introduce the initial conditions for z = 0

$$E_1(0) = 1$$

$$\text{and} \quad E_2(0) = 0$$

then the solutions are

$$E_1 = \frac{1}{2}\left(1 + \frac{\Delta\gamma}{g}\right)e^{m_1 z} + \frac{1}{2}\left(1 - \frac{\Delta\gamma}{g}\right)e^{m_2 z} \qquad (4)$$

$$E_2 = j\,\frac{c}{g}\left(e^{m_1 z} - e^{m_2 z}\right)$$

where

$$m_{1,2} = -\frac{1}{2}(\gamma_1 + \gamma_2) \pm \frac{1}{2} g$$

$$g = \sqrt{\Delta\gamma^2 - 4c^2} \qquad (5)$$

$$\Delta\gamma = \gamma_2 - \gamma_1$$

There is a very interesting case when $\gamma_1 = \gamma_2 = \gamma$ (degeneracy). Then without losses

$$\gamma = j\beta$$

From

$$m_{1,2} = -j\beta \pm jc$$

and from (4)

$$E_1 = e^{-j\beta z} \cos cz$$

$$E_2 = j\, e^{-j\beta z} \sin cz \qquad (6)$$

This means that there is a total transfer of energy from the excited transmission line (1) into the transmission line (2). The distance where this occurs is

$$z_1 = \frac{\pi}{2c}$$

Graphically, it can be represented in the following way :

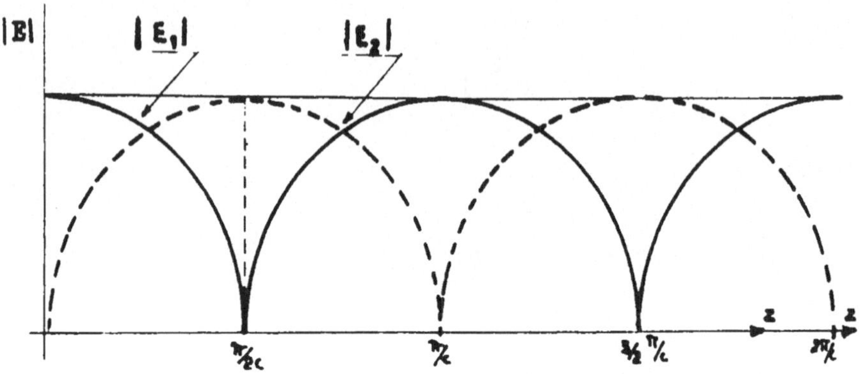

Energy flows periodically from one line into the other. In the absence of degeneracy, there is no transfer of energy and when the coupling is small

$$m_1 = - \gamma_1 - \frac{c^2}{\Delta\gamma}$$

$$m_2 = - \gamma_2 + \frac{c^2}{\Delta\gamma}$$

and graphically

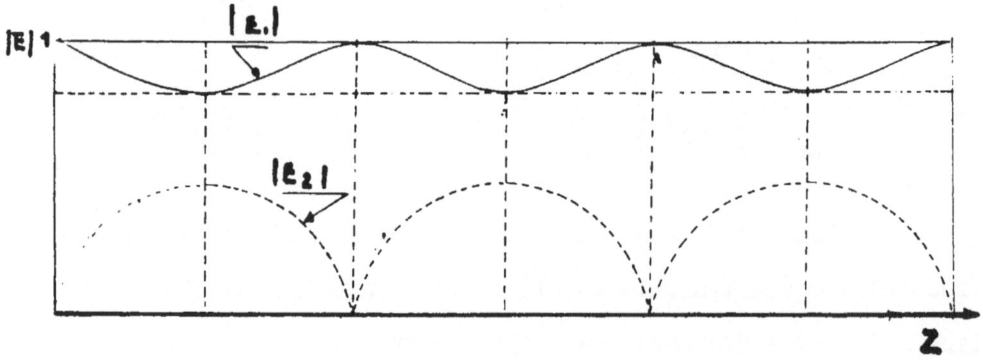

Unfortunately the coefficient c is not very easy to calculate in most cases, and each configuration needs a special treatment.

So for example it is possible to solve the problem of two thin infinite dieletric plates.

Let us mention here an approximative method of evaluating the coupling coefficient c of two identical lines by using thermodynamics.

As a consequence of the Ehrenfest theorem one can state that if the energy W of a lossless electromagnetic resonator oscillating at frequency f is increased slowly by δW: then the frequency increases by δf :

$$\frac{\delta f}{f} = \frac{\delta W}{W}$$

In particular

$$\delta f = \delta W = 0$$

Suppose that we form from one our line a resonator by placing the line between two perpendicular infinite metallic walls half a line-wavelength $\frac{\lambda_s}{2}$ apart.

If now another metallic plate (mirror) parallel to the line is being shifted from infinity towards the lines, we introduce a fictive image line.

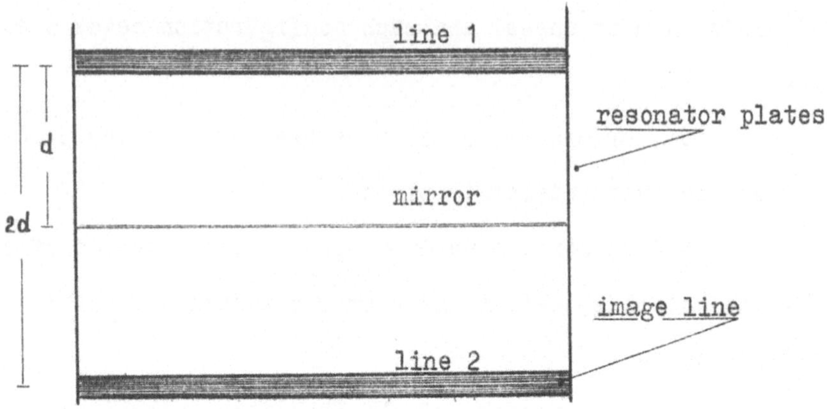

If we know the field at the mirror, we can easily calculate the radiation pressure ($p = \frac{1}{2} \mu_0 H_t^2 - \frac{1}{2} \varepsilon_0 E_n^2$) and from that the work done to bring the mirror from infinity (isolated line) to a certain distance d .

To satisfy $\delta W = 0$ we compensate by shifting one resonator plate by Δz . Then we are again in resonance at the same frequency.

But instead of $\beta = 2\pi / \lambda_s$, we have now

$$\beta' = \frac{\pi \beta}{\pi - \beta \Delta z}$$

and the variation of β caused by the coupling is

$$\frac{\Delta\beta}{\beta} = \pm \frac{2 \Delta z}{\lambda} .$$

The effect of coupling between two lines has been used with success in different endfire antennas.

In general to get a very extended field the surface wave must have a velocity very near to the velocity of light.

Thus a TEM line as feeding line is very advantageous.

We built a structure called "Saucisson antenna". It consists essentially of a two-wire line surrounded by a helix.

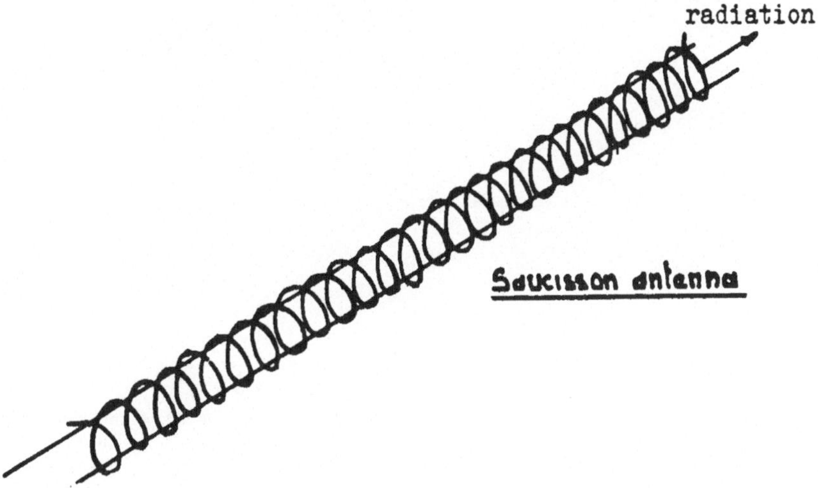

The energy of the two wire lines is progressively coupled to the helix supporting a surface wave. When all energy is in the helix ($z = z_1$), the structure is cut and radiation takes place in the endfire direction.

The effect of coupling can be used also for another system. We have seen that energy flows periodically from one line into the other. Thus if the two lines are not symmetrical to the outer space we can consider that the outer line supports an electrically amplitude modulated wave. If we arrange things in such a way that one of the new β is between $-k_0$ and k_0, ($k_0 = 2\pi/\lambda$) then radiation can take place in other direction than endfire (leaky wave).

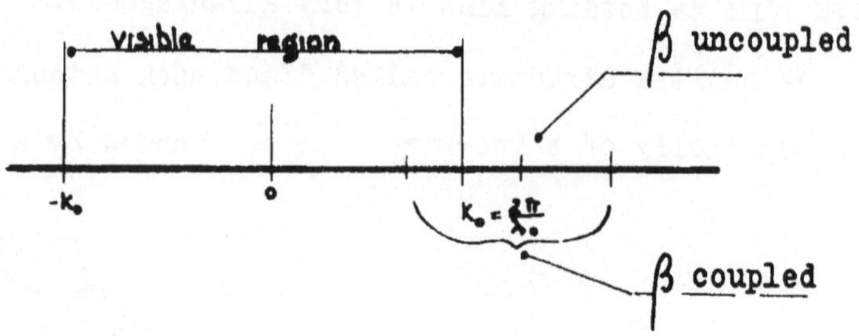
visible region
β uncoupled
-K₀
0
K₀ = 2π/λ₀
β coupled

CENTRO INTERNAZIONALE MATEMATICO ESTIVO

(C.I.M.E.)

FRANCIS J. ZUCKER

ELECTROMAGNETIC BOUNDARY WAVES

ROMA - Istituto Matematico dell'Università

ELECTROMAGNETIC BOUNDARY WAVES

(An Introduction)

Francis J. Zucker
Air Force Cambridge Research Laboratories
Bedford, Mass., USA

Contents:

F.J.Zucker

3. Surface Waveguides

   a) Sketch of the Marcuvitz-Schwinger formalism

   b) Surface Wave Excitation I: Source Radiation

   c) Surface Wave Excitation II: Influence of Source on Surface Wave Amplitude

   d) Surface Wave Excitation III: Orthogonality Relations

   e) Cylindrical Surface Waveguides

   f) Periodic Structures

4. Radiation of Surface and Leaky Waves

   a) Spatial Domains in which Surface and Leaky Waves are Dominant

   b) Radiation of Traveling Waves

   c) Surface-Wave Radiation: the Young and Fresnel Points of View

   d) Leaky-Wave Radiation: Pattern Control.

F.J.Zucker

ELECTROMAGNETIC BOUNDARY WAVES

(An Introduction)

To introduce the concepts of "surface" and "leaky" wave, the two types of electromagnetic boundary waves that will chiefly concern us, we begin with the concrete example of an electric line source exciting a grounded dielectric slab (Sec.1). By generalizing various features of this example, we can display the entire spectrum of boundary waves (Sec.2). This is followed by a discussion of surface waveguides (Sec.3). We conclude with a brief description of radiation from surface and leaky waves, with application to traveling-wave antennas (Sec.4).

1. INTRODUCTION : DIELECTRIC SLAB EXCITED BY AN ELECTRIC LINE SOURCE

a) Review of Maxwell's Equations

In the rationalized MKS (or Giorgi) system, Maxwell's equations are

$$\text{div}\ \underline{D} = \rho \tag{1a}$$

$$\text{div}\ \underline{B} = 0 \tag{1b}$$

$$\text{curl}\ \underline{E} = -\dot{B} \tag{1c}$$

$$\text{curl}\ \underline{H} = \underline{J} + \dot{\underline{D}}\ , \qquad (1\underline{d})$$

where the dots indicate partial derivatives with respect to time. The units of ρ are coulombs/meter^3, of $\underline{E}$ volts/meter, of $\underline{J}$ amperes/meter^2; from these, the units of the other quantities can be inferred, e.g., from (1d), $\underline{H}$ is in amperes/meter.

These equations are "induced" (guessed-at) from simpler laws known to hold at low frequencies : Coulomb's laws for electric charges and magnetic poles (1a and b), Faraday's law of induction (1c), and Ampère's (or Biot-Savart's) law (1d). The equations are verified by experimental checks of the totality of possible deductions from them.

Application of Maxwell's equations to the interface between two media results in the well-known boundary condition that the tangential components of $\underline{E}$ and $\underline{H}$ must be continuous across the interface. A third boundary condition is needed in the solution of radiation problems : at infinity, fields must travel outward and decay at least as fast as a spherical wave (Sommerfeld's "radiation condition").

Since all well-behaved time functions can be synthesized from a Fourier spectrum of simple harmonic eigenvalues, we shall restrict ourselves to the time dependence $e^{j\omega t}$, so that (1c) and (1d) become

$$\text{curl}\ \underline{E} = -\, j\omega \underline{B}\ ,\ \text{curl}\ \underline{H} = \underline{J} + j\omega \underline{D}\ .$$

In free space (air or vacuum), $\underline{D} = \varepsilon_0 \underline{E}$ = $(1/36\pi)\times 10^{-9}$farads/meter, and $\underline{B} = \mu_0 \underline{H}$, where $\mu_0 = 4\pi \times 10^{-7}$ henry/meter.

In general linear media, ε_0 and μ_0 are replaced by the constitutive parameters ε_{ij} and μ_{ij}, which are tensors of rank 2; in nonlinear media the tensor components are functions of the field quantities themselves. We restrict ourselves to simple media in which the constitutive parameters are scalar constants, and in which the total current $\underline{J}$ can be written as the sum of a current due to free sources, $\underline{J}_s$, and an induced current $\underline{J}_i$ that is directly proportional to the $\underline{E}$ field: $\underline{J}_i = \sigma \underline{E}$, where σ is the (scalar) conductivity. Maxwell's equations now read

$$\operatorname{div} \underline{E} = \rho/\varepsilon \qquad (2a)$$

$$\operatorname{div} \underline{H} = 0 \qquad (2b)$$

$$\operatorname{curl} \underline{E} = -j\omega\mu \underline{H} \qquad (2c)$$

$$\operatorname{curl} \underline{H} = \underline{J}_s + j\omega\underset{\sim}{\varepsilon} \underline{E}, \qquad (2d)$$

where $\underset{\sim}{\varepsilon} \equiv \varepsilon(1 - j\sigma/\omega\varepsilon)$. In a perfect conductor, $\sigma/\omega\varepsilon$ is infinite; in a good conductor, it is much larger than 1; in a dielectric with small loss, much smaller than 1; in a perfect dielectric, zero.

b) Integral Representation of the Field[1]

As shown in Fig.1, we assume an electric line source $\underline{J}_s = \hat{y}J_y$ ($\hat{y}$ is the unit vector out of the paper) at a height h in air (ε_0) above a two-dimensional loss-less dielectric slab of thickness d and dielectric constant ε_1, which in turn lies on a perfectly conducting ground plane. By taking the curl of (2c), substituting (2d), and noting that div $\underline{E} = 0$, we obtain

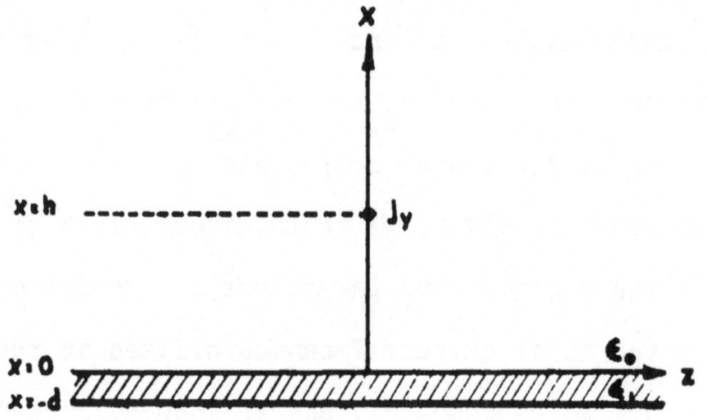

Fig. 1

the inhomogeneous Helmholtz (or reduced wave) equation

$$\nabla^2 \underline{E} + k^2 \underline{E} = j\omega\mu \underline{J}_s , \tag{3}$$

with $k^2 \equiv \omega^2 \mu \varepsilon$. In a two-dimensional problem, $\partial/\partial y = 0$, and since $\underline{E} = \hat{y}E_y$, (3) becomes

$$\nabla^2_{x,z} E_y + k^2 E_y = \delta(x-h)\,\delta(z) , \tag{4}$$

where for convenience we have set $J_y = 1/j\omega\mu$; the delta functions are defined to be zero everywhere except at $(x = h,\ z = 0)$, where they are discontinuous in such a way that $\int_{-\infty}^{+\infty} \delta(x-h)dx$ and $\int_{-\infty}^{+\infty} \delta(z)dz$ are both equal to 1.

With $E_0 \equiv E_y$ in air and $E_1 \equiv E_y$ in the dielectric, we have the two differential equations

$$(\nabla^2_{x,z} + k_0^2)\, E_0(x,z) = \delta(x-h)\,\delta(z) \qquad x \geqslant 0 \tag{5a}$$

$$(\nabla^2_{x,z} + k_1^2)\, E_1(x,z) = 0 . \qquad x \leqslant 0 \tag{5b}$$

We now resolve E_0 and E_1 into their spatial-frequency Fourier components β_z :

$$E_0(x,z) = \frac{1}{2\pi} \int_{-\infty}^{+\infty} e_0(x \mid \beta_z) e^{-j\beta_z z}\, d\beta_z \tag{6a}$$

$$e_0(x \mid \beta_z) = \int_{-\infty}^{+\infty} E_0(x,z) e^{j\beta_z z}\, dz , \tag{6b}$$

and analogously for $E_1(x,z)$ and $e_1(x \mid \beta_z)$. Each β_z is an eigenvalue or "mode", and e_0 and e_1 are the modal amplitudes. Integrating (5a),

$$\int_{-\infty}^{+\infty} (\nabla^2_{x,z} + k_0^2)\, E_0(x,z) e^{j\beta_z z} dz = \int_{-\infty}^{+\infty} \delta(x-h)\,\delta(z) e^{j\beta_z z} dz.$$

Separating out the x-dependent terms, and using (6b),

$$\left(\frac{\partial^2}{\partial x^2} + k_0^2\right) e_0 + \int_{-\infty}^{+\infty} \frac{\partial^2}{\partial z^2} E_0\, e^{j\beta_z z}\, dz = \delta(x-h).$$

The integral on the left-hand side is evaluated by integrating by parts twice, which results in the three terms

$$e^{j\beta_z z} \left(\frac{\partial}{\partial z} - j\beta_z \right) E_0 \Big|_{-\infty}^{+\infty} - \beta_z^2 e_0 .$$

If E_0 is to be a solution, it must satisfy the radiation condition and decay at least as fast at infinity as an outgoing spherical wave; its derivative with respect to z then decays even faster, and therefore in the limits the first two terms vanish. We are left with

$$\left(\frac{\partial^2}{\partial x^2} + k_{xo}^2\right) e_0(x|\beta_z) = \delta(x-h) \tag{7a}$$

and, analogously for e_1,

$$\left(\frac{\partial^2}{\partial x^2} + k_{x1}^2\right) e_1(x|\beta_z) = 0 , \tag{7b}$$

where

$$k_{xo}^2 \equiv k_0^2 - \beta_z^2 \quad \text{and} \quad k_{x1}^2 \equiv k_1^2 - \beta_z^2 . \tag{7c}$$

Equations (7) are reduced wave equations in the x-direction, and their solution can be written in the form of up-and downward traveling waves :

$$e_o(x|\beta_z) = A\, e^{-jk_{xo}x} \qquad x \geqslant h \qquad (8\underline{a})$$

$$= B_1\, e^{jk_{xo}x} + B_2\, e^{-jk_{xo}x} \qquad 0 \leqslant x \leqslant h \qquad (8\underline{b})$$

$$e_1(x|\beta_z) = C_1\, e^{jk_{x1}x} + C_2\, e^{-jk_{x1}x} \qquad -d \leqslant x \leqslant 0. \qquad (8\underline{c})$$

The constants in (8) are evaluated by satisfying the boundary conditions, which require that

$$\text{I} \quad e_o(h^+) = e_o(h^-)$$

$$\text{II} \quad e_o(0^+) = e_1(0^-)$$

$$\text{III} \quad e_1(-d) = 0 \, ,$$

and also, from (7a), that

$$\int_{h^-}^{h^+} \left(\frac{\partial^2}{\partial x^2} + k_{xo}^2 \right) e_o \, dx = 1,$$

or

$$\text{IV} \quad \left. \frac{\partial e_o}{\partial x} \right|_{h^-}^{h^+} = e'(h^+) - e'(h^-) = 1,$$

and from (7b), that

$$\text{V} \quad e'(0^+) - e'(0^-) = 0.$$

Using III, we obtain a relation between C_1 and C_2, from which

$$e_1 = 2jC_1\, e^{-jk_{x1}d} \sin k_{x1}(x+d). \qquad (9\underline{a})$$

From V,

$$B_1 - B_2 = \frac{2k_{x1}}{k_{xo}}\, C_1\, e^{-jk_{x1}d} \cos k_{x1}d \, , \qquad (9\underline{b})$$

which in combination with II gives

$$\frac{B_2}{B_1} = \frac{jk_{xo} - k_{x1} \cot k_{x1}d}{jk_{xo} + k_{x1} \cot k_{x1}d} \quad . \qquad (9\underline{c})$$

Since this is the amplitude ratio in (8$\underline{b}$) between the wave traveling downward from the source and the wave reflected upward from the surface we term it the surface reflection coefficient $\Gamma(k_{xo}, k_{x1})$ or $\Gamma(\beta_z)$ (both k_{xo} and k_{x1} being functions of β_z). Using I and IV, we find

$$B_1 = \frac{j}{2k_{xo}} e^{-jk_{xo}h}$$

$$A = \frac{j}{2k_{xo}} \left[e^{jk_{xo}h} + \Gamma e^{-jk_{xo}h} \right] , \qquad (9\underline{d})$$

so that

$$e_o(x|\beta_z) = \frac{j}{2k_{xo}} \left[e^{-jk_{xo}(x-h)} + \Gamma e^{-jk_{xo}(x+h)} \right] \qquad x \geqslant h \quad (10\underline{a})$$

$$= \frac{j}{2k_{xo}} \left[e^{jk_{xo}(x-h)} + \Gamma e^{-jk_{xo}(x+h)} \right] \qquad 0 \leqslant x \leqslant h \quad (10\underline{b})$$

Substituting (9$\underline{b}$) and (9$\underline{d}$) in (9$\underline{a}$),

$$e_1(x|\beta_z) = \frac{je^{-jk_{xo}h}}{2k_{xo}} \frac{(1+\Gamma)}{\sin k_{x1}d} \sin\left[k_{x1}(x+d)\right] \quad . \qquad (10\underline{c})$$

(10$\underline{a}$) represents a wave traveling upward from the source and a reflected wave traveling upward from the image of the source mirrored in the air-dielectric interface (at $x = -h$). (10$\underline{b}$) contains the same reflected wave, and a wave coming downward from the source. (10c) represents a standing wave that varies sinusoidally from the metal bottom.

This completes the modal analysis. Referring to (6a), and using a convenient notation to combine (10a) and (10b), we now sum over all modal amplitudes:

$$E_0(x,z) = \frac{j}{4\pi}\int_{-\infty}^{+\infty} \frac{1}{k_{xo}}\left[e^{-jk_{xo}|x-h|} + \Gamma(\beta_z)\, e^{-jk_{xo}(x+h)}\right] e^{-j\beta_z z}\, d\beta_z \qquad x \geq 0 \qquad (11)$$

An analogous expression holds for $E_1(x,z)$.

We have used the method of composite Green's function (e_0 and e_1). Other methods are available, but ours is best suited for the generalizations to be made in Secs.2-4.

It is useful, for later purposes, to visualize clearly the connection in (11) between wave travel in the z- and x-direction. As β_z passes through all values between $-\infty$ and $+\infty$, what are the corresponding values of k_{xo}? Let us focus on the second term in (11), which represents an upward traveling wave for all values of x. It can be viewed as the projection along x of a two-dimensional wave of the form $\exp(-jk_{xo}x-j\beta_z z)$. When $\beta_z = 0$, (7c) implies that $k_{xo} = k_o$, i.e., the wave travels broadside to the surface. When $\beta_z = \pm k_o$, (7c) makes $k_{xo} = 0$, and the wave travels parallel to the surface to the right or left, respectively. Inbetween, i.e., for $|\beta_z| < k_o$, k_{xo} is real, and $\exp(-j\beta_{xo}x-j\beta_z z)$ represents a wave traveling toward the upper right (β_z positive) or upper left (β_z negative). When $|\beta_z| > k_o$, (7c) forces k_{xo} to be imaginary; $\exp(-\alpha_{xo}x - j\beta_z z)$ now represents a wave traveling to the right or left with exponential amplitude decay away from the surface, i.e., the planes of constant phase and constant amplitude are orthogonal [see Sec.2(a)]. In waveguide language, the spectrum of waves along x in (8) and (11) corresponding to $-\infty \leq \beta_z \leq \infty$ thus includes all propagating (β_{xo}) as well as 'evanescent' ('below cutoff', α_{xo}) modes. According to (11), the sum of these appropriately-weighted modes satisfies the boundary conditions of Fig.1. (As we shall see below, an evanescent mode sometimes satisfies the boundary conditions all by itself, and this mode is then called a 'surface wave').

c) Surface- and Leaky-Wave Poles.

The integral (11) has branch point singularities at $\beta_z = \pm k_o$ ($k_{xo} = 0$), and additional singularities at all values of β_z for which the denominator of $\Gamma(\beta_z)$ is zero. Its analytic continuation in the $k_z = \beta_z - j\alpha_z$ plane is shown in Fig.2, in which $-j\alpha_z$ is drawn upward so that visually the location of contour, poles, branch cuts, etc. is the same as in the frequently-used $k_z = \beta_z + i\alpha_z$ plane corresponding to a choice of time dependence $e^{-i\omega t}$. The solid line is the original path for $-\infty \leq \beta_z \leq \infty$; it is drawn slightly displaced into the third (top left) and first (bottom right) quadrant to indicate that vanishingly small losses in air would move branch points and on-axis poles (if any) into the second and fourth quadrant, respectively. $k_{xo} = \beta_{xo} - j\alpha_{xo}$ is double valued; its imaginary part

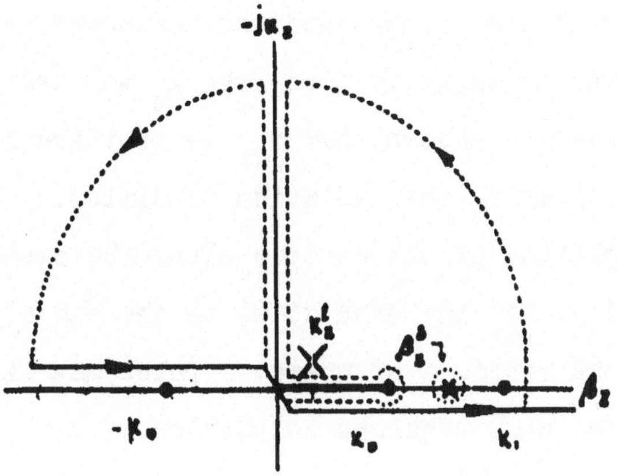

Fig. 2

changes sign when

$$\mathrm{Im}\left\{\sqrt{k_o^2 - k_z^2}\right\} = 0 \ , \quad \text{or } \alpha_z \beta_z = 0$$

The branch cut in the fourth quadrant is therefore drawn from the branch point to infinity along the $\alpha_z = 0$ and $\beta_z = 0$ axes, and the top sheet so chosen that α_{xo} is positive for outgoing waves, thus satisfying the radiation condition.

Completion of the contour along the dashed line allows us to write (11), for the case $z > 0$, as the sum of a branch cut integral and the residues of whatever poles may lie in the top sheet. (The case $z < 0$ requires completion of the contour around the lower half of Fig.2). The branch cut contribution is termed the 'continuous spectrum', and we defer discussion of it until Sec.3. The residues constitute the 'discrete spectrum', and as we shall now see there may be none or any finite number of them.

To locate the poles, we set the denominator of $\Gamma(k_z)$ in (9c) equal to zero:

$$j\, k_{xo} + k_{x1} \cot k_{x1} d = 0 \tag{12}$$

Let us first look for solutions of the form $k_z = \beta_z$ (no attenuation in the z-direction). It follows from (7c) that both k_{xo} and k_{x1} must be either purely real or purely imaginary. Assuming first that $k_{xo} = \beta_{xo}$, we find that neither $k_{x1} = \beta_{x1}$ nor $k_{x1} = -j\alpha_{x1}$ can satisfy (12). With $k_{xo} = -j\alpha_{xo}$ and $k_{x1} = \beta_{x1}$, a solution appears possible, provided the cotangent is negative. We therefore look for the roots of

$$\alpha_{xo} d = -\beta_{x1} d \cot \beta_{x1} d \tag{13a}$$

subject to (7c):

$$\left.\begin{aligned} k_o^2 &= \beta_z^2 - \alpha_{xo}^2 \\ k_1^2 = \varepsilon_r k_o^2 &= \beta_z^2 + \beta_{x1}^2 \end{aligned}\right\} \qquad (13\underline{b})$$

(where $\varepsilon_r \equiv \varepsilon_1/\varepsilon_o$), which upon subtraction gives

$$k_o^2\, d^2(\varepsilon_r - 1) = (\beta_{x1} d)^2 + (\alpha_{xo} d)^2. \qquad (13\underline{c})$$

Equations (13a) and (13c) can be solved graphically, as shown in Fig.3. The roots are labeled α_{xo}^s, β_{x1}^s, corresponding [via (13b)] to β_z^s, because they define a surface wave, that is, a wave which propagates along the interface without radiating, and attenuates vertically away from it. Since planes of constant amplitude are perpendicular to planes of constant phase, the surface wave is an example of an inhomogeneous plane wave (see Fig.4a). Equation (13b) implies that $|\beta_z^s|$ lies between $|k_o|$ and $|k_1|$, as indicated in Fig.2 for positive β_z^s; the surface wave velocity $v_z = \omega/\beta_z$ is therefore slower than $c/\sqrt{\varepsilon_r}$, the velocity in an infinite dielectric medium.

We find no surface wave until the radius in (13c) or Fig.3 is at least $\pi/2$, that is

$$k_o d \sqrt{\varepsilon_r - 1} = \frac{\pi}{2} \qquad (13\underline{d})$$

This is a cutoff condition, or perhaps it should be called 'kill-off' condition, because, unlike in shielded waveguide, the surface wave mode does not become evanescent below this point but simply ceases to exist. Equation (13d) says that, for a particular frequency and ε_r, the slab must be thick enough before the surface wave can be excited, or for a given thickness and ε_r, the frequency must be high enough. If the radius is larger than $3\pi/2$, two surface wave modes are excited, and for a very large radius (dielectric layer at optical frequencies, for example), an extre-

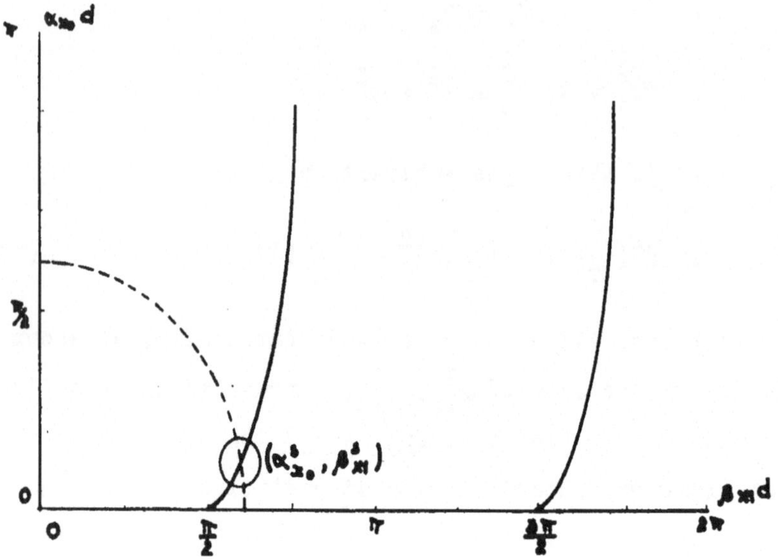

Fig. 3

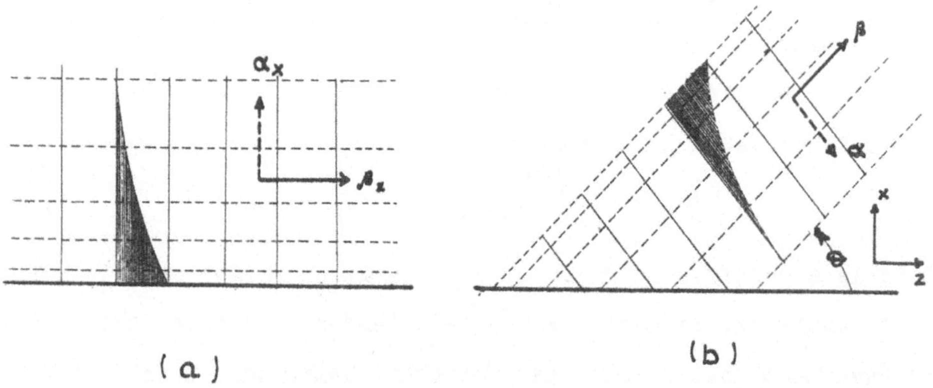

Fig. 4

mely large number of modes may exist, though never infinitely many of them. At each cutoff point, $\alpha_{xo}^{s} = 0$, so the wave extends throughout the upper half space. As the radius increases, α_{xo}^{s} increases also, the wave therefore becomes more tightly bound and [from (13b)] slower. Since we have just shown that β_{z}^{s} cannot increase beyond k_1, (13b) implies that the vertical attenuation reaches a limiting value of $\alpha_{xo} = k_o \sqrt{\varepsilon_r - 1}$.

As a final possibility for finding solutions of the form $k_z = \beta_z$, consider $k_{xo} = -j\alpha_{xo}$ and $k_{x1} = -j\alpha_{x1}$ in (12). Then instead of (13a), we have $\alpha_{xo}d = -(\alpha_{x1}d)\coth(\alpha_{x1}d)$, and instead of (13c), $k_o^2 d^2(\varepsilon_r - 1) = (\alpha_{xo}d)^2 - (\alpha_{x1}d)^2$. Plotting in the manner of Fig.3 shows that no real roots exist.

Let us now look for poles with complex k_z. The complete examination of (12) for complex roots would be tedious, and we restrict ourselves to the case of a very thin slab, so that (12) can be written to first order as

$$j(\beta_{xo} - j\alpha_{xo}) + \frac{1}{d} = 0, \tag{14a}$$

from which it follows that α_{xo} is negative. Let us first look for waves traveling out from the interface, i.e., β_{xo} positive. Equating the imaginary parts of (7c), we have

$$0 = \alpha_{xo}\,\beta_{xo} + \alpha_z\,\beta_z\ , \tag{14b}$$

and we conclude that α_z and β_z must be of equal sign, positive for travel to the right ($e^{-\alpha_z z - j\beta_z z}$),negative to the left

($e^{\alpha_z z+j\beta_z z}$). This is a <u>leaky wave</u>, which attenuates as it travels along the surface and increases exponentially (negative α_{xo}) away from it, as illustrated in Fig.4 b . Because they violate the radiation condition, leaky wave poles (such as the one indicated by the dashed cross marked k_z^{ℓ} in Fig.2) lie in the wrong Riemann sheet and contribute no residue if the contour is chosen as in Fig.2. In asymptotic evaluations of the integral, however, they may turn up within the modified contour, as we shall observe in Sec.4. When the leaky wave is the dominant part of the field above an interface, as it sometimes is in the case of plasma sheaths, and always in the case of leaky wave antennas, its residue in fact consitutes the major part of the total radiation field.

Both surface and leaky waves violate the radiation condition. Surface waves do so at $x = 0$, $z = \pm\infty$, but this is easily taken care of by noting that small losses in the medium above and/or below the interface introduce an imaginary part, however small, in k_z^s. The more serious case of violation by leaky waves is dealt with in Sec. 4.

We have not yet discussed the case of the complex pole corresponding to an incoming wave (β_{xo} negative). Equation (14a) still holds, and from (14b) it follows that a positive β_z is now paired with a negative α_z, that is, the wave increases exponentially in the direction of travel. The pole is therefore located on the wrong sheet in the second quadrant (lower right) of Fig.2, and it turns out that in the case of the dielectric slab no contour deformation ever includes it. [See Sec. 3(b)].

d) Evaluation of the Residue

It is instructive to perform at least one residue calculation. We restrict ourselves to the case of a surface wave near cutoff, that is, $\beta^s_{x1} d = \pi/2 + \kappa$, with $\kappa \ll 1$, so that (13a) becomes

$$\alpha^s_{xo} \cong \beta^s_{x1} \left(\beta^s_{x1} d - \frac{\pi}{2}\right) \cong \frac{\pi\kappa}{2d}$$

and

$$\beta^s_z = \sqrt{k_o^2 + (\alpha^s_{xo})^2} \cong k_o$$

(to first order, wave propagates with velocity of light).

To find the residue of (11),

$$E^s_o(x,z) = -2\pi j \text{ Res.} \left\{ \frac{j}{4\pi} \int_{-\infty}^{+\infty} \frac{1}{k_{xo}} \Gamma(k_z) e^{-jk_{xo}(x+h) - jk_z z} \right\}_{k_z = \beta^s_z}$$

we recall that

$$\text{Res.} \int \frac{N(t)}{M(t)} dt \Big|_{t=t'} = \frac{N(t')}{\left.\frac{dM}{dt}\right|_{t'}}$$

Therefore

$$E^s_o(x,z) = A\, e^{-\alpha^s_{xo}(x+h) - j\beta^s_z z} ,$$

with

$$A = \frac{1}{2} \frac{\alpha^s_{xo} - \beta^s_{x1} \cot \beta^s_{x1} d}{\frac{d}{dk_z} \left[k_{xo}(jk_{xo} + \beta_{x1} \cot \beta_{x1} d \right] \Big|_{k_z = \beta^s_z}}$$

With our approximations, the numerator of A simplifies to $\pi\kappa/d$, and the denominator turns out to be $-j2k_0(1+\pi\kappa/2)$, so that

$$A \cong \frac{j\pi\kappa}{2k_0 d} \cong j\,\frac{a^s_{xo}}{k_0}$$

The final result is best displayed in the form

$$E^s_0(x,z) = j\,\frac{a^s_{xo}}{k_0}\,e^{-a^s_{xo}h}\,e^{-a^s_{xo}x\,-j\beta^s_z z} \qquad (15)$$

valid for all z provided β^s_z is chosen negative for $z \lesssim 0$.
The last exponential represents the x and z behavior of the surface wave; the first exponential shows that the amplitude of excitation decreases sharply as the source is raised above the slab; and from the initial factor we see that, within the assumptions made, the surface wave is the more strongly excited the tighter it is bound. Note also the reciprocity in (15) between source and observation point: a source at height h produces the same field on the surface (x = 0) as a source on the surface (h = 0) at an observation point x = h.

From the curl equation (2c) and (15), we find the magnetic field components

$$H_{xo} = \frac{1}{j\omega\mu}\,\frac{\partial E_0}{\partial z}, \qquad H_{zo} = \frac{j}{\omega\mu}\,\frac{\partial E_0}{\partial x}$$

which are seen to be exponentials in x also. Fig.5 sketches these components separately and also the field lines of the combined H-components; we observe that the field lines start out from, and terminate on, the same interface.

We have not evaluated E_1 in detail, but it is clear from an examination of (10c) that the x-dependent term, which

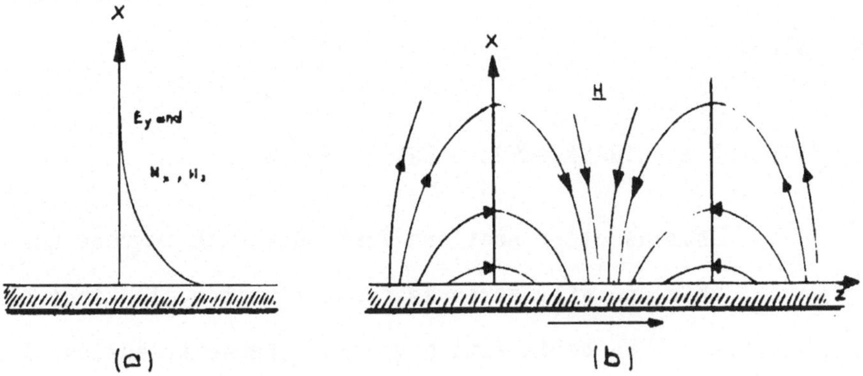

Fig. 5

is unaffected by the residue evaluation, is a sinusoid starting with zero on the metal bottom : $\sin\left[\beta^{s}_{x1}(x+d)\right]$. From the curl equations it follows therefore that H_{x1} is also a sine, and H_{z1} a cosine.

2. TYPES OF ELECTROMAGNETIC BOUNDARY WAVES.

Whereas other sections deal only with surface and leaky waves, this one will serve as bestiarium for all kinds of boundary waves. We begin with a general characterization of inhomogeneous plane waves, and with the help of "transverse impedance" language are then in a position to describe very rapidly all possible types of "pole" and "zero waves". This is followed by a brief discussion of other boundary waves. Our definition of a surface wave in Sec.1 was specifically restricted to pole waves of the nonradiating variety, but it should be pointed out that nomenolature in this area of electromagnetic research is loose, and that at one time or another almost every wave type listed in this section has been referred to as a "surface wave".

a) Inhomogeneous Plane Waves

In free space, the reduced wave equation (3) reads

$$(\nabla^2 + k^2)\,\underline{E} = 0 \qquad (16)$$

(and analogously for $\underline{H}$), with k in general complex :

$$k = \omega\sqrt{\mu\underset{\sim}{\varepsilon}} = k' - jk'' . \qquad (17)$$

A general solution in rectangular coordinates can be written in the form

$$\underline{E} + \underline{E}_0 \, e^{-j(k_x x + k_y y + k_z z)} , \tag{18}$$

where $\underline{E}_0$ is independent of the spatial coordinates. Substitution of (18) in (16) produces the "separability condition" of the wave equation :

$$k^2 = k_x^2 + k_y^2 + k_z^2 \; . \tag{19a}$$

If all k's are real, (19a) describes a <u>homogeneous</u> plane wave,

$$\underline{E} = \underline{E}_0 \, e^{-j \underline{\beta} \cdot \underline{r}} , \tag{20a}$$

where $\underline{\beta}$ and $\underline{r}$ are the vectors $(\beta_x, \beta_y, \beta_z)$ and (x, y, z), with $k = 2\pi/\lambda$ the norm of $\underline{\beta}$. Constant phase fronts are defined by $\underline{\beta} \cdot \underline{r}$ = const., and the wave travels in the direction of the unit vector $\hat{\beta}$ (collinear with $\underline{\beta}$). Since the wavelength λ and the phase velocity $v = 1/\sqrt{\mu\varepsilon}$ are inversely proportional to the phase constant, neither of them project like a vector. Thus, if $\hat{\beta}$ makes an angle θ with the z-axis, then $\beta_z = k \cos\theta$, but $v_z = v/\cos\theta$, i.e., the homogeneous wave is faster in any projection than in its direction of propagation.

We now write (19a) in the form

$$\begin{aligned} k^2 &= (\beta_x^2 + \beta_y^2 + \beta_z^2) - (\alpha_x^2 + \alpha_y^2 + \alpha_z^2) - 2j(\alpha_x\beta_x + \alpha_y\beta_y + \alpha_z\beta_z) \\ &= \beta^2 - \alpha^2 - 2j \, \underline{\alpha} \cdot \underline{\beta} \end{aligned} \tag{19b}$$

with β the norm of $\underline{\beta}$, and α the norm of $\underline{\alpha}$ $(\alpha_x, \alpha_y, \alpha_z)$. Equating real and imaginary parts,

$$(k')^2 - (k'')^2 = \beta^2 - \alpha^2 \tag{21a}$$

$$k'k'' = \underline{\alpha} \cdot \underline{\beta} \tag{21b}$$

If the medium is lossless, k" = 0, k' = k, and (21b) implies that the direction of propagation is perpendicular to the direction of amplitude decay : we are dealing with an <u>inhomogeneous</u> plane wave of the form

$$\underline{E} = \underline{E}_0 \, e^{-\underline{\alpha} \cdot \underline{r}} \, e^{-j\underline{\beta} \cdot \underline{r}} \tag{20b}$$

as illustrated, for two different orientations with respect to an interface, in Figs.4a and b. Equation (21a) now reads $k^2 = \beta^2 - \alpha^2$, or $\beta = k\sqrt{1+(\alpha/k)^2}$, which implies that in the direction of propagation the inhomogeneous wave is always slower than light, and this is just as true for leaky as for surface waves. In projection, however, the inhomogeneous wave may be either faster or slower than light. Consider first the two-dimensional surface wave in Fig.4a ; its $\hat{\beta} = \hat{z}$, hence $\beta_z = \beta$, and this wave is therefore always slow in the z-direction (as already noted in Sec.1).In a three-dimensional problem, however, the wave might be bouncing back and forth between the walls of a duct while propagating along the interface, as shown in Fig.11a; this has the effect of speeding the wave up, as always in a hollow waveguide, and the phase velocity along z might easily <u>exceed</u> that of light (see Sec. 2(c) for details). In Fig. 4b, $\beta_z = \beta \cos\theta = k\sqrt{1+(\alpha/k)^2}\cos\theta$, which is smaller or larger than k depending on the relative values of α and θ . In the case of leaky wave antennas, $\alpha/k \ll 1$ and θ is not very close to 0, so that along the interface leaky waves are usually, though not necessarily, faster than light. The assertion occasionally found in the literature that surface waves are always slow and leaky waves always fast is therefore not correct.

If the medium is lossy, $k' \neq 0$, and $\underline{\alpha}$ and $\underline{\beta}$ are no longer orthogonal, though (21b) implies that they will be nearly so if the losses are small. Examples of this type are treated in Sec.2(d).

Before passing on to the impedance aspects of inhomogeneous plane waves, we introduce two notations that will be useful to us later on. One is that of the complex wave vector $\underset{\sim}{\underline{k}} = \underline{\beta} - j\underline{\alpha}$, which is well known to workers in ionospheric propagation and (with a different meaning) in quantum mechanics. Its norm, defined in unitary space, is $\underset{\sim}{\underline{k}} \cdot \underset{\sim}{\underline{k}}^* = \beta^2 + \alpha^2$. Note that k is not the norm of $\underset{\sim}{\underline{k}}$, as can be seen by referring to (19b) which we can write in the form $k^2 = \underset{\sim}{\underline{k}} \cdot \underset{\sim}{\underline{k}}$.

The second useful notation is illustrated in Fig.6a, where the angle $j\eta$ is introduced to characterize the degree of inhomogeneity of a wave. (To avoid confusion, note that this triangle does not represent $\underset{\sim}{\underline{k}} = \underline{\beta} - j\underline{\alpha}$, but a relation in hyperbolic space between k and the norms β and α). We can now define the complex angle of reflection $\underset{\sim}{\xi} = \xi + j\eta$, Fig.6b, in terms of which

$$\begin{aligned} k_z &= k \sin \underset{\sim}{\xi} \\ k_x &= k \cos \underset{\sim}{\xi} \,. \end{aligned} \qquad (22)$$

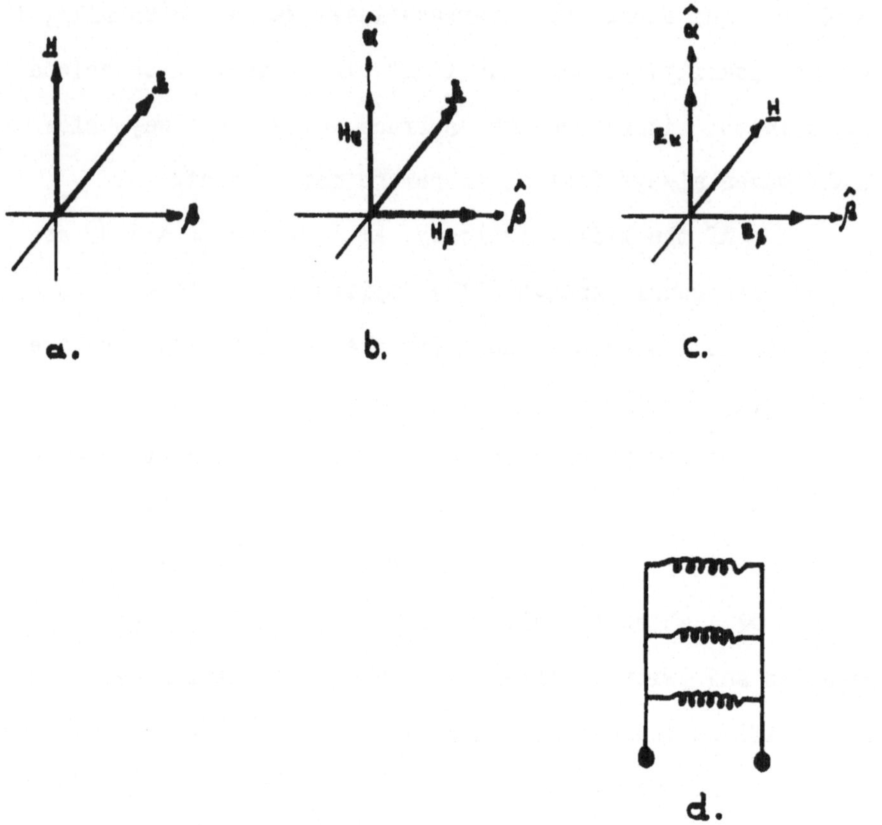

Fig. 7

we can now write in the form

$$\underline{E} = \underline{E}_o\, e^{-j\underset{\sim}{k} \cdot \underline{r}} ,$$

we find analogously from (2a) that $\hat{\underset{\sim}{k}} \cdot \underline{E} = 0$, from which it follows that $\hat{\beta} \cdot \underline{E} = 0$ and $\hat{\alpha} \cdot \underline{E} = 0$, as sketched in Fig.7b. From (2c)

$$\begin{aligned} \underline{H} &= \frac{1}{\omega\mu} \underset{\sim}{k} \times \underline{E} \\ &= \frac{\beta}{\omega\mu}\hat{\beta} \times \underline{E} - \frac{j\alpha}{\omega\mu}\hat{\alpha} \times \underline{E} \qquad (24a) \\ &= H_\alpha \hat{\alpha} + H_\beta \hat{\beta} ; \end{aligned}$$

thus $\underline{H}$ has two components, both perpendicular to $\underline{E}$, as shown. We term this wave TE, because in both the $\hat{\beta}$ and $\hat{\alpha}$ directions (and in fact in any direction in the $\hat{\beta}, \hat{\alpha}$ -plane) the $\underline{E}$-field is purely transverse. From (24a), TE vector wave admittances can be defined as

$$\frac{\beta}{\omega\mu}\hat{\beta} \quad \text{and} \quad \frac{-j\alpha}{\omega\mu}\hat{\alpha} ; \qquad (24b)$$

these are purely resistive in the direction $\hat{\beta}$ of wave propagation, and purely reactive (inductive) in the direction $\hat{\alpha}$ of evanescence.

Had we started with an $\underline{H}$-field of the form (20b), we would have found from (2d) that the wave is TM (Fig.7c):

$$\begin{aligned} \underline{E} &= \frac{1}{\omega\varepsilon} \underline{H} \times \underset{\sim}{k} \\ &= \frac{\beta}{\omega\varepsilon} \underline{H} \times \hat{\beta} - \frac{j\alpha}{\omega\varepsilon} \underline{H} \times \hat{\alpha} \qquad (25a) \\ &= E_\alpha \hat{\alpha} + E_\beta \hat{\beta} . \end{aligned}$$

The corresponding vector wave impedances are

$$\frac{\beta}{\omega\varepsilon}\hat{\beta} \text{ and } \frac{-j\alpha}{\omega\varepsilon}\hat{\alpha}\,; \qquad (25\underline{b})$$

the latter is capacitive.

We have used $\hat{\beta}$, $\hat{\alpha}$, and the directions of $\underline{E}$ or $\underline{H}$ as our natural coordinate system for discussing TE or TM waves, respectively. We will need, however, an expression for the wave impedance component in the directions of coordinate axes that make an arbitrary angle with $\hat{\beta}$ and $\hat{\alpha}$. From (24b), the admittance components along, say, x are

$$\frac{\beta}{\omega\mu}\hat{\beta}\cdot\hat{x} \quad \text{and} \quad \frac{-j\alpha}{\omega\mu}\hat{\alpha}\cdot\hat{x}\,,$$

and this can be combined into the complex scalar admittance $(\beta_x - j\alpha_x)/\omega\mu$, or

$$Z_{TE,x} = \frac{\omega\mu}{k_x} \qquad (26\underline{a})$$

From (25b)

$$Z_{TM,x} = \frac{k_x}{\omega\varepsilon}\,. \qquad (26\underline{b})$$

Note that since the vector admittances (24b) and impedances (25b) point in the direction of wave travel (and evanescence), their components (26a) and (26b) on a fixed coordinate axis change sign when the wave reverses its direction.

In terms of the Poynting vector, $\underline{S} = \frac{1}{2}\underline{E} \times \underline{H}^*$, we have from (24a) and (25a) :

$$\underline{S}_{TE} = \frac{1}{2}\frac{\beta}{\omega\mu}|E|^2\hat{\beta} + \frac{1}{2}\frac{j\alpha}{\omega\mu}|E|^2\hat{\alpha} \qquad (27)$$

$$\underline{S}_{TM} = \frac{1}{2}\frac{\beta}{\omega\varepsilon}|H|^2\,\hat{\beta} - \frac{1}{2}\frac{j\alpha}{\omega\varepsilon}|H|^2\,\hat{\alpha}$$

The real part, representing power flow, points in the direction of wave travel, while the reactive part, representing power storage, is transverse to it. The transmission line shown in cross section in Fig.7d, loaded transversely with a parallel bank of inductive coils (capacitors), is an anologue of a TE (TM) wave: here, too, the voltage and current would propagate without attenuation along the lossless line, and decay exponentially transversely to it.

We have now completed our characterization of the periodicity and electromagnetic aspects of the simplest spatial eigenfunctions, the exponential ones.

b) Pole Waves I: Transverse Resonance and the Impedance Sheet

Every specialized discipline within electromagnetic theory - such as circuitry and geometric optics - develops a language of its own in which problems are best formulated and solved. In the case of modal waves along interfaces, transmission line language furnishes a very convenient ansatz that saves us the labor of analyzing diverse cases as though each were a separate boundary value problem. The full justification for the use of this language is given by the Schwinger-Marcuvitz formalism (Sec.3); here it suffices to point to equations (8), which represent the incident and reflected wave of the transverse mode k_{xo}, and apply to them some well known results of transmission line theory.

Let V^+ and I^+ be the e_{yo} and h_{zo} components of an incident mode, as in (8), and V^- and I^- those of the mode reflected by an interface characterized by its surface impedance Z_s. Then in the coordinate system of Fig.1,

$$V(x) = V^+ e^{jk_{xo}x} + V^- e^{-jk_{xo}x}$$
$$I(x) = I^+ e^{jk_{xo}x} + I^- e^{-jk_{xo}x} \quad . \tag{28}$$

Since $V^+/I^+ = Z_o$, the wave impedance of the transverse transmission line, and $V^-/I^- = -Z_o$ [see remark following (26b)], we have

$$I(x) = \frac{V^+}{Z_o} e^{jk_{xo}x} - \frac{V^-}{Z_o} e^{-jk_{xo}x} \quad .$$

Now

$$\frac{V(0)}{I(0)} = Z_s = Z_o \frac{V^+ + V^-}{V^+ - V^-}$$
$$= Z_o \frac{1 + \Gamma}{1 - \Gamma}$$

or

$$\Gamma = \frac{Z_s - Z_o}{Z_s + Z_o} \tag{29}$$

where Γ, as in Sec.1, is the ratio V^-/V^+ of the reflected to the incident wave amplitude. When

$$Z_s + Z_o = 0 \ , \tag{30a}$$

Γ has an infinity, that is, the reflected amplitude $V^- = \Gamma V^+$ can be finite even though V^+ is zero; we therefore term (30a) the condition of <u>transverse resonance</u>. When, on the other hand,

$$Z_s - Z_o = 0 \,, \tag{30b}$$

the reflection coefficient has a zero. In view of the role played by Γ in the integrand of (11), we call a wave that satisfied (30a) a "pole" wave, and one that satisfied (30b) a "zero" (or "Brewster angle") wave; discussion of the latter is deferred to Sec.2(f).

We will need one further result from transmission line theory, namely the input impedance of a short- or open-circuited line of length d. The boundary condition on e_{yo} requires that $V = 0$ at a short circuit, and analogously $I(x) = 0$ at an open circuit. Therefore at the short circuit, the incident and reflected voltage amplitudes are equal in magnitude and opposite in sign, and from (28)

$$Z_{short} = \frac{V(d)}{I(d)} = \frac{V^+}{I^+} \frac{j\sin k_x d}{\cos k_x d}$$

$$= Z_o j \tan k_x d \,, \tag{31a}$$

where the subscript 0 has been dropped from k_x because we have in mind an arbitrary transmission line, not the region in air above the interface. For an open circuit, $I^+ = - I^-$, and

$$Z_{open} = -Z_o j \cot k_x d \,. \tag{31b}$$

We will now use our tools to determine the possible types of pole waves that can be excited in air over an impedance sheet, that is, an interface that is characterized by a surface impedance Z_s. We restrict ourselves to the scalar case; tensor surface impedances are fully treated in the accompanying paper by M.A.Miller and V.L.Talanov. Let us first assume that the waves are TM; then from transverse resonance (30a)

$$\frac{k_{xo}}{\omega\varepsilon_o} + Z_s = 0 ,$$

or

$$\frac{\beta_{xo}}{\omega\varepsilon_o} - \frac{j\alpha_{xo}}{\omega\varepsilon_o} = -(R_s + jX_s) , \qquad (32)$$

where X_s positive is an inductance, and X_s negative a capacitance. We begin with X_s inductive; from (32), $\alpha_{xo} = \omega\varepsilon_o X_s$ is then also positive, so that $\underline{\alpha}$ is directed into the upper half space at some angle with the interface. In Fig.8a - c we show three angles for $\underline{\alpha}$, and also the corresponding $\underline{\beta}$ directions $[\perp\underline{\alpha}$ from (21b)$]$. In Fig.8a, $\beta_{xo} = 0$; (32) implies that $R_s = 0$ in that case, and we have a surface wave of the type encountered in Sec.1. If $R_s > 0$, the surface is lossy, β_{xo} from (32) is negative and the surface wave is inclined toward the interface, Fig.8b. Fig.8c corresponds to $R_s < 0$ (β_{xo} positive), and we will just call this a complex pole wave : it radiates out of the interface, but it is not the leaky wave of Fig.4b. While no example comes to mind of such waves on a scalar impedance sheet (to which we are restricting ourselves), this type does occur over anisotropic and gyrotropic media. (Waves of the type Fig.8b also occur over such media, even in the absence of loss). Cases a-c are called "proper" modes, because with α_{xo} positive these poles lie in the top sheet of the k_z-plane in Fig.2, on which the radiation

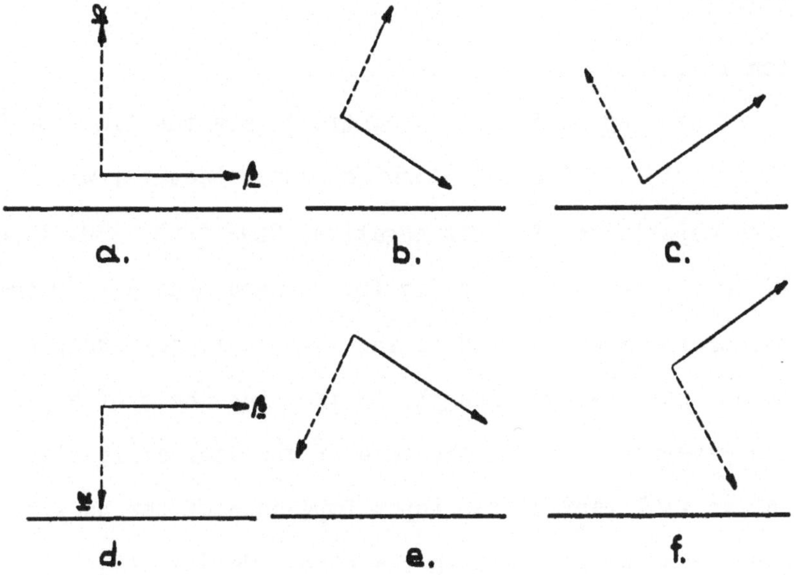

Fig. 8

condition is satisfied.

Cases d-f in Fig.8, by contrast, are the "improper" poles, which violate the radiation condition because, with X_s now chosen capacitive, α_{xo} is negative. Case f ($R_s < 0$, thus $\beta_{xo} > 0$) is the leaky wave of Fig.4b. We note that $R_s < 0$ implies the presence of sources in the interface, for example small holes (many per wavelength) in the stop sheet of a parallel-plate waveguide, or, in the case of the dielectric slab, the negative real part of the input impedance of the complex transverse standing wave within the slab. Case e ($R_s > 0$) corresponds to a pole in Fig.2 which, like the leaky wave pole k_z^{ℓ}, is located on the wrong Riemann sheet. Since β_z is positive and α_z negative for this wave type, the pole is located in the lower-right quadrant of the k_z-plane and thus, unlike k_z^{ℓ}, completely outside the contour drawn in Fig.2. As we shall see in Sec.3(b), this implies that the pole contributes no residue even when, as the result of a branch cut deformation, the leaky wave does so. The same observation holds for case d ; therefore neither wave is ever observed as the result of a resonance phenomenon, although a wave of type d is of interest in quite a different context (as part of a continuous source spectrum, see Sec.3(c)).

Had we started with a TE wave, the roles of X_s inductive and capacitive would merely have been reversed. In case a, for example, the TM surface wave, known from (25b) to be capacitive, needed an inductive surface to satisfy transverse resonance, whereas a TE wave, inductive from (24b), resonates with a capacitive surface.

We will now, in Secs.2(c), (d), and (e) below, examine several physical structures that give rise to pole waves : Sec. 2(c) treats surface waves of type Fig.8a, Sec. 2(d) surface waves of type Fig. 8b, and Sec. 2(e) briefly mentions pole waves in non-rectangular coordinate systems, and leacky waves of type Fig. 8f. For Secs. 2(c) and (d), we need only the following simple formulas :

(19a) : $$k^2 = k_x^2 + k_y^2 + k_z^2 \qquad (33a)$$

(30a) : $$Z_s + Z_o = 0 \qquad (33b)$$

(23) : $$Z_{TEM} = \sqrt{\mu/\varepsilon} \quad (\equiv R_c \text{ in air}) \qquad (33c)$$

(26) : $$Z_{TE,x} = \frac{\omega\mu}{k_x}, \quad Z_{TM,x} = \frac{k_x}{\omega\varepsilon}, \qquad (33d)$$

and equations (31) for short- and open-circuited transmission lines.

c) Pole Waves II : Surface Waves on Lossless Interfaces

We will look for surface waves of the type Fig. 8a, so that $k_z = \beta_z$ in (33a) and, in the region above the interface, $k_x = -j\alpha_{xo}$ in (33a) and (d). Let us begin with interfaces that extend indefinitely in the y-direction, so $k_y = 0$ in (33a) and (d).

Corrugated Surface : The input impedance of the dominant TEM wave in each of the grooves in Fig. 9 is that of a pa-

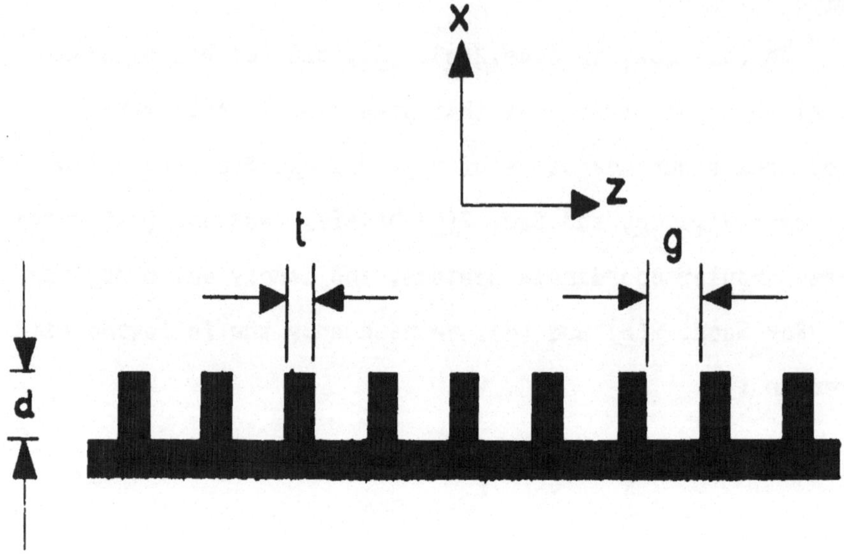

Fig. 9

rallel plate waveguide short-circuited at depth d: $j\sqrt{\mu_0/\varepsilon_0}\,\tan k_0 d$ [using (33 c) and (31)] . Across the teeth, the impedance is zero. Assuming many grooves per wavelength, the average surface impedance is therefore

$$Z_s \cong \frac{g}{t+g}\; j\sqrt{\frac{\mu_0}{\varepsilon_0}}\;\tan k_0 d \qquad (34)$$

In the first quadrant of the tangent (and its odd multiples), Z_s is inductive and will resonate a TM wave :

$$\frac{\alpha_{x0}}{\omega\varepsilon_0} = \frac{g}{t+g}\sqrt{\frac{\mu_0}{\varepsilon_0}}\;\tan k_0 d \;.$$

(Rigorous analysis requires consideration of the higher-order TM waves in the grooves. [2]) A TE surface wave cannot exist, as the following consideration shows. To excite the dominant TEM wave in the grooves, an E_z - or an H_y-component must be provided. The TM surface wave provides both, the TE surface wave neither; the latter can therefore excite only higher-order evanescent ($g \ll \lambda$) TE modes in the grooves, and the wave impedance of these modes is inductive. In view of (31a), Z_s is then real, and cannot resonate a TE surface wave.

Dielectric Slab-on-Metal : The TE surface wave modes of a conductor-backed dielectric sheet have already been discussed in Sec.1. From the point of view of transverse resonance, the impedance looking downward from the interface in Fig. 10a is that of a short-circuited transmission line with characteristic impedance $\omega\mu_0/\beta_{x1}$, as given by (33d). Thus

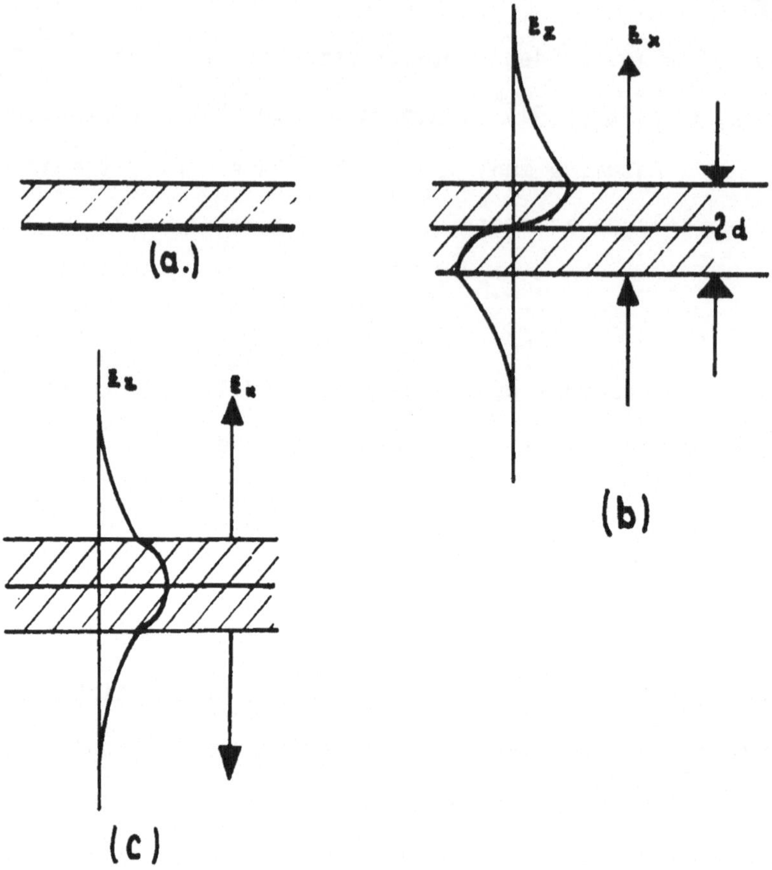

Fig. 10

$$Z_s = \frac{j\omega\mu_o}{\beta_{x1}} \tan \beta_{x1} d ,$$

and with $Z_o = \omega\mu_o/-j\alpha_{xo}$, (33b) becomes

$$\frac{1}{\alpha_{xo}} + \frac{1}{\beta_{x1}} \tan \beta_{x1} d = 0 , \qquad (35a)$$

which is of course identical with (13a) in Sec. 1. (33a) applied to the air and dielectric medium yields (13b) in Sec. 1, and the graphical solution of Fig. 3 follows. (Note that, whereas the surface impedance Z_s on the impedance sheets of Sec. 2(b) was a constant, it here is a function of β_{x1} and thus, through (33a), of the wavenumber β_z ; Z_s is now said to be "spatially dispersive". In the case of the corrugated surface, the approximation (34) is spatially nondispersive, but higher-order terms do show dependence on β_z : spatially dispersion is, in fact, a characteristic of all physical surfaces). In the TM case, (33b) becomes

$$-\frac{\alpha_{xo}}{\varepsilon_o} + \frac{\beta_{x1}}{\varepsilon_1} \tan \beta_{x1} d = 0 , \qquad (35b)$$

and (33a) again yields (13b); a graphical solution analogous to Fig. 3 immediately shows that the lowest-order mode has no cut-off, i.e., the surface wave exists for arbitrarily thin layers.

Dielectric Slab in Air (Panel): In Figs. 10b and c, the TM wave on a conductor-backed slab is shown with its mirror image. One would expect such a dielectric panel to support the same modes as the slab-on-metal. If the mirroring wall is thought of as a metal ground plane, the field distribution is anti-symmetric, Fig. 10b, and this is indeed the case. But symmetric field distributions, Fig. 10c, can also exist on the panel; these correspond to mirror imaging the slab through a perfect "magnetic conductor" (an open circuit, as contrasted with the short circuit in b). (31) now requires us to replace j tan $\beta_{x1}d$ in (35a) and (35b) by -j cot $\beta_{x1}d$, and a wholly new set of modes is produced; the lowest-order TE mode in this case turns out to have no cutoff, while the lowest-order TM mode has.

It is usually convenient to display the results of modal calculations as a set of curves plotting $\beta_z/k_o = \lambda_o/\lambda_z$ as a function of the geometric constants, here d/λ_o, with dielectric constant as a parameter. This is done in Fig. 10d for the case of Fig. 10a; the range of λ_o/λ_z shown is that of principal interest in surface wave antenna design (waves only slightly slower than light, see Sec. 4). Had we plotted larger values of λ_o/λ_z, all curves would eventually flatten out and approach $\sqrt{\varepsilon_r}$ $(\lambda_o/\lambda_z = k_1/k_o)$ asymptotically, as proved in the discussion following (13c) in Sec. 1.

Trough guide with dielectric bottom : We have so far assumed $k_y = 0$ in (33a). Let us now investigate the modal properties of the trough guide, Fig. 11a. If the wave is TE, the side-walls are everywhere perpendicular to the E-field and might as well not be there; the modal properties are those of the infinite slab-on-metal. If the wave is TM, the side-walls act as short circuits : only integral multiples of half-wavelengths are therefore allowed in the y-direction, or $\lambda_y = 2w/n$, $k_y = n\pi/w$, and for the lowest-order mode (33a) reads

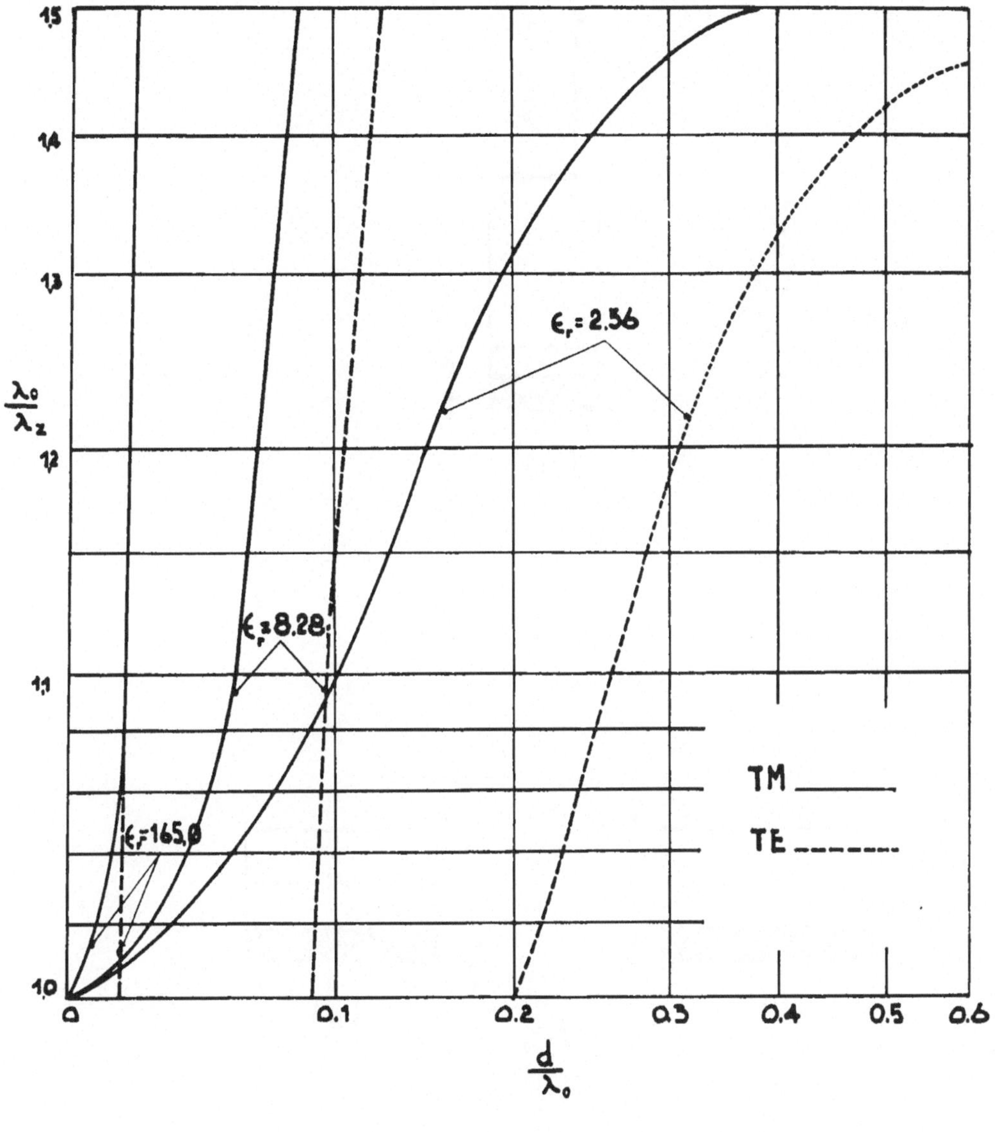
1.5
1.4
1.3
1.2
1.1
1.0
$\frac{\lambda_0}{\lambda_z}$
0
0.1
0.2
0.3
0.4
0.5
0.6
$\frac{d}{\lambda_0}$
$\epsilon_r = 2.56$
$\epsilon_r = 8.28$
$\epsilon_r = 165.0$
TM
TE

Fig. 10(d)

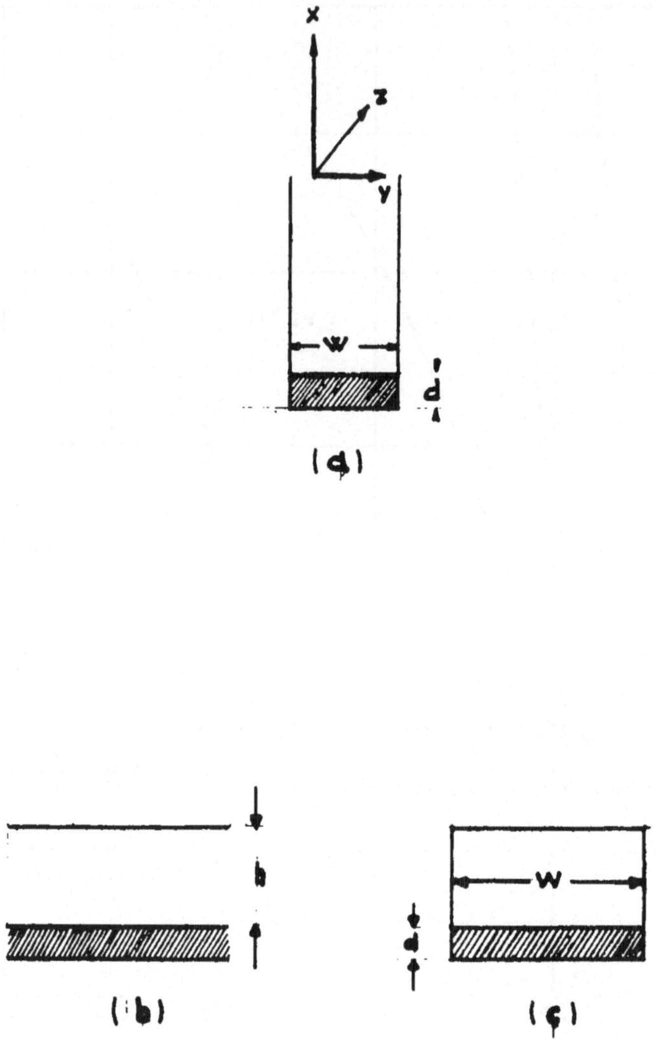

Fig. 11

$$\beta_z = \sqrt{k_0^2 + \alpha_{xo}^2 - \left(\frac{\pi}{w}\right)^2}$$

which, as already mentioned in the discussion following (20b), can represent a slow or fast wave, depending on whether $\alpha_{xo} > \pi/w$ or vice versa.

Shielded waveguides with dielectric slab : We can show quickly that structures of the type Figs. 11b and c (which differ from each other only in that $\beta_y = 0$ in b and $n\pi/w$ in c) support two types of modes : in one, the field distribution in the air region is hyperbolic, in the other, trigonometric; the field distribution for both is trigonometric in the dielectric region. In Fig.11b, a dielectric-loaded parallel plate waveguide, assume a TM surface wave and apply transverse resonance with the air-dielectric interface as reference plane:

$$\frac{j\beta_{x1}}{\omega\varepsilon_1} \tan\beta_{x1}d + \frac{\alpha_{xo}}{\omega\varepsilon_0} \tan(-j\alpha_{xo}h) = 0 ;$$

the second term is negative imaginary and, with (33a) in the form of (13c), a solution is readily found. The effect of the top plate is evidently to short circuit the surface wave $\exp(-\alpha_{xo}x)$ and add a reflected (upside-down) surface wave $\exp(\alpha_{xo}x)$, which in combination give a $\sinh\left[\alpha_{xo}(h-x)\right]$ distribution for E_z, a $\cosh\left[\alpha_{xo}(h-x)\right]$ distribution for H_y, and thus the upward-looking impedance $(-j\,\alpha_{xo}/\omega\varepsilon_o)\tanh(\alpha_{xo}h)$ in the above transverse resonance equation. If we assume, on the other hand, that the transverse wavenumber in air is β_{xo}, then resonance requires

$$\frac{j\beta_{x1}}{\omega\varepsilon_1} \tan\beta_{x1}d + \frac{j\beta_{xo}}{\omega\varepsilon_o} \tan\beta_{xo}h = 0 ,$$

which can be satisfied if, for example, $\beta_{x1}d$ is in the first quadrant and $\beta_{xo}h$ in the second. A real tranverse wave-number in air implies a trigonometric field distribution there $[\exp(-j\beta_{xo}x) \pm \exp(j\beta_{xo}x)]$, just as β_{x1} does in the dielectric. The hyperbolic modes in shielded waveguide are appropriately called surface waves because, when the walls are removed, they become surface waves on an infinite slab; the trigonometric modes, on the other hand, cannot exist on a single interface (because $\exp(j\beta_{xo}x)$ violates the radiation condition) and are therefore "ordinary" waveguide modes.

d) Pole Waves III : Surface Waves on Lossy Interfaces

We now consider surface waves of the type Fig. 8b. Losses in the structures shown in Figs. 9, 10 and 11 will produce such waves, but what we shall concentrate on are interfaces between two semi-infinite media, as in Fig. 12. Formulas (33) are still we need, with $k_y = 0$. Transverse resonance requires, for a TM wave, that

$$\frac{k_{xo}}{\omega\varepsilon_0} + \frac{k_{x1}}{\omega\underset{\sim}{\varepsilon}_1} = 0\ , \tag{36a}$$

where medium 0 is assumed lossless and medium 1 lossy, $\underset{\sim}{\varepsilon}_1 = \varepsilon_1(1-jh)$, with the power factor $h \equiv \sigma_1/\omega\varepsilon_1$ [see eqn. (2d)] ; also, we let $\mu_0 = \mu_1$. We can calculate k_z immediately by noting from (36a) that

$$\frac{k_{x1}}{k_{xo}} = -\frac{\underset{\sim}{\varepsilon}_1}{\varepsilon_0} = -\frac{k_1^2}{k_0^2}\ , \tag{36b}$$

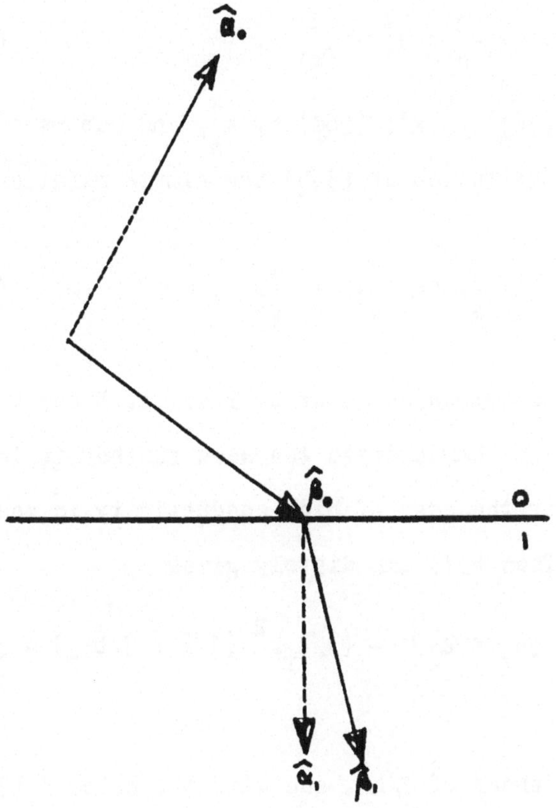

Fig. 12

and by writing (33<u>a</u>) as the pair of equations

$$k_z^2 = k_o^2 - k_{xo}^2 \tag{36\underline{c}}$$

$$k_z^2 = k_1^2 - k_{x1}^2 \ . \tag{36\underline{d}}$$

On multiplying (36<u>c</u>) by k_1^4, (36<u>d</u>) by k_o^4, and subtracting, one obtains upon substitution of (36<u>b</u>) the simple relation

$$\frac{1}{k_z^2} = \frac{1}{k_o^2} + \frac{1}{k_1^2} \ , \tag{37}$$

also cited in the companion paper by Bremmer. Since k_1 is complex, so is k_z. To characterize the wave further it is simplest to specialize for the case of high conductivity in medium 1, i.e., $h \gg 1$. Then (37) immediately gives

$$k_z \equiv \beta_z - j\alpha_z \cong k_o \left[1 - (1/\epsilon_r h^2)(1/2 + 3/8\epsilon_r) - j/2\epsilon_r h \right] \tag{38\underline{a}}$$

to terms of the order of $1/h^2$, and with the help of (36<u>c</u>) and (36<u>d</u>) we find

$$k_{xo} \equiv \beta_{xo} - j\alpha_{xo} \cong k_o \sqrt{1/2 \ \epsilon_r h} \ (-1-j) \tag{38\underline{b}}$$

$$k_{x1} \equiv \beta_{x1} - j\alpha_{x1} \cong k_o \sqrt{\epsilon_r h/2} \ (1-j) \ . \tag{38\underline{c}}$$

We see from (38<u>a</u>) that the wave is very slightly faster than light along the surface, and from (38<u>b</u>) and (38<u>c</u>) that α_{xo} as well as α_{x1} are positive (decay away from the interface in both media); β_{xo} is negative, which implies an incoming wave in medium 0, while β_{x1} is outgoing. The result is sketched in Fig. 12 : from (38a) and (38b) it is evident that the surface wave in medium 0 is of the form Fig. 8b, and from (38a) and (38c) it follows that since $\beta_{x1} \gg \beta_z$, the net phase constant β_1 is nearly perpendicular to the interface, and since $\alpha_{x1} \ggg \alpha_z$, $\underline{\alpha}_1$ is even more so. (In at least one well-known text the problem of the skin effect, which we have in effect just treated, is tackled by assuming explicitly that β_1 and $\underline{\alpha}_1$ are collinear, a numerically trivial but, in view of (21b), conceptually nontrivial error).

The surface wave that satisfies (36a) is often referred to as the "Zenneck" wave. An extensive literature is devoted to the question of its existence. We have here demonstrated that (36a) does have a solution in the case of high conductivity (it turns out the same in the case of arbitrary conductivity), and thus the Zenneck wave is indeed a pole in the radiation spectrum of an appropriate exciting source (say a vertical dipole, since the TM nature of the wave requires an E_x-component). Since α_{xo} is positive, this pole lies on the top sheet in the integral representation shown in Fig. 2, and thus contributes a residue; in this sense, therefore, the Zenneck wave can be said to "exist". There is, however, another side to this story, and we will refer to it in Sec. 3(b).

Under the assumptions made, the transverse resonance relation for a TE wave is found to have no solution. If the roles of ε and μ are interchanged, i.e., if $\underline{\mu}$ is now assumed to be complex, a TE but no TM solution can be shown to exist.

Such "two-sided" surface waves can also propagate along the interface between two lossless half spaces, provided the ε or μ in one medium is a tensor. The wavenumbers k_{xo} and k_{x1} in that case are purely imaginary, and the phase fronts in both media are perpendicular to the interface (unlike in Fig. 12).

e) Pole Waves IV : Non-Cartesian Coordinate Systems

In all coordinate systems in which the vector wave equation is separable, relations between the wavenumber components can be written that are analogous to (33a). Spatial eigenfunctions analogous to the exponentials discussed in Sec. 2(a) lead to wave impedances that correspond to (33c) and (33d), and boundary conditions at the interface can be formulated in terms of transverse resonance as in (33b). In radial cylindrical coordinates --- used in the analysis, say, of a dipole center-feeding a dielectric disc or a flat surface with annular corrugations -- one finds

$$k^2 = k_x^2 + k_\rho^2 \tag{39}$$

where ρ is the radial direction in the y-z plane. Pure surface waves are possible, with $k_{xo} = -j\alpha_{xo}$ and $k_\rho = \beta_\rho$. The eigenfunctions are zero-order Hankel functions of the second kind (for outgoing waves), and for $k_o\rho \gg 1$ the wave behaves like $(1/\sqrt{k_o\rho})\exp(-\alpha_{xo}x - j\beta_\rho\rho)$; note that the $1/\sqrt{\rho}$ dependence implies constant-amplitude fronts that are not parallel to the interface. The axial cylindrical case -- waves propagating on open cylindrical waveguides -- will be described in Sec. 3(e); true surface waves exist, with constant-amplitude fronts that are concentric with the cylindrical interface. In the azimuthal cylindrical or spherical case, by contrast -- waves propagating circumferentially -- the wavenumber components in the radial and azimuthal directions are necessarily complex, and completely-trapped surface waves are no longer possible.[3] If $k_0 \rho_0 \gg 1$, with ρ_0 the radius of the interface, the problem can be trated as a perturbation of the flat surface : curvature is then found to slow down the surface wave and introduce a slight amount of leakage. Rigorous solution of the problem for arbitrary $k_0 \rho_0$ leads to an infinite set of poles that satisfy the radiation condition, and there is no branch cut integral. As $k_0 \rho_0$ increases, a finite number (zero, one, or more) of these poles approach their limiting position for $k_0 \rho_0 = \infty$; e.g., the pole marked β_z^s in Fig.2 can be thought of as the limiting position of such a pole on a dielectric-coated sphere. The remaining poles become infinitely dense as $k_0 \rho_0$ increases, and finally coalesce into the branch cut in Fig.2. Only poles of the former kind are therefore "quasi-surface" waves; poles of the latter kind are a representation of the radiation field (continuous spectrum) on the cylinder or sphere.

Another case of a quasi-surface wave is of particular interest because it involves a variable surface impedance : in Fig. 13, $R_s = 0$ and

$$X_s(z) = 0 \qquad -\infty < z < 0$$
$$= a/z \,, \qquad 0 \leqslant z < \infty$$

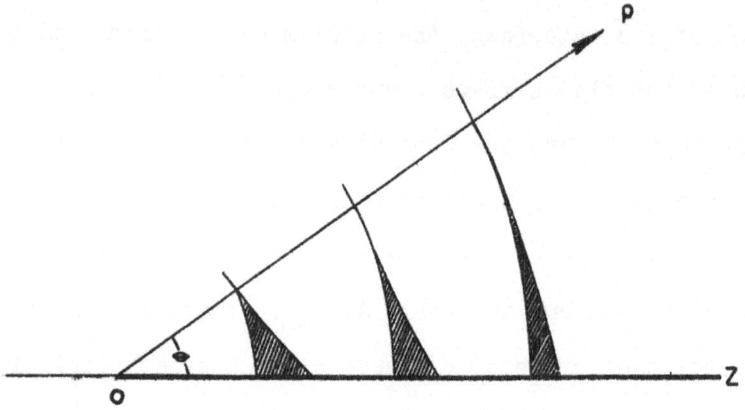

Fig. 13

where a is a positive constant. The wave equation in this instance was shown by Felsen[4] and also by Talanov[5] to be separable for a TM wave in the cylindrical coordinate system indicated in the figure; the result is an inhomogeneous cylindrical wave which propagates at the velocity of light and decays exponentially along the phase fronts, i.e., the magnetic field for $k_o \rho \gg 1$ is given by

$$H_y'(\rho, \vartheta) \sim \frac{1}{\sqrt{k_o \rho}} e^{-b\theta} . e^{-jk_o \rho}, \qquad (40)$$

where $b = a \, \varepsilon_o \omega$. We will derive (40) asymptotically in Sec. 4(c).

We have restricted our discussion of specific pole wave examples to surface waves of types Fig. 8a and 8b and their variants in other coordinate systems. Examples of type Fig. 8c are given in the companion paper by Miller and Talanov. Types d and e, we have said, do not occur within the context of the usual asymptotic field representations [see Sec. 3(b)] . An example of type Fig. 8f, the leaky wave, was calculated in Sec. 1. Of greater interest would be the example of a leaky waveguide, e.g., a rectangular waveguide with an infinite slot in one of the narrow walls. Calculations of this sort involve a complicated discontinuity in the transverse cross section of the structure -- as complicated as the exact calculation of the Z_s of the corrugated surface except that fortuitously simple approximations of the type (34) cannot now be found. Transverse resonance can be applied only after the effect of the discontinuity has been formulated, with the help of the Waveguide Handbook[6] or other sources, in equivalent lumped-circuit terms.

Oliner[7] has worked out a perturbation procedure that greatly facilitates calculation of the complex leaky wave parameters once the transverse resonance equation has been set up, but even so the detailed presentation of a leaky waveguide problem is too lengthy for inclusion here.

f) Boundary Waves in Refraction I : Zero (Brewster-angle) Waves

Pole waves are the resonant response of an interface, that is, they are reflected waves whose x-dependence is of the form $\exp(-jk_{xo}x)$, where a positive k_{xo} signifies that the wave is outgoing. Incident waves, by contrast, are of the form $\exp(jk_{xo}x)$, and a positive k_{xo} now signifies a wave traveling in towards the surface. (The z-dependence of both reflected and incident wave is $\exp(-jk_z z)$ for a wave traveling to the right). We found in (30b) that an incident wave will be entirely transmitted (zero reflection coefficient) if

$$Z_s = Z_o \, . \qquad (30b)$$

To examine these "zero" waves, assume, to begin with, that the lower half-space is filled with a lossless dielectric, with air above it. Let a TEM wave be incident at an angle ξ_o (Fig. 6b), with the H-field out of the paper. Viewed in the x-direction, this wave is TM, and $Z_o = \beta_{xo}/\omega\varepsilon_o = k_o \cos\xi_o/\omega\varepsilon_o$. In the lower medium, the refracted wave travels at an angle ξ_1 with the normal, so $Z_s = k_1 \cos\xi_1/\omega\varepsilon_1 = k_o \cos\xi_1/\omega\varepsilon_o\sqrt{\varepsilon_r}$, where we have assumed that $\mu_1 = \mu_o$. (30b) now reads

$$\cos \xi_1 = \sqrt{\varepsilon_r} \cos \xi_0 . \tag{41a}$$

As in the case of pole waves, $\beta_{zo} = \beta_{z1}$ (periodicity of fields above and below interface must match), or

$$k_o \sin \xi_o = k_1 \sin \xi_1, \qquad \sin \xi_1 / \sin \xi_o = 1/\sqrt{\varepsilon_r}, \tag{41b}$$

which is Snell's law in optics. Combining (41a) and (41b),

$$\sin^2 \xi_o = \varepsilon_r/(1 + \varepsilon_r), \tag{42a}$$

thus: $$\tan \xi_o = \sqrt{\varepsilon_r} \tag{42b}$$

and ξ_o is recognized as Brewster's angle in optics. In the analogous TE case ($\underline{E}$-field of the incident TEM wave out of the paper), a Brewster angle can be found only if the media differ in μ.

What if the lower medium is lossy? Then ε_r is complex, and the angle of incidence likewise : $\tan \underset{\sim}{\xi}_o = \sqrt{\underset{\sim}{\varepsilon}_r}$, with $\underset{\sim}{\xi}_o$ the complex Brewster angle. The zero-reflection-coefficient wave is now inhomogeneous, its parameters from equations (22) and (42) are

$$k_z = k_o \sin \underset{\sim}{\xi}_o = k_o \sqrt{\underset{\sim}{\varepsilon}_r/(1+ \underset{\sim}{\varepsilon}_r)},$$

$$k_{xo} = k_o \cos \underset{\sim}{\xi}_o = 1/\sqrt{1+ \underset{\sim}{\varepsilon}_r} .$$

If we assume, as in the case of the Zenneck wave, that the lower medium is a good conductor, then k_z turns out to be identical with (38 a), the k_z of the Zenneck wave, and

$$k_{xo} \cong k_o \sqrt{1/2\ \varepsilon_r h}\ (1+j),$$

which is numerically the same as (38b) but of opposite sign. Now in view of the $\exp(k_{xo}x)$ dependence assumed for the incident wave, a positive β_{xo} is also incoming and thus corresponds to the negative β_{xo} of the pole wave; and vice versa for α_{xo}. We have therefore shown that the complex Brewster angle wave is of precisely the same form in medium 0 as the Zenneck pole wave, and we could in fact easily verify this for medium 1 as well. Indeed, had we written the zero equation (30b) in the form

$$\frac{k_{xo}}{\omega \varepsilon_o} - \frac{k_{x1}}{\omega \varepsilon_1} = 0 \qquad (43)$$

instead of using angle-of-incidence notation, we would have noticed immediately that (43) and (36a) differ only in the sign of k_{xo} and hence lead to identical results if k_{xo} in (36a) is associated with a reflected wave and in (43) with an incoming wave (k_{x1} is associated with a transmitted wave in both cases). Fig. 12 evidently illustrates both a zero and a pole wave.

Does this mean that zero and pole waves are one and the same thing? The answer is "yes" insofar as the wave parameters are concerned : zero and pole equations are in fact merely two ways of satisfying the same boundary conditions, and cannot but lead to identical wave solutions. In Fig. 14, Z_s is the same in a and b, and because no sources are present on the interface,

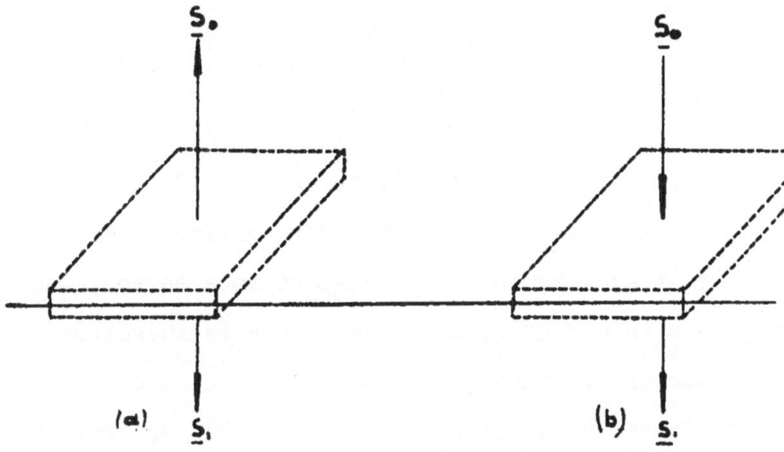

Fig. 14

the net power outflow through the unit area boxes must be zero. If, as in a, we assume that the upper half space contains only a reflected wave, then

$$\underline{S}_0 + \underline{S}_1 = 0$$

implies, via (26) and (27), the transverse resonance condition $Z_0 + Z_s = 0$; if, as in b, we assume that it contains only an incident wave, then

$$\underline{S}_0 - \underline{S}_1 = 0$$

implies the zero-reflection conditions $Z_0 - Z_s = 0$.

Zero and pole waves do, however, differ essentially when their roles are examined with respect to a source. Incident waves originate in a source, and a zero wave is therefore part of a source spectrum. Since the radiation field of a source is a continuous function of direction in space, the spectrum of plane waves that produces this radiation field is a continuous function of the wave number [see Sec. 3 (c) for an example]. The zero wave is only one line of infinitesimal width in this continuous source spectrum, and the fraction of source power it carries is therefore zero. The pole wave, by contrast, is a discrete spectral line β_z^s in the reflected wave spectrum, with no β_z values in its immediate neighborhood, and it carries an amount of power that is proportional to the square of the absolute magnitude of the residue wave which, as we shall see in Sec. 3(c), may constitute an appreciable fraction of the total available source power.

One further point needs clarification : in the limiting case of a lossless lower medium, (42b) defines the zero-wave parameters; the corresponding pole wave, however, does not exist since $\beta_z = k_o \sin \xi_o < k_o$ falls right on the branch cut shown in Fig.2.

g) Boundary Waves in Refraction II : Total Internal Reflection

Two lossless semi-infinite media are assumed in Fig.15, and a plane wave incident at an angle ξ_1 in the denser medium is reflected by the interface. The angle of refraction ξ_0, given by Snell's law (41b)

$$\sin \xi_o = \sqrt{\varepsilon_r} \sin \xi_1 ,$$

becomes complex when $\sqrt{\varepsilon_r} \sin \xi_1$ exceeds 1. At the "critical" angle of incidence, ξ_o equals 90°, for larger ξ_1 we write $\sin \underset{\sim}{\xi}_o = \sin(90° + j\eta) = \sqrt{\varepsilon_r} \sin \xi_1$, or

$$k_z = k_o \sin \underset{\sim}{\xi}_o = k_o \sqrt{\varepsilon_r} \sin \xi_1 , \qquad (44)$$

which is a slow wave in medium 0. The associated $\alpha_{xo} = \sqrt{\beta_z^2 - k_o^2}$ is seen to be zero at the critical angle and increases to $\sqrt{k_1^2 - k_o^2}$ at glancing incidence, $\xi_1 = 90°$; the latter value is identical to the maximum vertical attenuation found on a metal-backed dielectric slab, derived in the discussion following (13d) in Sec. 1.

The surface wave in total internal reflection is clearly not a zero wave, since the amplitude of the reflection coefficient is 1, not 0. Nor is it a pole wave -- in fact, if we refer to a pole wave as a "free" oscillation of the surface, we can now

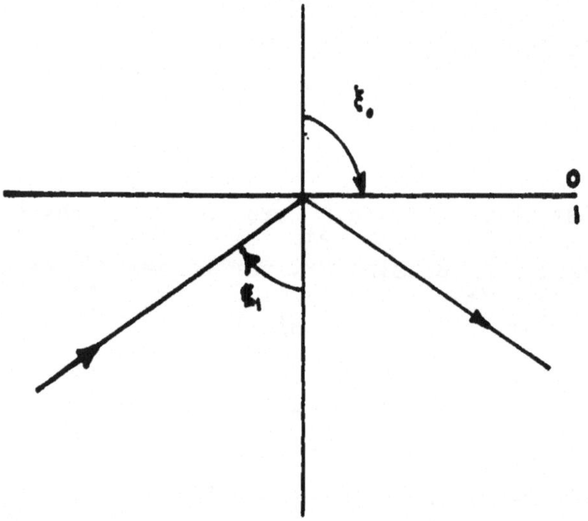

Fig. 15

call the total-reflection surface wave a 'forced' oscillation: the β_z of free oscillations depends only on the geometric and constitutive parameters of the structure, whereas β_z in (44) depends also on the angular location ξ_1 of the driving source. Examination of total internal reflection in a dielectric slab with metal bottom, Fig. 1, does however lead to a physically suggestive relation with pole waves. Consider a totally reflected ray incident at such an angle ξ_1 that in successive bounces between the slab faces its phase remains in step, i.e., differs from one complete bounce to the next by integral multiples of 2π. The reflection coefficient of, say, a TE wave at the ground plane is -1, hence the phase shift there is π ; at the dielectric interface, $\Gamma = |\Gamma| \exp j\phi$ is given by (23), with $Z_s = \omega\mu/-j\alpha_{xo}$ and $Z_o = \omega\mu/\beta_{x1}$, so that $\phi = \text{arc tan}\left[2\alpha_{xo}\beta_{x1}/(\beta_{x1}^2 - \alpha_{xo}^2)\right]$. Therefore

$$- 2\beta_{x1}d + \pi + \phi = 2n\pi , \qquad n = 0, 1,... \qquad (45)$$

[This expression can also be derived from equations (13a) or (35a) for the TE surface wave on a slab, by writing $\tan 2\beta_{x1}d$ in terms of $\tan \beta_{x1}d = -\beta_{x1}/\alpha_{xo}$.] The surface wave is now seen to be equivalent to the superposition of two of these totally reflected rays, criss-crossing each other symmetrically with respect to the median plane of the slab so as to produce a standing wave in cross section; each permissible mode corresponds to a discrete value of β_{x1} in (45), and thus of the angle of incidence $\cos \xi_1 = \beta_{x1}/k_1$. The resolution of waveguide modes into criss-crossing plane waves is familiar in connection with shielded rectangular waveguides, and it can also be applied to leaky waves : if ξ_1 is smaller than the critical angle, a refracted ray emerges from the slab at each bounce off the air-dielectric interface, and the amplitude of the reflected ray is correspondingly reduced. To satifsfy (45), ϕ must be complex, i.e., $j\alpha_{xo}$ becomes k_{xo} and β_{x1} becomes k_{x1}, and we are in effect solving (12) for the complex modes.

If two dielectric half spaces are separated by a narrow ($\leq \lambda$) air gap, the surface wave associated with a totally reflected ray incident in one medium can be shown to couple energy into the other -- a phenomenon that underlies several well-known effects in optics, e.g., Newton's rings.

h) Boundary Waves in Diffraction and Propagation

The electromagnetic field diffracted by a sphere can be expanded in terms of the poles mentioned in Sec. 2(e). To perform actual calculations, however, diverse transformations and asymptotic evaluations are called for, and each gives rise to a recognizable type of boundary wave. Prominent among these are the waves in Fock's representation of the field at the light-shadow boundary, and the "shedding ray" in Keller's quasi-geometric theory of diffraction. Involving as they do the vast field of diffraction, these topics cannot be treated in the present context.

Two boundary waves that play important roles in propagation -- the subject to which the companion lectures by H.Bremmer are devoted -- are the "head" or "lateral" wave, and the Norton surface wave, and we shall describe these at least qualitatively. A line source located at the origin on the interface between two lossless media, Fig. 16, produces cylindrical waves that travel more slowly in the denser medium 1 than in medium 0; the ratio of the radii is $\sqrt{\varepsilon_r}$. A plane wave front extending from the footpoint at the interface of the faster wave to the point of tangency on the slower, mediates between the two. This "lateral" wave can be clearly seen on schlieren photographs of acoustic fields, and mathematically it arises from the deformation of the path of integration occasioned by the presence of a branch cut starting at k_1 in Fig. 2 (which is a branch point whenever medium 1 is semi-infinite).[8] Inspection of Fig. 16 leads to the immediate result that sin $\xi_1 = 1/\sqrt{\varepsilon_r}$, i.e., the lateral wave emerges from the interface at the critical angle defined in Sec. 2(g). Although both phase and amplitude fronts are plane and at right angles to each other, the lateral wave is not an inhomogeneous plane wave of the type discussed at the beginning of this section : instead of decaying exponentially along the phase front, the amplitude decays as a power of the footpoint distance from the source, i.e., at P it is proportional to $1/z^2$.

Certain portions of the total field arising in propagation problems are often referred to as "surface waves" by writers on that subject, but these surface waves are not necessarily related to any that we have described. The best known example is the "Norton surface wave" at the interface between air and a conducting half space : it represents the difference, at large distances from the exciting dipole, between the total field and the "space" wave [i.e., the direct plus the surface-reflected

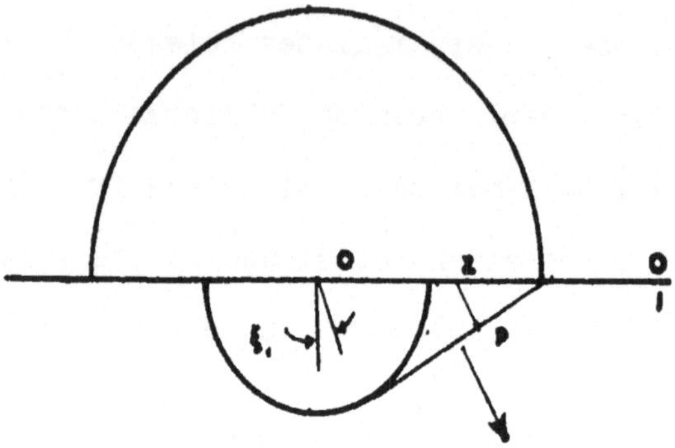

Fig. 16

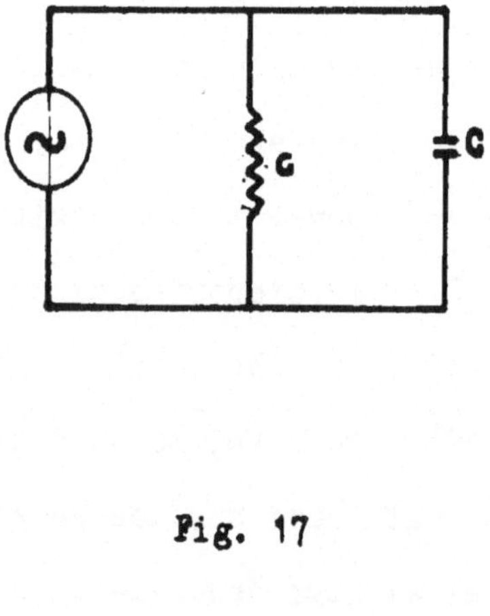

Fig. 17

far-field radiation from the dipole, see Sec. 3(b)] . J.Wait [9] has suggested that a surface wave can be defined in the most gegeral case -- at any distance from the source, for interfaces of any shape and with any kind of boundary condition -- as the remainder field after the saddle point contribution in the radiation integral [cf. Sec. 3(b)] has been split off the total field. Although in certain instances, as we shall see, this remainder field does closely resemble our surface wave type Fig. 8a or b, it more usually resembles a superposition of surface and leaky waves, or cannot be related to our wave types at all. Unlike all other boundary waves discussed by us (except for the lateral wave), the Norton (or any remainder-field) surface wave does not by itself satisfy the boundary conditions at the interface, and is not a solution of the homogeneous wave equation.

We have now concluded our survey of boundary wave types. If every bestiary must have a moral it is perhaps this : a clear understanding of the context in which a boundary wave arises, and of its characteristics, is more important than terminological consistency and unanimity; as past experience has shown, the latter is in any case unachievable. From here on we confine ourselves to pole waves, and among these largely to surface (Secs. 3 and 4) and leaky waves (Sec. 4).

3. SURFACE WAVEGUIDES

We will in this section sketch the theory of open waveguides, starting in 3(a) with an introduction to the Marcuvitz-Schwinger formalism. Sec. 3(b) outlines the asymptotic evaluation of the source radiation field, 3(c) details the influence of the source on the amplitude of the residue wave, and 3(d) proves that the power flow in the radiation field and in the residue waves are independent of each other. In Sec. 3(e) the varieties of cylindrical open waveguides are displayed, and 3(f) introduces the periodic surface waveguide.

a) Sketch of the Marcuvitz-Schwinger Formalism

The method used in Sec. 1 to calculate the total field of a dipole above a grounded slab can be generalized so as to apply to an arbitrary source configuration in any coordinate system in which the wave equation is separable in at least one direction. Because it leads to a systematic approach to all waveguide mode and discontinuity problems, this generalized formalism is used, explicitly or implicitly, in a large body of literature. The following sketch is intended as an introduction to that literature and to the original paper on the subject by Marcuvitz and Schwinger,[10] which the uninitiated will not find easy to read.

We begin with an analogy in the time domain, due to Marcuvitz himself. Consider the simple circuit in Fig. 17 : voltage and current satisfy Kirchhoff's differential equation

$$i(t) = y\frac{d}{dt}v(t) \tag{46a}$$

where the admittance operator for our circuit is

$$y\frac{d}{dt} = g + c\frac{d}{dt}.$$

Our problem is to find v(t) when the generator current i(t) is prescribed. The spectral representation

$$\begin{aligned} v(t) &= \frac{1}{2\pi}\int_{-\infty}^{\infty} V(\omega)e^{j\omega t}\,d\omega \\ i(t) &= \frac{1}{2\pi}\int_{-\infty}^{\infty} I(\omega)e^{j\omega t}\,d\omega \end{aligned} \tag{47a}$$

analyzes voltage and current in terms of exponential eigenfunctions (modes), with ω the radian frequency and $V(\omega)$, $I(\omega)$ the corresponding modal amplitudes. The eigenfunctions are orthogonal, i.e.

$$\int_{-\infty}^{\infty} e^{j\omega_m t}\, e^{-j\omega_m t}\, dt = \delta(\omega_m - \omega_m) \tag{48a}$$

where δ is the delta function. Substitution of (47a) in (46a), subject to (48a), results in Ohm's law valid at each frequency:

$$I(\omega_m) = Y(\omega_m)V(\omega_m) \tag{49a}$$

or

$$V(\omega_m) = I(\omega_m)/Y(\omega_m), \tag{50a}$$

where $Y(\omega_m)$ is the circuit admittance. v(t) is now resynthesized by integrating over all ω

$$v(t) = \frac{1}{2\pi}\int_{-\infty}^{\infty} \frac{I(\omega)}{Y(\omega)}\, e^{j\omega t}\, d\omega, \tag{51a}$$

and by analytic continuation in the ω plane, we obtain the residue

$$v(t) = -j \frac{I(\omega_p)}{\frac{d}{d\omega} Y(\omega)\Big|_{\omega_p}} e^{j\omega_p t} \tag{52}$$

More complicated circuits give rise to several poles, the "natural" frequencies or "free" oscillations of the system; branch point singularities never occur.

We now translate this procedure into the spatial domain, where we assume a waveguide with cross sections that are arbitrary in shape but uniform along ξ, the direction of wave travel -- a restriction made for our convenience only since the formalism also applies to nonuniform structures, e.g., the conical waveguide. (ξ is not to be confused with the angle in Fig.6). The basic idea is to analyze the field into cross-sectional modes, then discover what equation each modal amplitude obeys in the ξ-direction, solve it, and resynthesize the total field. To begin with, we use the existence of a preferred direction to eliminate the ξ-components of the electromagnetic field by dot and cross multiplying in turn the two Maxwell equations (2c) and (2d) :

$$j\omega\mu \underline{H} \times \hat{\xi} = \hat{\xi} \times \nabla \times \underline{E} = \hat{\xi} \times \left[(\nabla \times \underline{E})_t \hat{t} \right]$$

$$= -\frac{\partial}{\partial \xi} \underline{E}_t + \nabla_t E_\xi \, , \tag{53a}$$

where $\hat{t}$ is the unit vector in the transverse cross section, and

$$j\omega\mu \underline{H} \cdot \hat{\xi} = -\hat{\xi} \cdot (\nabla \times \underline{E}) = -\hat{\xi} \cdot \left[(\nabla_t + \hat{\xi} \frac{\partial}{\partial \xi}) \times \underline{E} \right]$$

or

$$j\omega\mu H_{\xi} = -(\hat{\xi} \times \nabla_t)\cdot\underline{E} = -\nabla_t\cdot\underline{E}\times\hat{\xi}$$
$$= \nabla_t\cdot\hat{\xi}\times\underline{E} \ . \qquad (53\underline{b})$$

Similarly,

$$j\omega\underset{\sim}{\varepsilon}\hat{\xi}\times\underline{E} + \hat{\xi}\times\underline{J}_s = -\frac{\partial}{\partial\xi}\underline{H}_t + \nabla_t H_{\xi} \qquad (53\underline{c})$$

$$j\omega\underset{\sim}{\varepsilon}E_{\xi} + J_{s\xi} = \nabla_t\cdot\underline{H}\times\hat{\xi} \ . \qquad (53\underline{d})$$

Substituting in turn (53d) into (53a) and (53c) into (53b), we obtain

$$-\frac{\partial}{\partial\xi}\underline{E}_t = j\omega\mu\underline{H}_t\times\hat{\xi} - \frac{1}{j\omega\underset{\sim}{\varepsilon}}\left[\nabla_t\,\nabla_t\cdot\underline{H}_t\times\hat{\xi} - \nabla_t J_{s\xi}\right]$$

$$= j\omega\mu\left[\underset{\approx}{1} + \frac{1}{k^2}\nabla_t\,\nabla_t\right]\cdot\left[\underline{H}_t\times\hat{\xi}\right] + \frac{1}{j\omega\underset{\sim}{\varepsilon}}\nabla_t J_{s\xi}$$

$$-\frac{\partial}{\partial\xi}\underline{H}_t = j\omega\underset{\sim}{\varepsilon}\left[\underset{\approx}{1} + \frac{1}{k^2}\nabla_t\,\nabla_t\right]\cdot\left[\hat{\xi}\times\underline{E}_t\right] . \qquad (46\underline{b})$$

These are the differential equations whose time-domain analogue is Kirchhoff's equation (46a). We can now write a spectral representation involving only the transverse modes :

$$\underline{E}_t(\underline{t},\xi) = \frac{1}{2\pi}\int_{-\infty}^{\infty} V(\xi|\beta_t)\,\underline{e}_t(\underline{t}|\beta_t)\,d\beta_t$$
$$\underline{H}_t(\underline{t},\xi) = \frac{1}{2\pi}\int_{-\infty}^{\infty} I(\xi|\beta_t)\,\underline{h}_t(\underline{t}|\beta_t)\,d\beta_t \qquad (47\underline{b})$$

in direct analogy with (47a). Comparison should also be made with (6a) in Sec. 1 : there x corresponds to our waveguides direction ξ, β_z is the scalar transverse wavenumber corresponding to our vector β_t, our modal amplitude V was e_o, and the transverse mode functions $\underline{e}_t$ were the exponential eigenfunctions $\exp(-j\beta_z z)$; in other words, the waveguide in Fig.1 is viewed as a rectangular guide of infinite cross section in the y-z plane. (Were the exciting source a dipole, our waveguide would be a cylindrical one of infinite cross section). We could also, in the problem of Sec.1, choose z as the waveguide direction and expand in terms of the x-y plane modes; these alternatives are thoroughly explored in the companion lectures by L.Felsen. The vector eigenfunctions are defined so as to be orthogonal:

$$\int_{-\infty}^{+\infty} \underline{e}_{tm}(\underline{t}|\underline{\beta}_{tm})\cdot\underline{e}^{*}_{tn}(\underline{t}|\underline{\beta}_{tn})d\underline{t} = \text{const.}\,\delta(\underline{\beta}_{tm} - \underline{\beta}_{tn}) \qquad (48\underline{b})$$

Without given details,[10] we state that substitution of (47b) in (46b), subject to (48b) with an appropriate constant, results in a pair of differential equations for each mode:

$$-\frac{d}{d\xi} V(\xi|\underline{\beta}_{tm}) = j\underline{\beta}_{tm} Z_m I(\xi|\underline{\beta}_{tm})$$

$$-\frac{d}{d\xi} I(\xi|\underline{\beta}_{tm}) = j\underline{\beta}_{tm} Y_m V(\xi|\underline{\beta}_{tm}) + I_{sm}, \qquad (49\underline{b})$$

where $Z_m(Y_m)$ is the modal wave impedance (admittance) which, in the special case of plane wave modes, is given by (23) or (26), and $I_{sm} = \int_S \underline{J}_s\cdot\underline{e}^{*}_{tm}dS$, where S is the total cross section. Although not as simple as the algebraic Ohm's law (49a) to which they correspond, these scalar ordinary differential equations are very much easier to solve than the vector partial differential equations (46b). The electrical engineer recognizes in them the transmission line -- with the modal amplitudes as voltage and current, and with I_{sm} the equivalent source (generator) current for the m-th mode -- and can instantly write the solution

$$V(\xi \mid \underline{\beta}_{tm}) = Z_m I_{sm} \left[e^{-j\beta_{\xi m}\xi} + \Gamma(\beta_{\xi m}) e^{j\beta_{\xi m}\xi} \right] \qquad (50\underline{b})$$

(with $\beta_{\xi m}^2 + \beta_{tn}^2 = k^2$), and analogously for I, which corresponds to (50a) in the time domain, and to (10) in the special case of Sec.1.

[Alternatively, one can differentiate (49b) once more with respect to ξ and obtain wave equations in V and I that correspond to (7) in Sec.1; these are then solved in terms of the boundary conditions, paralleling (8) and (9), to yield (50b)]. Synthesis, as in (51a) and (11), sums over the modes (50b):

$$\underline{E}_t(\underline{t},\xi) = \frac{1}{2\pi} \int_{-\infty}^{\infty} Z I_s \left[e^{-j\beta_\xi \xi} + \Gamma(\beta_\xi) e^{j\beta_\xi \xi} \right] \underline{e}_t(t \mid \underline{\beta}_t) d\underline{\beta}_t, \qquad (51\underline{b})$$

and analytic continuation, as in Fig.2, into the complex $\underline{k}_t$ plane produces the analogue of (52), except that branch cuts will now be present whenever a portion of the waveguide cross section extends to infinity.

The formalism is seen to reduce all electromagnetic problems in which at least one variable is separable to the solution of a standard vector eigenfunction problem in the cross section, and to the solution (50b) of a simple scalar differential equation for each modal amplitude. The solution (11) in Sec.1 can be written down on sight from the general formula (51b) by choosing ξ and $\underline{e}_t$ as already indicated, and by noting that $Z = Z_{TE_0 x} = \omega\mu/k_{xo}$, and $I_s = J_y/2 = 1/2j\omega\mu$ (because only one-half of J_y flows toward the slab, the other half flows in the positive x-direction).

b) Surface Wave Excitation I : Source Radiation

We now return to our example in Sec.1 to discuss the branch cut contribution, which represents the radiation field of the source. To simplify the problem, let us assume that the interface at x = 0 is a capacitive impedance sheet, $Z_s = -jX_s$ (X_s positive). Analytic continuation of (11) or (15b) into the complex k_z plane and completion of the contour as indicated in Fig.2 leads to the radiation field (total field less residues) in the form of a branch cut integral

$$E^{rad}(x,z) = \frac{1}{4\pi j}\int_b \frac{1}{k_x}\left[e^{-jk_x|x-h|} + \Gamma(k_z)e^{-jk_x(x+h)}\right]e^{-jk_z z}\,dk_z, \tag{54}$$

where we have dropped the now unneeded subscripts 0, and where E, as in (11) is E_y, all other components being zero. The path b extends around the branch cut in Fig.2. This integral cannot be directly evaluated; it can however be transformed through two changes in coordinates, and evaluated asymptotically by the saddle point method. We set $k_z = k_0 \sin\xi$, where ξ is the complex angle of reflection defined in (22) and in Fig.6a and b, and introduce the cylindrical coordinate system shown in Fig.18, where $z = r\sin\phi$, $x = r\cos\phi$. Substitution in (54) immediately gives

$$E^{rad}(r,\phi) = \frac{1}{4\pi j}\int_c \Big[e^{-jk_0(r-h\cos\phi)\cos(\xi-\phi)} \tag{55}$$

$$+\Gamma(\xi)e^{-jk_0(r+h\cos\phi)\cos(\xi-\phi)}\Big]d\xi,$$

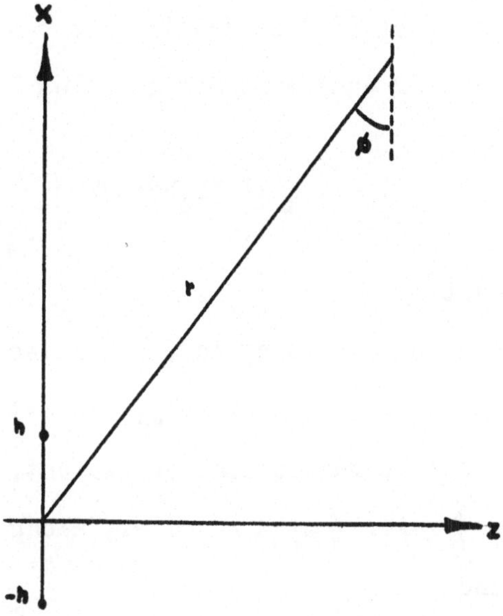

Fig. 18

where the path c in the ξ-plane is so chosen as to keep the phase stationary. Because the saddle point method is carefully treated in the companion lectures by L. Felsen, we do not need to discuss the evaluation of (55) in detail. The result, valid for $k_0 r \gg 1$ and $\phi \neq \pm \pi/2$ (and assuming no poles lie on or very near the contour), is

$$E^{rad}(r,\phi) \cong \frac{1}{\sqrt{8\pi k_0 r}} e^{-j(\pi/4+k_0 r)} \left[e^{jk_0 h\cos\phi} + \Gamma(\phi) e^{-jk_0 h\cos\phi} \right], \tag{56}$$

which is a combination of two cylindrical waves originating in the source and its image, respectively; the latter is modified by the geometric-optics reflection coefficient, which is $\Gamma(\xi)$ evaluated at $\xi = \phi$. For $\phi = \pm \pi/2$, i.e., along the interface at $|k_0 z| \gg 1$, we find

$$E^{rad}(z) \cong \frac{1}{\sqrt{2\pi |k_0 z|^3}} e^{j(\frac{\pi}{4} - |k_0 z|)} \left[(X_s/R_c)^2 + jk_0 h X_s/R_c - (k_0 h)^2/2 \right], \tag{57}$$

where the magnitude signs are needed to ensure that the wave travels away from the source for all z, and R_c is defined in (33c). Expressions (56) and (57) are referred to as the "space wave", which always accompanies the excitation of surface waves on open structures; if the structure is shielded, the branch cut integral in (54) becomes a sum over infinitely many poles -- the discrete set of "evanescent waves" that surround the source in a conventional waveguide.

The contour deformation involved in the asymptotic evaluation of the space wave affects the excitation of pole waves, as we shall now show. In the $\underset{\sim}{\xi}$-plane, Fig. 19, both Riemann sheets of the k_z-plane, the proper top sheet T as well as the improper bottom sheet B, are mapped as adjacent semi-infinite strips. The strip $(0 \leq \xi \leq \pi/2)$, η negative), for example, is marked B because (22) implies $\alpha_x = -\mathrm{Im}\left\{k_0 \cos(\xi + j\eta)\right\}$, so that α_x is negative (amplitude increasing away from surface). The original path along the β_z axis in Fig. 2, mapped by (22) into the heavy portions of the lines $\xi = -\pi/2$, $\eta = 0$, and $\xi = \pi/2$, is deformed into the steepest descent path c along which the imaginary part of the exponent in (55) is constant :

$$\cos(\xi - \phi_0)\cosh \eta = \text{const.}, \tag{58a}$$

where the subscript O emphasizes that the observation angle is held fixed. The saddle point occurs when

$$\frac{d}{d\underset{\sim}{\xi}} \cos(\underset{\sim}{\xi} - \phi_0) = 0 ,$$

which has the solution $\underset{\sim}{\xi} = \phi_0$. In words : at an angle of observation ϕ_0 in Fig. 18, the maximum contribution to the integral $E^{rad}(r, \phi_0)$ comes from the homogenous plane waves that leave the source and its image at the angle $\underset{\sim}{\xi} = \phi_0$. Since the contour must pass through its saddle point, (58a) becomes

$$\cos(\xi - \phi_0)\cosh \eta = 1 . \tag{58b}$$

As ϕ_0 varies from $-\pi/2$ to $\pi/2$, the contour is displaced parallel to itself from the dashed line on the left to that on the right. The region within the closed contour in Fig. 2 is now

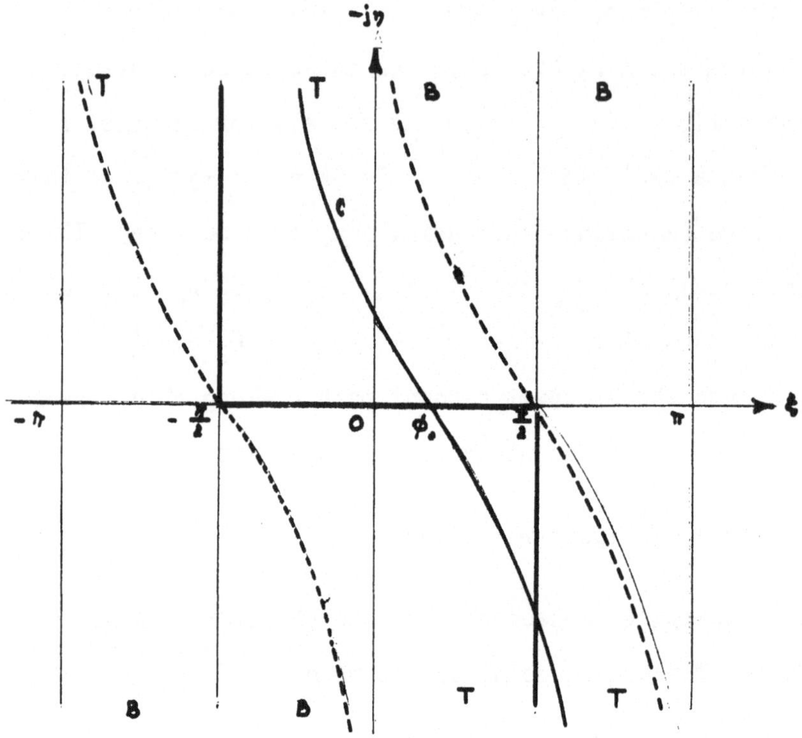

Fig. 19

bounded in Fig. 19 on the left by the heavy double step that maps the β_z axis, and on the right by the steepest descent path c . Poles that lie in this region contribute residues -- even the improper poles that lie on the included portion of the B strip -- and poles that lie to the right of c do not. This circumstance, as we shall see in Sec. 4(a), leads to the appearance, in a limited spatial domain, of leaky wave residues, and to the disappearance, from a limited spatial domain, of surface wave residues. We also recognize that, in the steepest descent representation, no poles that lie outside the region between the left- and right-hand dashed paths can ever contribute residues. This immediately excludes the improper pole types shown in Fig. 8d and e which, using (22), lie respectively on the line ($\xi = \pi/2$, η negative) and on the B strip to the right of that line. It also excludes the Zenneck wave (38) : using (22), this pole lies just to the right of the point $(\pi/2, 0)$ and, almost imperceptibly, below the line $\xi = \pi/2$ in Fig. 19; but (58b) has a slope of 45 degrees at $\xi = \phi_0$, and the right-hand dashed contour therefore passes to the left of the pole.

It is worth examining the effect of contour deformation in the k_z plane : using (22) and (58b) with $\phi_0 = \pi/2$, the right-hand dashed contour becomes the branch cut ($\beta_z = k_0$, α_z negative) shown in Fig. 20a. In deforming the branch cut from its original position in Fig. 2 to its present one, the shaded area has to be swept over, and as indicated in the cross section through the two-sheeted k_z plane, Fig. 20b, any pole (marked x) on the

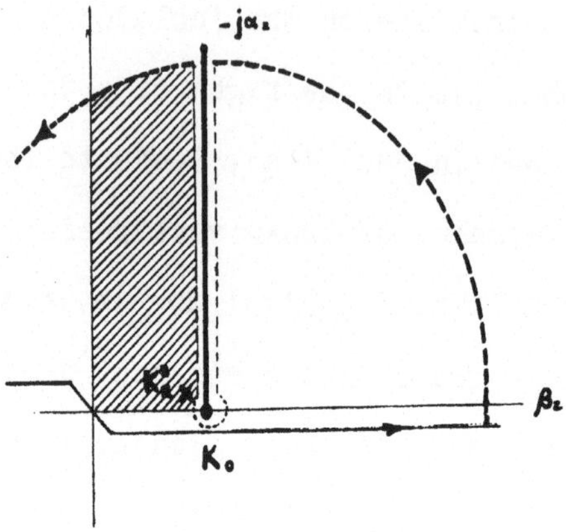

(a)

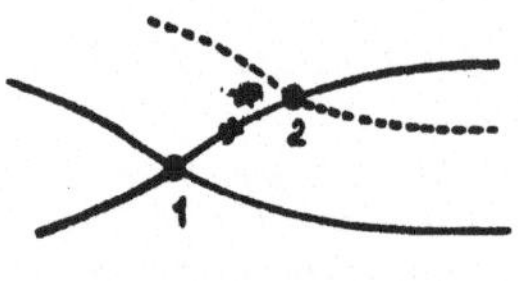

(b)

Fig. 20

top sheet when the branch cut is in its initial position 1 will find itself on the bottom sheet when the branch cut has moved to position 2, This clearly applies to the Zenneck wave pole k_z^z in Fig. 20a; because of its close proximity to the contour expressions (56) and (57) require additional terms, which lead to the Norton surface wave of Sec. 2(h) -- but no Zenneck wave residue shows up in this representation. In terms of the "spectral" representation of Fig. 2 and equation (54), the original branch cut contribution completely masks the Zenneck surface wave; in the case of a purely reactive interface, by contrast, the surface wave is by far the dominant part of the total field near the interface [at sufficiently far distances from the source -- see Sec. 4(a)] .

c) Surface Wave Excitation II : Influence of Source on Surface Wave Amplitude

Assuming the existence of a pole that lies within the modified contour of Fig. (19), we must add its residue $E^s(x,z)$ to the space wave (56) and (57) in order to represent the total field. On the capacitive impedance sheet, the pole is located, using (33b) and (33d), at $\alpha_x^s/k_o = R_c/X_s$, and we can write the residue (15) in the form

$$E^s(x,z) \cong (R_c/X_s)\, e^{-\alpha_x^s h}\, e^{-\alpha_x^s x + j(\frac{\pi}{2} - \beta_z^s z)} \tag{59}$$

where we assume, as in Sec. 1(d), that $(\alpha_x^s/k_o)^2 = (R_c/X_s)^2 \ll 1$.

The line source in Sec. 1 is represented by the delta

function $\delta(x-h)\,\delta(z)$ in (4) and (5a). What if we have a more complicated source, say one that extends over a finite distance in the z-direction, $\delta(x-h)f(z)$? Then its Fourier transform, which in the case of $\delta(z)$ is unity, will show up on the right-hand side of (7a) as the factor $F(\beta_z)$. This factor reappears in equations (9),(10), and (11), and in the form $F(k_z)$ (by analytic continuation) it multiplies the entire integrand in (54), and thus also the integrand in the residue calculation of Sec.1(d). Evaluated at the pole, the factor $F(\beta_z^s)$ multiplies the right-hand side of the surface wave (15) or (59). A particular f(z) is illustrated in Fig.21(a), which shows two line sources separated by approximately $\lambda/4$ and fed 90 degrees out of phase so as to produce, in the absence of the impedance sheet, a cardioid pattern. The lower half of this cardioid is incident on the surface at angles $-\pi/2 \leq \xi \leq \pi/2$, and the corresponding incident plave wave spectrum is obtained by extending ξ analytically beyond $\pm\,\pi/2$. The pattern itself, and therefore also the plane wave spectrum, turn out to be complex in the case of our two line-sources combination, i.e., the amplitude F of the spectrum as well as its phase are functions of β_z. Only the amplitude is sketched in Fig.21b and its value F_2 at $\beta_z = \beta_z^s$ multiplies the surface wave amplitude. The single line-source pattern is omnidirectional in ξ and its spectrum is therefore the dashed horizontal line with amplitude F_1.

The sharper we peak the source spectrum at β_z^s, the larger the amplitude of excitation; since β_z^s is in most applications close to k_o, F_2 is approximately the same as $F(k_o)$, and for all prac-

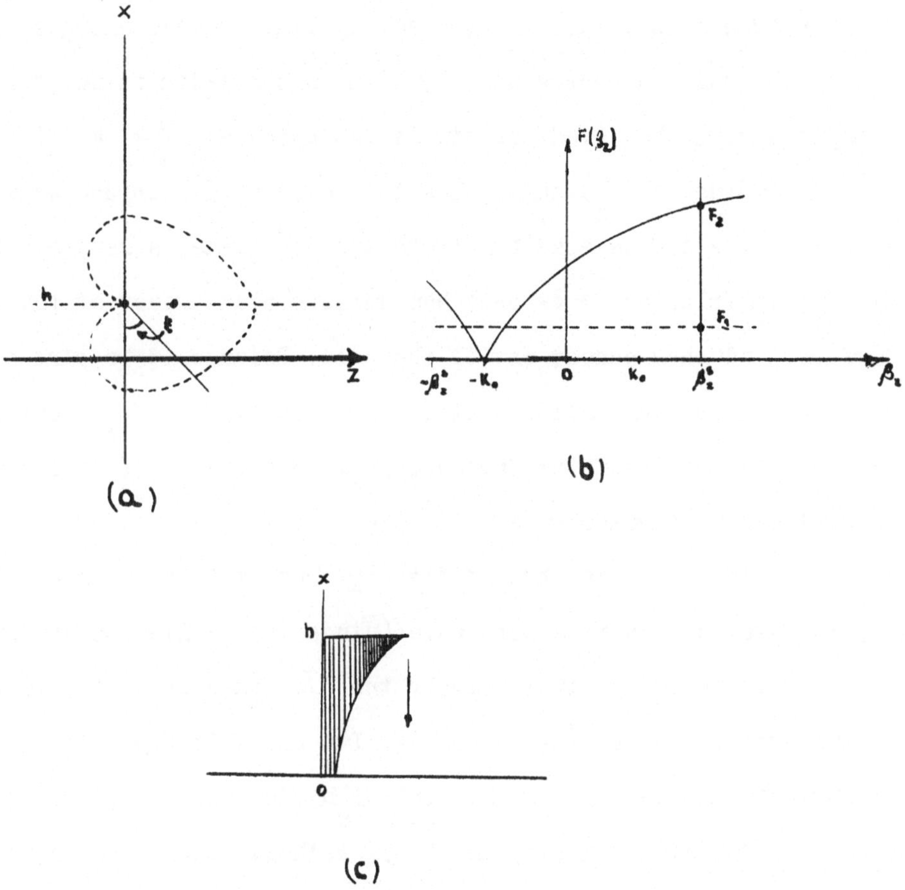

Fig. 21

tical sources the amplitude of the surface wave is therefore proportional to the endfire amplitude ($\zeta = \pi/2$) of the source pattern in free space. Instead of stacking line sources along z, we could also build a source distribution in the x-direction, for example a horn that produces strong radiation at $\zeta = \pi/2$.

Since $F(\beta_z^s)$ multiplies the residue, no surface wave exists if the source spectrum lacks the β_z^s line; a remote source, for example, produces only homogeneous plane waves at the interface, and thus cannot excite the pole. Because it is part of the source spectrum incident at the interface, the β_z^s wave in the x-direction is of the form $\exp(jk_x x) = \exp(\alpha_x^s x)$, which is an upside-down surface wave, Fig. 21c.

Its shape is that of Fig. 8d, but it is of course not a pole wave, nor is it a zero wave (since the reflection coefficient is infinite) -- it is simply the β_z^s wave prior to its resonance reflection by the interface. Its shape in Fig. 21c clearly suggests the reason for the $\exp(-\alpha_x^s h)$ dependence of the surface wave amplitude in (15) and (59). Although as a line in the continuous source spectrum it carries zero energy, the pole-exciting wave evidently couples, or funnels, energy from the source into the surface wave.

In launching a surface wave on a transmission line or antenna, we are usually interested in transferring maximum power from the source into the surface wave. The "efficiency of excitation" is defined as the ratio of the power in the surface wave to that in the total field, which in turn is the sum of the surface wave power and the power in the entire radiation field (56); calculation of the latter requires a laborious numerical integration. (The definition implies that the power flows in surface wave and radiation field are separable, and we shall prove in Sec. 3(d) below, that on a lossless structure this is in fact the case). Cullen[11] found for a line source that it excites most efficiently at a height $h = 1/\alpha_x^s$, an astonishing result in view of the exponential decrease with h of the surface wave amplitude in (59). The result can be interpreted as follows. Let $P^{rad} = \frac{1}{2} R^{rad} |J_y|^2$ be the power in the radiation field, with R^{rad} the radiation resistance of the source, and let P^s be the power in the surface wave, which is proportional to the square of the amplitude in (59) and thus to $\exp(-2 \alpha_x^s h)$. Then the excitation efficiency $P^s/(P^{rad} + P^s) = P^s/(\frac{1}{2} R^{rad} |J_y|^2 + P^s)$, and if now the radiation resistance decreases faster than the square of the surface wave amplitude, the efficiency increases. This in fact occurs as the line source is raised to its optimum height, but it should be pointed out that the effect is very frequency-sensitive. Source arrangements ordinarily used (open-mouthed waveguide, horn, or the configuration of Fig. 21a) are more broadband, i.e., R^{rad} is fairly constant, and the excitation efficiency therefore increases monotonically with the residue amplitude. Maximum efficiencies achievable in practice are around 90 percent, but in the usual antenna applications 70 percent are rarely exceeded.

A thorough treatment of the surface wave launching

problem can be found in the new book by Barlow and Brown.[12]

d) Surface Wave Excitation III : Orthogonality Relations

We will now show that, on a lossless structure, the power flow in the radiation field of the source is independent of that in the surface wave (no cross-coupling), and that if several surface waves are excited, their respective power flows are also independent of each other. We need the reciprocity theorem and apply it to an axially uniform waveguide of inhomogeneous and arbitrarily shaped cross section. Assume the guide contains a source $\underline{J}_s$, and two fields $\underline{E}$, $\underline{H}$ and $\underline{E}'$, $\underline{H}'$. We form the product

$$\nabla\cdot(\underline{E}'\times\underline{H}^*) = \underline{H}^*\cdot\nabla\times\underline{E}' - \underline{E}'\cdot\nabla\times\underline{H}^*$$

$$= -j\omega\mu\underline{H}'\cdot\underline{H}^* + j\omega\underset{\sim}{\varepsilon}^*\underline{E}'\cdot\underline{E}^* - \underline{J}_s^*\cdot\underline{E}' ,$$

where we have used Maxwell's equations (2c) and (2d). In the same manner,

$$\nabla\cdot(\underline{E}^*\times\underline{H}') = j\omega\mu^*\underline{H}'\cdot\underline{H}^* - j\omega\underset{\sim}{\varepsilon}\underline{E}'\cdot\underline{E}^* - \underline{J}_s'\cdot\underline{E}^*$$

Since the media are lossless $\underset{\sim}{\varepsilon} = \varepsilon = \varepsilon^*$ and $\mu = \mu^*$, and on addition we are left with

$$-\nabla\cdot(\underline{E}'\times\underline{H}^* + \underline{E}^*\times\underline{H}') = \underline{J}_s^*\cdot\underline{E}' + \underline{J}_s'\cdot\underline{E}^* , \qquad (60)$$

the differential form of the reciprocity theorem in a lossless medium.

Now let $\underline{E}$ and $\underline{E}'$ represent two surface wave modes m and n whose transverse components are of the form

$$\underline{E}_{tm}(\underline{t},z) = \underline{e}_{tm}(\underline{t})e^{-j\beta_{zm}z}$$
$$\underline{H}_{tm}(\underline{t},z) = \underline{h}_{tm}(\underline{t})e^{-j\beta_{zm}z} \quad (\text{with } \underline{h}_{tm} = Y_m\underline{e}_{tm}), \tag{61}$$

and likewise for n. Fig. 22<u>a</u> shows a cross section of particularly simple form : the interior region S_i represents, say, a dielectric rod, and the exterior region S_e extends to infinity or else is bounded by a lossless circular waveguide wall; is is assumed that $\underline{J}_s$ and $\underline{J}'_s$ are zero over the entire cross section. Integration of (60) over S_i produces 0 on the right-hand side. The left-hand side, with (61), can be written as

$$(\nabla_t \cdot + \hat{z}\cdot\frac{\partial}{\partial z})(\underline{E}_m x\underline{H}_n^* + \underline{E}_n^* x\underline{H}_m) =$$
$$= \nabla_t\cdot(\underline{E}_m x\underline{H}_n^* + \underline{E}_n^* x\underline{H}_m) - j(\beta_{zm} - \beta_{zn})\hat{z}\cdot(\underline{E}_{tm} x\underline{H}_{tn}^* + \underline{E}_{tn}^* x\underline{H}_{tm}) \tag{62}$$

(because the dot product removes the longitudinal components). In view of the two-dimensional divergence theorem, the integral over the first term is the line integral over the boundary s_i of S_i :

$$\int_{S_i} \nabla_t\cdot(\underline{E}_m x\underline{H}_n^* + \underline{E}_n^* x\underline{H}_m)dS_i = \oint \hat{n}\cdot(\underline{E}_m x\underline{H}_n^* + \underline{E}_n^* x\underline{H}_m)ds_i$$
$$= \oint (\hat{n}x\underline{E}_m\cdot\underline{H}_n^* + \hat{n}x\underline{E}_n^* x\underline{H}_m)ds_i \quad .$$

The $\hat{n}x\underline{E}$ products select the $\underline{E}$ components tangential to s_i, and the dot products with $\underline{H}$ then leave only the tangential $\underline{H}$ components. These are continuous across s_i, and if now we add the analogous integral over S_e, the contour s_i will be traversed again, but in the opposite direction, Over the exterior contour,

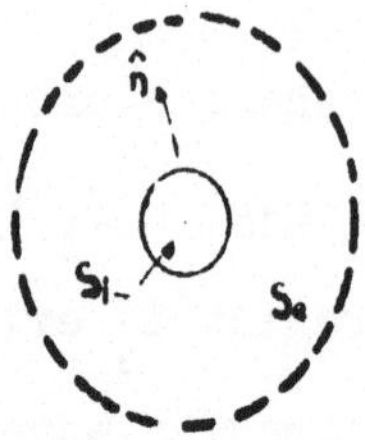

Fig. 22a)

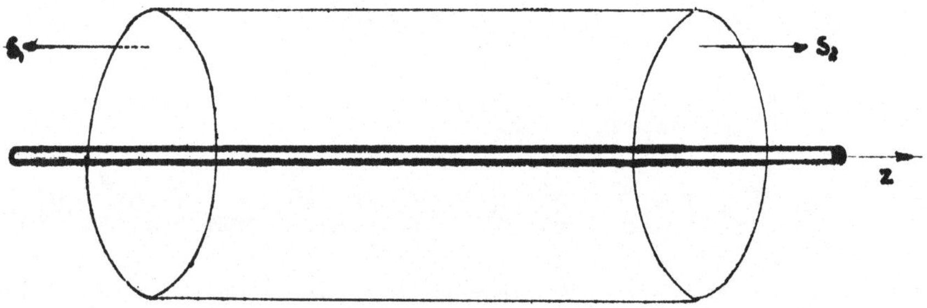

Fig. 22b)

the cross products $\hat{n}x\underline{E}$ vanish on the lossless waveguide wall or, if the cross section is infinite, they vanish because of the radiation condition. We are therefore left with only the second term of (62), so that over the entire cross section $S = S_i + S_e$

$$(\beta_{zm} - \beta_{zn}) \int_S (\underline{e}_{tm} x \underline{h}^*_{tn} + \underline{e}^*_{tn} x \underline{h}_{tm}) \cdot d\underline{S} = 0 \,, \qquad (63\underline{a})$$

where the common exponential factor in (61) has been cancelled. Because of reflection symmetry, we can also postulate a pair of modes m and n of which only the first depends on z as in (61), while the second depends on z as

$$\begin{aligned} \underline{E}_{tn}(\underline{t},z) &= \underline{e}_{tn}(\underline{t}) e^{j\beta_{zn} z} \\ \underline{H}_{tn}(\underline{t},z) &= -\underline{h}_{tn}(\underline{t}) e^{j\beta_{zn} z} , \end{aligned} \qquad (64)$$

where the minus sign is required because, as we showed in connection with equations (26), Y_m in (61) changes sign on reversal of the direction of wave travel. In lieu of (63a) we now obtain

$$(\beta_{zm} + \beta_{zn}) \int_S (-\underline{e}_{tm} x \underline{h}^*_{tn} + \underline{e}^*_{tn} x \underline{h}_{tm}) \cdot d\underline{S} = 0 \qquad (63\underline{b})$$

The factors multiplying the integrals (63) can be cancelled provided $\beta_{zm} \neq \beta_{zn}$, i.e., the modes must not be degenerate. Subtracting (63b) from (63a) gives the desired orthogonality relation between surface wave modes :

$$\int_S \underline{e}_{tm} x \underline{h}^*_{tn} \cdot d\underline{S} = 0 \,. \qquad (65)$$

Our derivation holds equally well for any number of contours s_j of arbitrary shape. Since the axial power flow of the two surface wave modes is given by one-half of the real part of $(\underline{e}_{tm}+\underline{e}_{tn}) \times (\underline{h}^*_{tm}+\underline{h}^*_{tn})$, relation (65) (in which m and n are of course interchangeable) eliminates all terms containing both m and n; the total power flow is therefore simply the sum of that carried by each mode separately.

To prove orthogonality between the radiation field and any surface wave mode m, we assume a source $\underline{J}_s$ located in the cross section z = 0 somewhere within the cylinder, Fig. 22b, and write the field over the cross section S_2 at $z = z_2$ in the form

$$\underline{E}^+ = \underline{E}^{rad+} + \sum_n a_n^+ \underline{E}_n^+$$
$$\underline{H}^+ = \underline{H}^{rad+} + \sum_n a_n^+ \underline{H}_n^+ \,, \qquad (66a)$$

where $\underline{E}^+$ refers to the total field propagating in the z-positive half space, consisting of the radiation field [i.e., the continuous spectrum represented by a branch cut integral, for example (54)] and the finite sum of all possible surface wave modes. For the cross section S_1 at $z = z_1$, expressions (66a) hold with the plus signs replaced by minus signs to indicate propagation in the z-negative half space :

$$\underline{E}^- = \underline{E}^{rad-} + \sum_n a_n^- \underline{E}_n^- \,, \qquad (66b)$$

and analogously for $\underline{H}^-$. We use the reciprocity relation (60) with $\underline{E}$, $\underline{H}$ again representing a surface wave mode $\underline{E}_m^+$ of the form (61) but with $\underline{E}'$, $\underline{H}'$ now the total field (66) of the source. Since $\underline{E}$, $\underline{H}$

is a solution of the homogeneous (source-free) wave equation, we set $\underline{J}_s = 0$ and can thus drop the prime in $\underline{J}'_s \neq 0$. Application of the three-dimensional reciprocity theorem to (60) produces

$$- \oint (\underline{E}' x \underline{H}_m^{+*} + \underline{E}_m^{+*} x \underline{H}') \cdot d\underline{S} = \int \underline{J}_s \cdot \underline{E}_m^{+*} \, dV \qquad (67)$$

over the cylinder. The integration over the cylinder mantle at infinity is zero because of the radiation condition; or if the cylinder radius is finite, the integral is zero because of the boundary conditions on the perfectly-conducting waveguide wall. The left-hand side of (67) therefore consists of the integrals over S_1 and S_2 only, which are

$$-e^{j\beta_{zm}z_1} \int_{S_1} (\underline{E}_t^{rad-} \times \underline{h}_{tm}^{*} + \underline{e}_{tm}^{*} \times \underline{H}_t^{rad-}) \cdot d\underline{S}$$

$$-e^{j\beta_{zm}z_2} \int_{S_2} (\underline{E}_t^{rad+} x \underline{h}_{tm}^{*} + \underline{e}_{tm}^{*} \, x \underline{H}_t^{rad+}) \cdot d\underline{S}$$

$$- 2a_m^{+} \int_{S_2} \underline{e}_{tm} \, x \underline{h}_{tm}^{*} \cdot d\underline{S}),$$

where we have made repeated use of (61),(64),(65) and (66), and of the fact that the dot product with $d\underline{S}$ takes out the axial field components; also, in connection with the second term (missing because it turns out to be zero) and the fourth, of the fact that the surface wave admittance is real in the direction of propagation. Since z_1 and z_2 are arbitrary, we can, following Goubau [13], let $z_1 = -\infty$ and $z_2 = \infty$, where the radiation condition renders the integrands of the first and third term zero, and thus for $z \geq 0$

$$\int_{S_2} (\underline{E}_t^{rad+} x\underline{h}_{tm}^{*} + e_{tm}^{*} x\underline{H}_t^{rad+}).d\underline{S} = 0 \, , \qquad (68\underline{a})$$

and similarly for $z \leqslant 0$. Only the second term now remains to be equated to the right-hand side of (67) :

$$-2a_m^{+} \int_{S_1} \underline{e}_{tm} x\underline{h}_{tm}^{*} \cdot d\underline{S} = \int \underline{J}_s \cdot \underline{E}_m^{+*} dV \, , \qquad (68\underline{b})$$

from which the amplitude a_m^{+} of the surface wave traveling to the right can be determined without recourse to a residue calculation.

Had we chosen $\underline{E}$, $\underline{H}$ in (60) to represent the leftward traveling surface wave $\underline{E}_m^{-}$ of the form (64), we would have obtained, in lieu of (68a),

$$\int_{S_2} (\underline{E}_t^{rad+} x\underline{h}_{tm}^{*} - \underline{e}_{tm}^{*} x\underline{H}_t^{rad+}).d\underline{S} = 0 \, , \qquad (68\underline{c})$$

and also an expression for a_m^{-} analogous to (68b). Addition of (68a) and (68c) produces the desired orthogonality relation between radiation field and surface wave mode for $z \geqslant 0$:

$$\int_{S_2} \underline{E}_t^{rad+} x\underline{h}_{tm}^{*} .d\underline{S} = 0 \, ; \qquad (69)$$

analogous integrals over S_1 produce the orthogonality condition for $z \leqslant 0$.

If the structure has losses, (65) and (69) no longer hold; in their stead, one obtains expressions identical with (65) and (69) except that the asterisks are removed, so that independence of power flow in the several surface wave modes and the radiation field is no longer implied. For a proof of these modified orthogonality relations, see Collin.[14]

e) Cylindrical Surface Waveguides

Open cylindrical waveguides support the axial cylindrical modes briefly mentioned in Sec. 2(e). We use the separability condition (39), but interchange k_x with k_z in order to retain z as the direction of propagation

$$k^2 = k_z^2 + k_\rho^2 \tag{70}$$

(Note that (39) or (70) do not involve the azimuthal wavenumber k_θ, whereas the separability condition in rectangular coordinates involves all three wavenumber components). With ε_0 the dielectric constant of the outside and ε_1 that of the inside medium, and $\mu_0 = \mu_1$, successive application of (70) to both media produces, upon subtraction,

$$k_0^2(\varepsilon_r - 1) = k_{\rho 1}^2 - k_{\rho o}^2 \tag{71}$$

where in the special case of a surface wave $k_{\rho 1} = \beta_{\rho 1}$ and $k_{\rho o} = -j\,\alpha_{\rho o}$, so that (71) becomes formally identical with (13c) for a dielectric slab. Relations (13c) and (71) imply that ε_1 must be larger than ε_0 for a surface wave to exist; a plasma column in air (without magnetic field), or an empty tube in a dielectric bloc (through which electrons can be injected that excite Čerenkov radiation in the surrounding dielectric) therefore support no surface waves.

The functional form of the surface wave in the region $\rho > \rho_0$ (ρ_0 is the radius of the structure) is

$$H_n^{(2)}(-j\alpha_{\rho 0}\rho)e^{-j(n\theta+\beta_z z)}$$

for the E_z and H_z components; the other components depend also on derivatives of the Hankel function.[6,14] We have written n in lieu of k_θ to indicate that, as in shielded waveguide, the azimuthal periodicity must be an integer. For $\alpha_{\rho 0}\rho \gg 1$, the Hankel function approximates $\exp(-\alpha_{\rho 0}\rho)/\sqrt{k_0\rho}$. (If the surface waveguide is surrounded by an outside cylindrical shield with a radius larger than ρ_0, a Hankel function of the first kind must be added, corresponding to the superposition of two exponentials in the rectangular shielded waveguide with slab, as explained in conjunction with Fig. 11c). In the region $\rho < \rho_0$, the appropriate functions are the cylindrical equivalents of the sine and cosine, i.e., the Bessel and Neumann functions. In the case of a homogeneous dielectric or metal wire, only the former can be used since the latter has a singularity at $\rho = 0$. In the case of a dielectric mantle surrounding a metal core, the two are superimposed. Matching of the fields at ρ_0 is equivalent to writing the transverse resonance relation in the cylindrical cross section, as described in the Waveguide Handbook.[6] The result[14] shows that for n = 0 (no circumferential field varations), either E_z or H_z are zero so that purely TE and TM waves are possible; the electric field lines of the latter are shown in Fig. 23a. On a dielectric rod, the lowest TE and TM modes are both cut off at $2\rho_0/\lambda_0 = 0.613$; on the dielectric-coated wire, the lowest TE mode is also cut off below a minimum thickness of dielectric, while the lowest TM mode has no cutoff -- just as in the correspon-

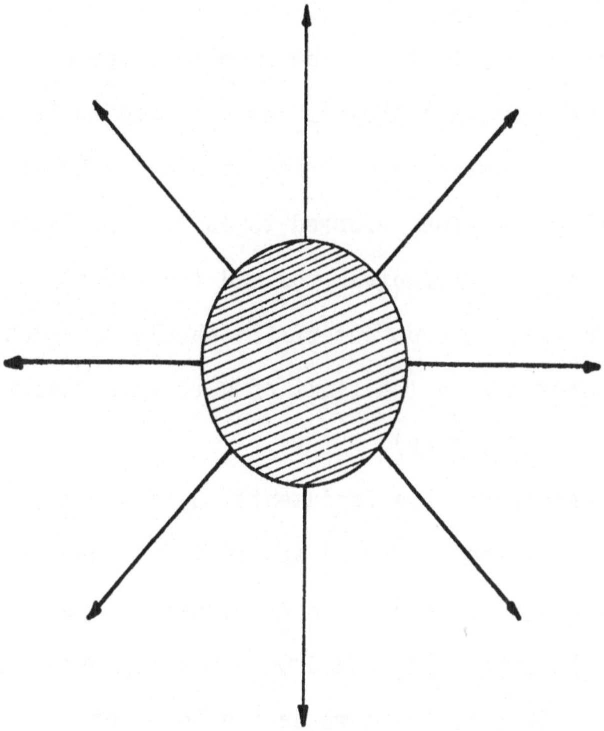

(a)

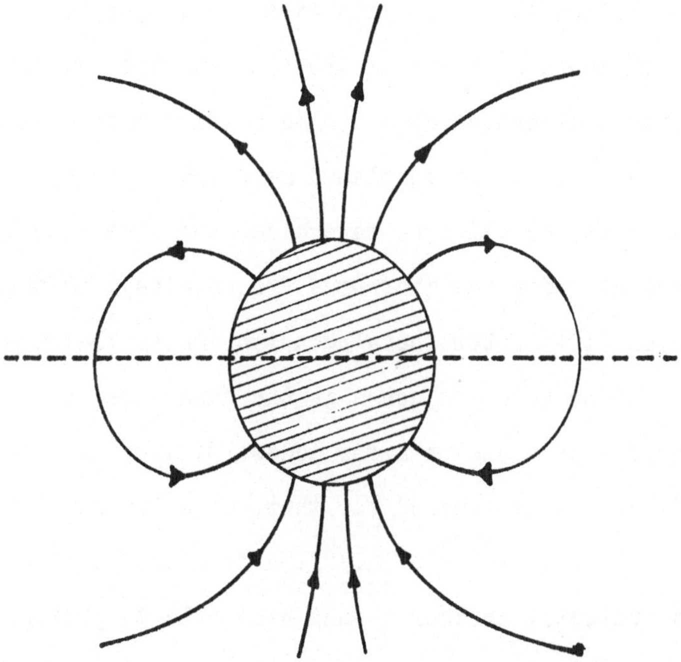

(b)

ding case of the dielectric slab on metal. For n = 1, both E_z and H_z are nonzero, and these modes are said to be 'hybrid'; because of their transverse eletric field distribution, shown in Fig. 23b, they are also referred to as "dipole" modes. The lowest hybrid mode on the dielectric rod and the lowest on the dielectric-covered wire have no cutoff (it should be pointed out that the latter does not in the limit of zero wire radius become identical with the former).

Operated in its fundamental hybrid mode, the dielectric rod has important antenna applications (Sec. 4(c)); sliced in half along the plane that is everywhere perpendicular to the electric field lines and placed on a metal ground sheet (dashed line in Fig. 23b), it is known as the "dielectric image line". The thin dielectric mantle with wire core, operated in its fundamental TM mode, is known as the "Goubau line", which has found specialized application as a single-conductor transmission line.[13] (Sharp bends, or obstacles placed near the surface, cause transmission losses on all open waveguides and generally limit their usefulness at radio and microwave frequencies). In the optical range, dielectric fibers have been fabricated that model the dimensions of the rods and cones in the human eye; they are so thin as to support only one or a few modes. By illuminating one end of the fiber with a focused ray that, upon entering the fiber, strikes the air-dielectric interface from within at an angle larger than critical, Snitzer[15] has been able to photograph at the other end the typical cross sectional energy distribution of surface-wave modes singly and in combination. When the angle of incidence was less than critical, he obtained photographs of leaky waves, as might be expected from our discussion in Sec. 2(g). Flexible fibers thick enough to support hundreds or thousands of modes, which in the aggregate produce a single light beam trapped inside the dielectric, are used for carrying light spots to inaccessible places (e.g., the stomach), for scrambling or unscrambling optical images, and as components in optical computers.

By examining the poles and branch cuts in the complex k_z plane, we can quickly survey the modal properties of a large variety of surface waveguides. Consider first the dielectric rod in Fig. 24a, excited in a single TM mode marked by the cross located somewhere on the real axis between the wavenumbers k_0 and k_1 of the outside and inside medium, respectively. In the left-hand column, the inside medium is allowed to become at first slightly lossy (Fig. 24b), then very lossy (i.e., a good conductor, Fig. 24c), and finally a perfect conductor, Fig. 24d. The corresponding power factor, defined as in Sec. 2(d) by $h_1 = \sigma_1/\omega\varepsilon_1$, increases from 0 to ∞, while $k_1 = \omega\sqrt{\mu_1\varepsilon_1(1-jh_1)}$ [see eqn. (2d)] moves off the real axis to infinity via the 45 degree line [because for $h_1 \gg 1$, $k_1 \cong (1-j)\sqrt{\omega\mu_1\sigma_1/2}$] In the right-hand column, the outside medium becomes slightly lossy (Fig. 24b'), then very lossy (i.e., a good-conductor wall, Fig. 24c'), and finally a perfect-conductor wall (Fig. 24d'), with h_0 and k_0 changing correspondingly. The branch cut is associated with k_0 and therefore stays put in the left-hand column in the position of Fig. 2, while in the right-hand column it gradually disappears toward the upper right, contributing less and less of a source radiation

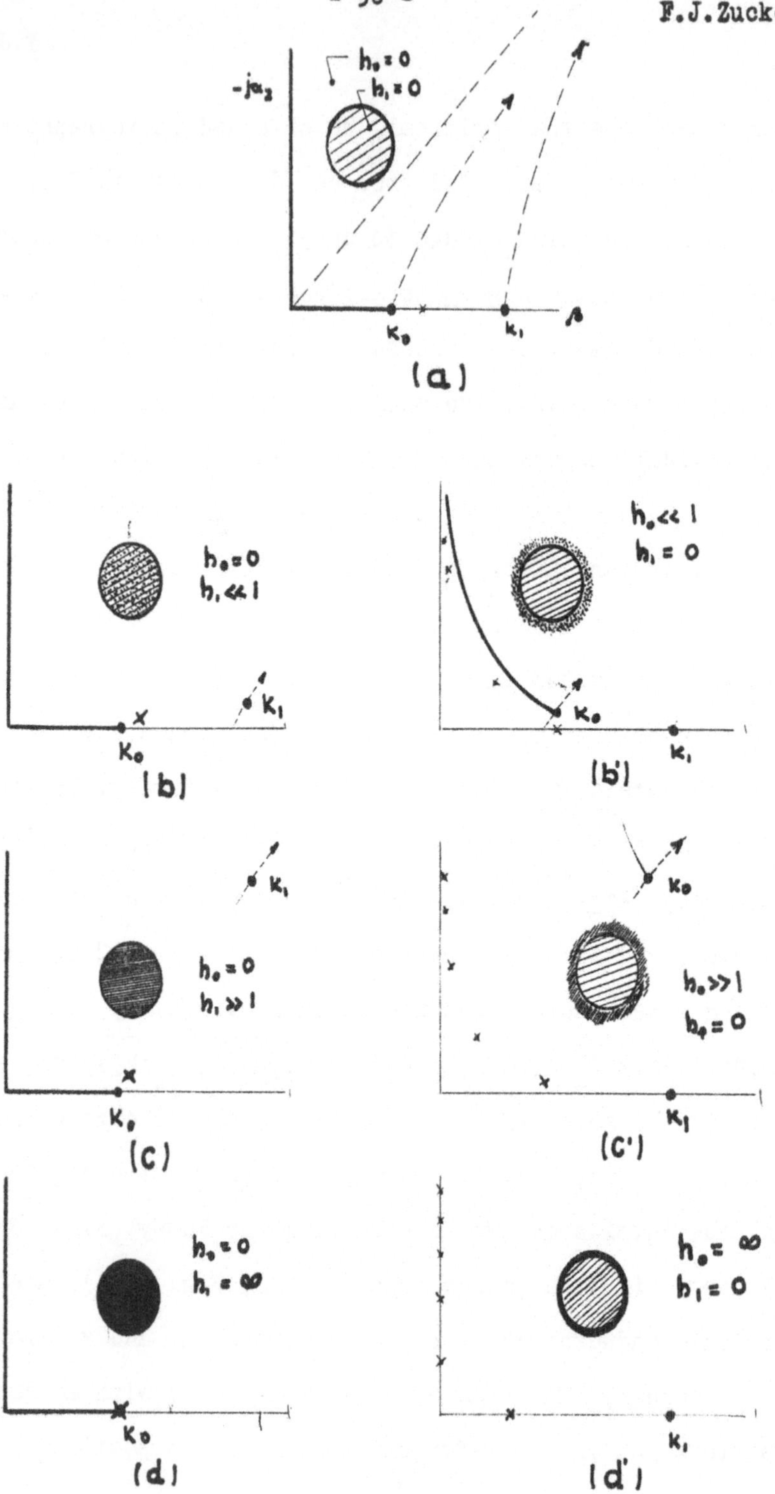
-jα₂
h₀ = 0
h₁ = 0
k₀
k₁
(a)
h₀ = 0
h₁ << 1
K₀
K₁
(b)
h₀ << 1
h₁ = 0
K₀
K₁
(b')
h₀ = 0
h₁ >> 1
K₀
K₁
(c)
h₀ >> 1
h₁ = 0
K₀
K₁
(c')
h₀ = 0
h₁ = ∞
K₀
(d)
h₀ = ∞
h₁ = 0
K₁
(d')

Fig. 24

field as it does so (because of increasing α_z), until in Fig.24d' the continuous spectrum has vanished altogether, leaving only a discrete one.

The TM surface wave pole of Fig. 24a moves slightly off the real axis as the dielectric rod becomes lossy in b, but is found to approach it again in c as the increasing σ_1 turns the rod into a conducting wire. Known as the Sommerfeld wire wave,[16] this mode is the analogue of the Zenneck wave over a plane conductor. Note, however, that in contrast to the Zenneck pole, which is seen in Fig. 20a to be located to the left of k_0, the Sommerfeld pole in Fig. 24c is located to its right; this slowing down of the surface wave on a wire can be interpreted in terms of the additional reactance contributed by the cylindrical curvature of the interface. Because of its location, the Sommerfeld pole contributes a residue even when the branch cut is drawn as in Fig. 20a for the asymptotic evaluation of the field along the surface, and it is in fact easy to observe the Sommerfeld wave experimentally.[17] (Since the pole is very close to k_0, $\alpha_{\rho o}$ is small, and in (15) we showed -- for a TE wave, but the relation holds also for a TM wave -- that loose binding of the surface wave in turn implies a low residue amplitude. The excitation efficiency is correspondingly low, and the direct source radiation therefore high; to suppress it, one might place a second wire in close proximity to the first and excite it 180 degrees out of phase : then the radiation fields of the two sources cancel, while the two Sommerfeld waves combine into a single coupled mode -- but this, as Goubau pointed out [13], is of course the ordinary two-wire wave!) When the wire at last becomes a perfect condictor, Fig. 24d, the pole becomes a TEM wave that coincides with k_o, thus $\alpha_{\rho o}$ is zero and the residue amplitude is zero as well. (The two-wire mode, whose residue is not of the form (15), of course does not vanish on perfect conductors). In the right-hand column, losses in the outer medium also move the pole off the real axis, and at the same time a finite set of new poles appears in the vicinity of the branch cut,[18] Fig. 24b'. As the outer medium becomes a good conductor in c', the number of new poles increases rapidly while the branch cut contribution gradually vanishes, until in d', the dielectric-filled shielded waveguide, the new poles form an infinite set of evanescent waves while the original pole has again become a purely propagating wave (which is faster than on the unshielded rod).

f) Periodic Structures

Because surface waves on periodic structures are frequently of practical importance -- on Yagis and other endfire antennas, along optical gratings, in traveling wave tubes and linear accelerators -- we now give a brief introduction to this topic. Our starting point, Floquet's theorem, states that the fields associated with any wave in a uniform waveguide with periodic boundary conditions differ from one cross section to another located any number of periods p away by only a complex constant. If we assume a modal wave k_z of the form

$$\underline{E}(x,y,z|k_z) = \underline{E}(x,y,z)e^{-jk_z z} \qquad (72\underline{a})$$

in the cross section at z, then in a cross section at z' any number n periods away the field is

$$\underline{E}(x,y,z'|k_z) = \underline{E}(x,y,z')e^{-jk_z z'} \quad . \qquad (72\underline{b})$$

With $z' = z + np$, and noting that the invariance of an infinite periodic structure under displacement along its length by integral multiples of p requires $E(x,y,z) = \underline{E}(x,y,z + np)$, (72b) becomes

$$\underline{E}(x,y,z+np|k_z) = \underline{E}(x,y,z)e^{-jk_z z - jnpk_z} \quad ,$$

which on comparison with (72a) proves the theorem, (As Bouwkamp has pointed out, Floquet's theorem does not imply that solutions of the exponential form (72a) are the only possible ones).

We can write (72) in terms of its Fourier components :

$$\underline{E}(x,y,z)e^{-jk_z z} = \sum_n \underline{e}_n(x,y)e^{-j(k_z^o + \frac{2\pi n}{p})z} \quad , \qquad (73)$$

where $n = 0, \pm 1, \pm 2, \ldots$, and k_z^o is the propagation constant of the dominant term ($n = 0$), k_z^o and the coefficients $\underline{e}_n$ are determined by the boundary conditions. Each of the space harmonics $k_{zn} = k_z^o + 2\pi n/p$ have their own phase velocity ω/β_{zn}, where β_{zn} is the real part of k_{zn}, but the group velocity $d\omega/d\beta_{zn} = d\omega/d\beta_z^o$ is identical for all of them; were it not so, no signal however narrow-band could be transmitted intact through a periodic structure, nor could energy be transferred from a charged particle to an r.f. field or vice versa.

In Sec. 2(c) we calculated the α_{xo}^{s} (and thus the β_z^s) of the surface wave on an interface characterized by the impedance (34), which approximates the surface impedance of a corrugated sheet. Were we to choose a physically more realistic interface than (34), with a finite number of grooves per wavelength in which higher-order modes are excited, then our field representation above the interface would be of the form (73), with $p = t+g$ (see Fig. 9). k_z^o and $\underline{e}_n$ are then evaluated by matching (73) to the groove fields and the tooth surfaces. So long as the structure is lossless, and p is small enough (see below), k_z^o turns out to be pure real, β_z^o, and in fact quite close to the β_z^s value based on the average-impedance approximation. Fig. 25a shows a spectrum of this type, with the "carrier" β_z^o and some of its "sidebands" (usually of much lower amplitude) β_{zn}. The $n = 1$ sideband (harmonic) at the extreme right corresponds to a very slow (and thus very tightly bound) inhomogeneous wave; since its β_z is positive, the wave travels toward the right. The $n = -1$ harmonic (usually higher in amplitude than $n = 1$) is located to the left of $-k_o$ provided $2\pi/p > \beta_z^o + k_o$ (see the figure), and in that case corresponds to an inhomogeneous wave traveling toward the left. The total wave in Fig. 25a is a surface wave; the term as introduced in Sec. 1 and used since explicitly refers to a wave that meets the boundary conditions, and we therefore do not individually apply it to β_z^o and the harmonics. This total surface wave is orthogonal, in the sense of Sec. 3(d), to the radiation field of a source exciting the periodic interface,

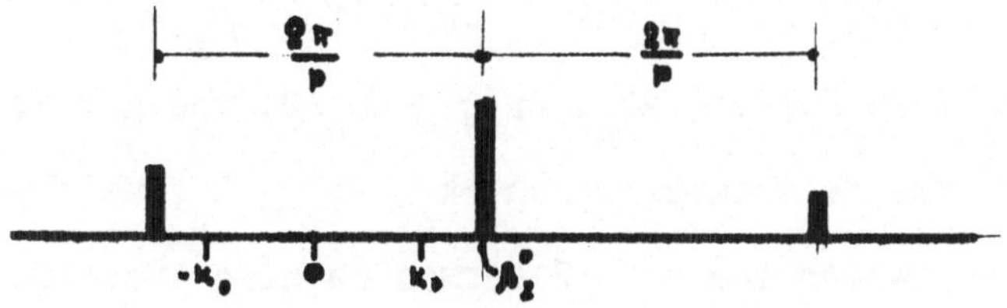

(a)

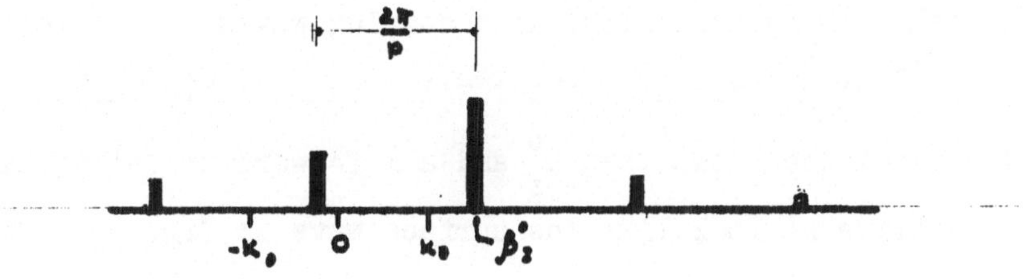

(b)

Fig. 25

but β_z^o and the harmonics are not orthogonal among each other.

When $2\pi/p < \beta_z^o + k_o$, which for $\beta_z^o \cong k_o$ can be written in the form $2\pi/p < 2k_o$, or $p > \lambda_o/2$, the n = -1 harmonic moves into the fast-wave region $-k_o \leqslant \beta_z \leqslant k_o$, as shown in Fig. 25b (in which the n = ±2 harmonics also appear). If the corrugated surface is shielded by means of a metal plate placed at some height above it, the spectrum of Fig. 25b applies directly, and the fast-wave harmonics in fact play an important role in certain traveling wave tubes. If, on the other hand, the structure is open, then the fast waves radiate energy off the surface and this converts all of the spectral terms, the fundamental as well as the harmonics, into complex waves. The detailed analysis [19] of an impedance sheet with sinusoidally varying inductive reactance shows that k_z^o and all forward-traveling harmonics (positive n) look like the surface wave in Fig. 8b on a lossy interface (even though our sheet is purely reactive); the backward-traveling harmonics to the left of $-k_o$ (n = -2 in Fig. 25b) also look like Fig. 8b, only reversed in direction (i.e., imaged with respect to the normal to the surface). The n = -1 harmonic as it passes $-k_o$ at $p \cong \lambda_o/2$ into the fast-wave range, looks like the complex-pole wave in Fig. 8c with reversed direction, i.e., radiating toward the left. With p increasing further (or, alternatively, the frequency increasing so that $-k_o$ and k_o move further out from $\beta_z = 0$), the angle of radiation with the surface increases until at $\beta_z = 0$ the radiation is broadside. As the n =-1 harmonic moves into the range $0 < \beta_z \leqslant k_o$, it now radiates toward the upper right, and by visualizing the clockwise rotation of the Fig. 8c wave from its original backward-radiating position we find that after passage through broadside it will now look like the leaky wave in Fig. 8f; at the same time, the n =-2 harmonic now also passes into the fast-wave region. This analysis has been used to explain the "Wood's anomalies" on certain optical diffraction gratings, and the radiation mechanism of "modulated" surface wave antennas [Sec. 4(c)] .

4. RADIATION OF SURFACE AND LEAKY WAVES

To prepare the way for a discussion of surface- and leaky-wave antennas in Secs. 4(c) and (d), respectively, we first examine in 4(a) the detailed field behavior in the vicinity of an interface that supports a surface or leaky wave, and in 4(b) derive a general formula for the radiation of traveling wave antennas.

a) Spatial Domains in which Surface and Leaky Waves are Dominant

We return to the problem of the line source exciting a dielectric slab on metal, and recall that the "spectral" representation of the field, developed in Sec. 1, consists in a finite number (or zero) of poles and a branch cut contribution. Let the slab be just thick enough to support a single pole; its residue gives rise to a surface wave defined in one entire space quadrant [Sec. 1(c)], while the branch cut integral represents a "remainder" field that cannot be directly calculated. By deforming the branch cut into a steepest descent path, we found in Sec. 3(b) that asymptotic evaluation of the integral results in a direct and a reflected space wave from the source. At the same time, however, the shift in contour captured some poles not contained in the spectral representation, and excluded others that were; the implications of this situation for surface- and leaky-wave poles must now be examined.

A surface wave pole β_z^s located as shown in Fig. 2 is transformed by (22) into $\underset{\sim}{\xi}^s = \xi^s + j\eta^s$ in Fig. 26a, where $\xi^s = \pi/2$, and η^s is positive [as explained in the lines immediately following (22)] . The corresponding saddle point ϕ_0^s is found by inserting in (58b) the values of ξ^s and η^s :

$$\cos\left(\frac{\pi}{2} - \phi_0^s\right) \cosh \eta^s = 1 \,, \tag{74a}$$

or, using Fig. 6a and remembering that for a surface wave $\beta = \beta_z^s$,

$$\sin \phi_0^s = k_0 / \beta_z^s \;. \tag{74b}$$

Only when the steepest descent path is to the right of the one shown in Fig. 26a will the pole $\underset{\sim}{\xi}^s$ contribute a residue; in this asymptotic representation, therefore, the surface wave exists only in the wedge-shaped region $\phi_0^s \leq \phi_0 \leq \pi/2$, as shown in Fig. 26b. The line c is labeled thus for two disparate reasons : because it marks the critical angle separating the spatial domains without and with surface wave, and because the surface wave phase velocity v_c along c is

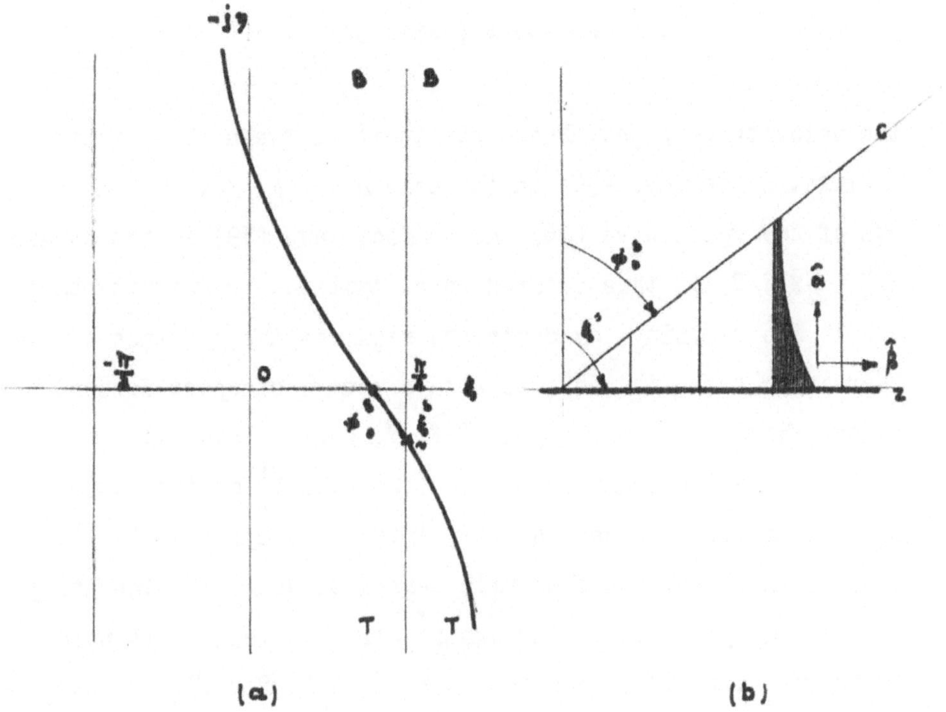

Fig. 26

$$v_c = \omega/\beta_c = \omega/\beta_z^s \cos(\tfrac{\pi}{2} - \phi_0^s)$$

$$= \left[\text{using (74b)}\right] \omega/k_0 = c,$$

the velocity of light. Phase continuity between the velocity-of-light space wave (56) in the domain $0 \leqslant \phi_0 \leqslant \phi_0^s$, and the sum of the space wave (56) and surface wave (59) in the domain $\phi_0^s \leqslant \phi_0 \leqslant \pi/2$, is thus assured on c. Amplitude continuity on c and in its vicinity, by contrast, requires the presence of higher-order terms in (56), as always when a steepest descent path passes through or near a pole (cf. Felsen's lectures).

In the leaky wave case, the pole k_z^{ℓ} on the bottom sheet in Fig.2 is mapped by (22) into the complex angle ξ^{ℓ}, located in Fig.27a on the strip marked B, Not until the steepest descent path passes through ϕ_0^{ℓ} will the leaky wave pole contribute a residue, and the leaky wave in Fig.27b is therefore also restricted to a wedge-shaped region bounded by c. Instead of (74a) we now have

$$\cos(\xi^{\ell} - \phi_0^{\ell})\cosh\eta^{\ell} = 1,$$

and, with the help of Fig.6a,

$$\cos(\xi^{\ell} - \phi_0^{\ell}) = k_0/\beta^{\ell}, \qquad (75)$$

where β^{ℓ} is the net phase constant β of the leaky wave [see sec.2(a)]. The leaky wave phase velocity along c is then $v_c = \omega/\beta_c = \omega/\beta^{\ell}\cos(\phi_0^{\ell} - \xi^{\ell}) = \omega/k_0 = c$, as before. In this representation, the leaky wave nowhere violates the radiation condition.

The field along a reactive interface that supports a

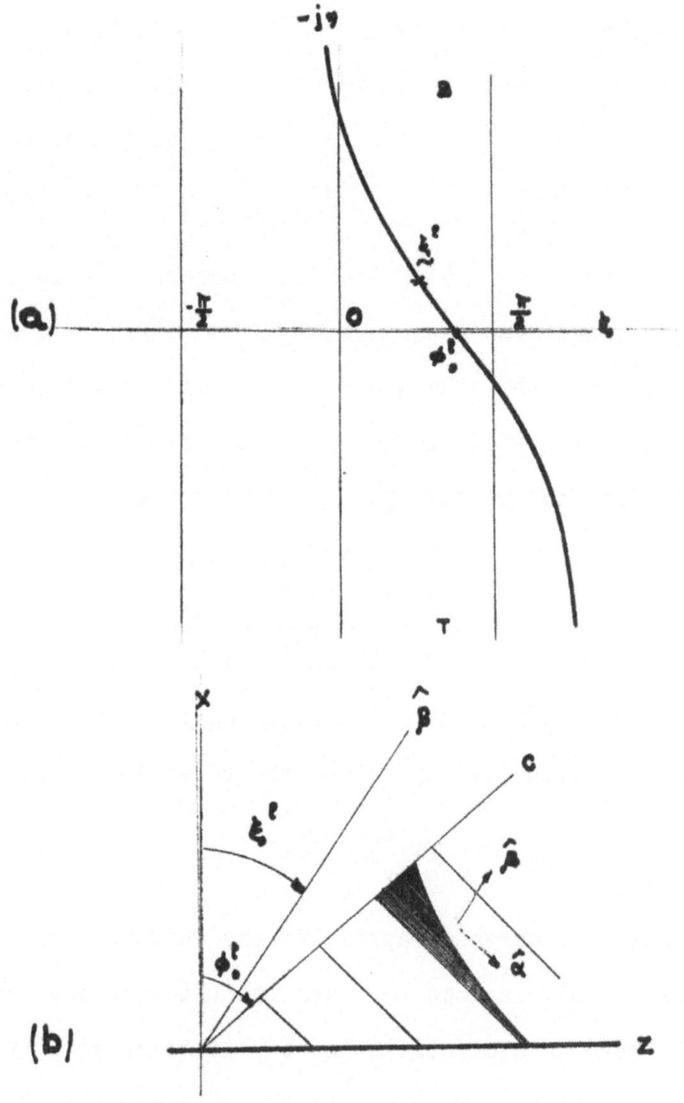

Fig. 27

single surface wave consists in the rapidly-decaying near-surface space wave (57) and the surface wave (59). The ratio of surface-wave to space-wave amplitude at x = 0 is therefore

$$\frac{|E^s(z)|}{|E^{rad}(z)|} = \left(\frac{R_c}{X_s}\right)^3 \sqrt{2\pi |k_o z|^3} \; , \qquad (76\underline{a})$$

where we assume, as in (59), that the surface wave is loosely bound. From transverse resonance, $R_c/X_s = \alpha_x^s/k_o < 1$, and (76a) therefore implies that the space wave dominates the surface wave until the normalized distance $k_o z$ is large enough to reverse the situation. What is the "peel out" distance of the surface wave? We write (76a) in the form

$$k_o z = \left[\frac{k_o}{\alpha_x^s}\right]^2 \sqrt[3]{\frac{|E^s(z)|^2}{2\pi\, |E^{rad}(z)|^2}} \; , \qquad (76\underline{b})$$

and if we now look for the distance along the surface at which the amplitude ratio is, say, 10 (power ratio 100), (76b) gives

$$k_o z \cong 2.5(k_o/\alpha_x^s)^2 \; , \qquad (76\underline{c})$$

which is seen to increase rapidly as the surface wave is bound more loosely. [Although we have derived (76) for the case of a TE surface wave on a capacitive sheet, it holds equally well for a TM wave on an inductive sheet (with the tangential E components replaced by tangential H components), and it holds even for the fundamental hybrid mode on a dielectric rod or a Yagi.] Beyond the distance (76c) the space wave is evidently quite negligible -- unless, that is, the interface is lossy, causing the surface wave to decay exponentially along z, so that eventually the more slowly-decaying space wave regains the upper hand. Since all physical structures have some loss, the space wave must in the end win out on any open waveguide --- even on the two-wire line, as Goubau pointed out.[13]

We saw in Sec. 1 that a source above a dielectric slab can also excite one or more leaky waves. Barone [20] has shown that their attenuation α_z^{ℓ} is normally quite large, so that they will noticeably modify the field only in the immediate vicinity of the source, and there they have in fact been observed.[21]

Structures exist that support a pure leaky wave, so that the field on and near the interface is entirely that in Fig. 27b. An example is the long slot in waveguide, Fig. 28, or rather, since we are here working with two-dimensional interfaces, a plane array of these slotted waveguides with common sidewalls; if the waveguides are excited in a TE mode with electric field lines parallel to the top and bottom walls, we can remove the sidewalls and are left with a parallel plate waveguide whose top wall is furrowed by z-directed parallel slots starting at z = 0. The discontinuity at z = 0 constitutes the source, and its radiation can be made entirely negligible by flaring the slots gradually from zero to their final width in the region $0 \leq z < \lambda_0$. As the constant slot width beyond $z \cong \lambda_0$ is decreased, the leakage α_z^{ℓ} becomes smaller and the pole $\underset{\sim}{\xi}^{\ell}$ in Fig. 27a moves closer to the saddle point ϕ_0^{ℓ} on the ξ-axis; in Fig. 27b, therefore, the lines c and $\hat{\beta}$ approach one another. The limit case $\alpha_z = 0$ is an ordinary plane wave emerging at $\xi = \phi_0$, i.e., the lines c and $\hat{\beta} = \hat{k}_0$ are identical. (This case can be physical-

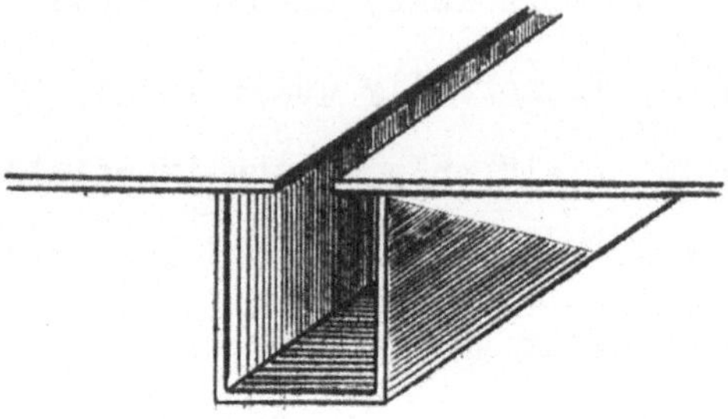

Fig. 28

ly realized by widening the slot with progressive distance from the origin at such a rate as to keep the ratiated power constant per unit length -- until the point is reached at which all of the power has leaked out).

The leaky wave sketch in Fig. 4b strongly suggested that this wave radiates into the far field along the direction of β. We know from Fig. 27b, however, that β lies outside the wedge in which the leaky wave is defined, and that within the wedge the leaky wave decays exponentially in all directions (with the exception of the limit case $\alpha_z = 0$). The leaky wave therefore cannot directly transfer energy from the vicinity of the interface to the radiation field, and its radiation mechanism needs further explaining (see below).

We now turn to the calculation of the radiation field of finite-length interfaces on which either a single surface or a single leaky wave is dominantly excited.

b) Radiation of Traveling Waves

We derive a formula well known in optics and in large aperture antenna theory: the radiation pattern of a finite length aperture along which the current, or a component of the electromagnetic field, is prescribed. Let the region $0 \leq z \leq \ell$ in Fig. 29 represent the aperture, and A(z) the current (or any 'Huygens source') distribution in the plane x = 0, with A(z) = 0 outside $0 \leq z \leq \ell$. We assume that $k_0 \ell \gg 1$, i.e., the aperture is at least two wavelengths long, and we wish to calculate the far field radiation pattern $F(\phi)$, i.e., the angular dependence of the electromagnetic field at points P for which $k_0 r' \gg k_0 \ell$. The simplified derivation usual in physical optics and antenna

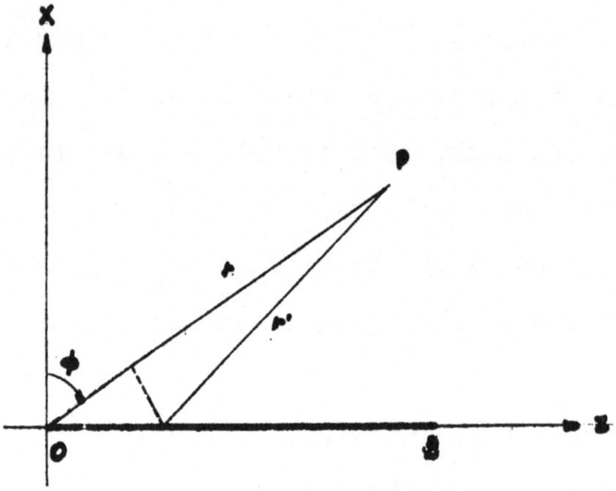

Fig. 29

theory assigns a pattern $I(\phi)$ to each individual current element, and notes that for $k_0 r' \gg k_0 \ell$, $\underline{r}$ and $\underline{r}'(z)$ can be taken parallel, so that $r' = r - z \sin\phi$. Then

$$F(\phi) = k_0 \int_0^{\ell} I(\phi) A(z) e^{jk_0 z \sin\phi} dz = I(\phi) \cdot R(\phi), \tag{77a}$$

where the aperture radiation integral $R(\phi)$ is defined by

$$R(\phi) = k_0 \int_0^{\ell} A(z) e^{jk_0 z \sin\phi} dz \; . \tag{77b}$$

[The k_0 factor is needed to make $R(\phi)$ dimensionless; see also (79b) below.] As we shall shortly observe, $I(\phi)$ is a slowly-varying function of ϕ, while for $k_0 r \gg k_0 \ell$, $R(\phi)$ is usually a rapidly-varying function of ϕ; if so, it is sufficient to examine the $R(\phi)$ term alone in discussing pattern characteristics.

To derive (77b) rigorously for a particular case, and at the same time illustrate the calculation of $I(\phi)$, assume that $A(z)$ represents the $E_y(0,z)$ field of a TE wave along an impedance-sheet 'aperture' $0 \leq z \leq \ell$ embedded in a metal ground plane extending over $z < 0$ and $z > \ell$. Green's theorem states that the field at point P in Fig. 29 is given by

$$E_y(r,\phi) = \int_{-\infty}^{+\infty} \left[G(r') \frac{\partial}{\partial x} E_y(0,z) - E_y(0,z) \frac{\partial}{\partial x} G(r') \right] dz, \tag{78}$$

where $G(r')$ is a two-dimensional half-space Green's function which we shall now construct from the free-space Green's function. The latter is a solution of the two-dimensional wave equation (4) with a delta-function source, and this we have already obtained in (54)-(56), where the first term was seen to represent free-space radiation from the source, while the second represented the modification due to the interface. In (56), the first term is recognized to be 1/4j times the asymptotic form for $k_o r \gg 1$ of $H_o^{(2)}(k_o r)$ [and the first terms in (54) and (55) are in fact integral representations of that Hankel function] . We can combine two of these free-space Green's functions in such a way as to make G(r') vanish at the interface, and with it the entire first term in the integrand of (78):

$$G(r') = \frac{1}{4j}\left[H_o^{(2)}(k_o r'_+) - H_o^{(2)}(k_o r'_-)\right], \qquad (79\underline{a})$$

where r'_+ and r'_- are the distances from the point P to two points located symmetrically with respect to the interface, separated by an infinitesimal distance along x. Then

$$\frac{\partial}{\partial x}G(r') = -\frac{k_o}{2j}\, H_1^{(2)}(k_o r')\frac{\partial r'}{\partial x},$$

and on substitution in (78),

$$E_y(r,\phi) = \frac{k_o}{2j}\int_{-\infty}^{+\infty} H_1^{(2)}(k_o r')\, E_y(0,z)\frac{\partial r'}{\partial x}dz .$$

We have already noted that in the far field of a finite-length aperture ($k_o r \gg k_o \ell \gg 1$), $r'(z) \cong r - z\sin\phi$, and we now add that also $r'(x) \cong r' - x\cos\phi$; then the asymptotic form of $H_1^{(2)}(k_o r')$ is $\sqrt{2/\pi k_o r}\,\exp\left[j(3\pi/4 - k_o r + k_o z\sin\phi)\right]$, and $\partial r'/\partial x =$ $= -\cos\phi$, so that

$$E_y^{rad}(r,\phi) \cong \frac{k_0 \cos\phi}{\sqrt{2\pi k_0 r}}\, e^{-j(3\pi/4 + k_0 r)} \int_0^{\ell} E_y(0,z)\, e^{jk_0 z \sin\phi}\, dz. \qquad (79\underline{b})$$

Apart from a constant-amplitude and -phase factor, which do not affect the angular pattern, this result agrees with (77a) if we identify the element pattern in this instance with $\cos\phi$. Had we chosen a TM wave, with prescribed $H_y(0,z)$, we would first of all observe that this field component cannot be set equal to zero on the metal ground plane outside $0 \leq z \leq \ell$; we must therefore work with $E_z(0,z) = (j/\omega\varepsilon_0)\,\partial H_y(0,z)/\partial x$ [from the second curl equation, (2d)], which does vanish there. In (78), with H_y replacing E_y throughout, the entire second term will be zero if $\partial G(r')/\partial x = 0$, i.e., $G(r')$ must have a maximum on the surface, which it will if two free-space Green's functions are combined as in (79a), but with a plus sign instead of the minus. In this instance, no $\cos\phi$ term appears in the analogue of (79b), and the element pattern is therefore omnidirectional [$I(\phi) = 1$ in (77a)]. If $A(z)$ is due to a wave type other than the simple TE or TM mode, the element pattern might be a more complicated function of ϕ, but never one that varies rapidly.

We now examine the aperture radiation pattern $R(\phi)$, given by 77b, which we recognize as the Fourier transform of the aperture distribution $A(z)$. As an illustration, consider the simple aperture distribution $A(z) = \exp(-j\beta_z z)$ produced by a traveling wave with constant amplitude and constant phase progression. For $0 \leq \beta_z \leq k_0$, this is an ordinary plane wave emerging at the angle $\sin\xi = \beta_z/k_0$ [i.e., the 'degenerate' leaky wave discussed in Sec. 4(a)], while for $\beta_z > k_0$ it is a surface wave. Substitution in (77b) gives

$$R(\phi) = k_0 \int_0^{\ell} e^{jz(k_0 \sin\phi - \beta_z)} dz \qquad (80a)$$

$$= k_0 \ell \frac{\sin\psi}{\psi}, \qquad (80b)$$

where

$$\psi = \frac{\ell}{2}(k_0 \sin\phi - \beta_z) \qquad (80c)$$

(an absolute phase factor having been omitted). The $\sin\psi/\psi$ function has its maximum at $\psi = 0$, i.e., when

$$\sin\phi_m = \beta_z/k_0; \qquad (80d)$$

the angle of maximum radiation ϕ_m is therefore identical with the angle of emergence ξ of the wave in the aperture, as one would expect. The magnitude of the radiation pattern (80b), normalized with respect to $k_0\ell$, is shown in Fig. 30. When the beam is broadside, $\beta_z = 0$ in (80c), and $\psi = 0$ therefore corresponds to $\phi = 0$, and $\phi = \pm\pi/2$ to $\psi = \pm k_0\ell/2$; any ψ-values outside the range $|\psi| \leq k_0\ell/2$ correspond to complex angles $\tilde{\phi}$ and consequently represent the analytic continuation of the pattern. Since the real-angle range is thus proportional to ℓ, more and more sidelobes appear, and the beamwidth of main lobe as well as sidelobes narrows progressively with increasing aperture length. Thus for $k_0\ell \gg 1$, $R(\phi)$ varies rapidly in the region of the main beam and the closest sidelobes, and we are justified in neglecting the effect of $I(\phi)$ on the total pattern. When the beam is endfire, $\beta_z = k_0$ in (80c) and $\psi = 0$

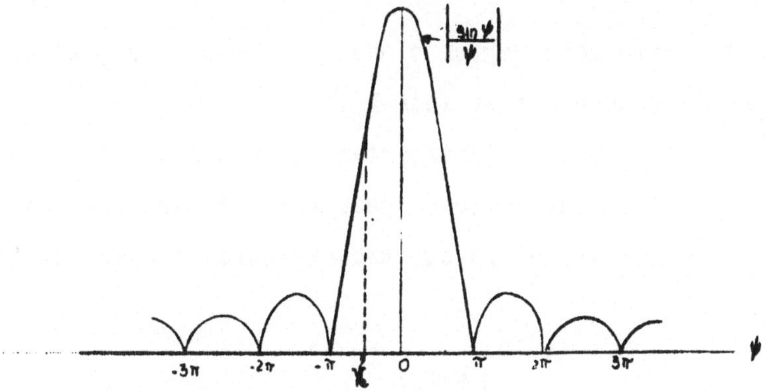

Fig. 30

corresponds to $\phi = \pi/2$, $\phi = 0$ to $\psi = -k_0 \ell/2$, and $\phi = -\pi/2$ to $\psi = -k_0\ell$; ψ values outside this range again represent complex angles of radiation, and in this instance all ψ-values to the right of $\psi = 0$ in Fig. 30 are thus invisible. If $\beta_z > k_0$, (80c) shows that even the point $\psi = 0$ is pushed into the complex-angle region, i.e., the pattern is real only to the left of a point such as the one marked ψ_e. We note that the radiation will still be predominantly endfire so long as ψ_e remains in the main lobe region close to $\psi = 0$, i.e., so long as β_z is not too much larger than k_0.

A pattern criterion important in antenna design is the 'directivity' D, the ratio of peak power to average power, both per unit solid angle; in our two-dimensional case, it is

$$D = \frac{|R(\phi_m)|^2}{\frac{1}{2\pi}\int_0^{2\pi} |R(\phi)|^2 d\phi} \tag{81}$$

The directivity is obviously larger the narrower the main beam and the lower the sidelobe level (and thus in turn, as we have seen, the longer the antenna). It can be shown, subject to certain restrictions,[22] that among all possible amplitude and phase distributions, our $\exp(-j\beta_z z)$ produces the highest possible directivity. The choice of β_z for a prescribed angle of maximum radiation is forced by (80d) -- except at endfire, where (80d) implies only that $\beta_z \geqslant k_0$. It is natural to ask, therefore, which β_z produces maximum endfire directivity. Increasing β_z beyond k_0, as we saw, moves ψ_e to the left of $\psi = 0$, and it is clear from Fig. 30 that this results in narrowing the main beam and, at the same time, in decreasing the peak radiation relative to the sidelobes. These two effects increase and decrease the directivity, respectively. Numerical evaluation of the maximum of (81) with respect to variations in β_z shows that ψ_e is optimum at $\psi_e \cong -\pi/2$, and (80c) therefore gives

$$\beta_z \ell - k_o \ell \cong \pi \quad , \tag{82}$$

which is well known to antenna engineers as the 'Hansen-Woodyard condition' for maximum endfire directivity.[22]

Because of the inevitable presence along the interface of the space wave (57), the surface-wave antenna aperture does not properly fall into the class of constant-amplitude, linear-phase distributions (80a). A pure leaky wave-execited aperture can, on the other hand, be described by (80a) provided we replace β_z by the complex leaky wave-number $k_z = \beta_z - j\alpha_z$, so that $\underset{\sim}{\psi} = \psi + j\alpha_z \ell/2$. The result (80b) still holds, but $R(\phi)$ is now complex and the amplitude pattern is therefore given by $|R(\phi)| = k_o \ell |(\sin \underset{\sim}{\psi})/\underset{\sim}{\psi}|$. At what angle ϕ_m does this function have its maximum? Differentiation with respect to ϕ of $|R(\phi)|$ results in $\sin\phi_m = \beta_z/k_o$, which is precisely (80d) again. This angle can easily be shown to lie between ξ^ℓ and ϕ_o^ℓ in Fig.27b. [If the element pattern $I(\phi)$ is not a constant, the peak direction of the total pattern (77a) will differ slightly from that just calculated for (77b).]

The foregoing discussion of the far field patterns (77) remains valid if the aperture $0 \leq z \leq \ell$ is in free space. Consider, for example, a sheet of length ℓ with identical impedance conditions on its top ($x = 0^+$) and bottom ($x = 0^-$): symmetry between the field configuration above and below the interface always allows transformation into a half-space problem. If the symmetry is of the type shown in Fig. 10b, the lower half space can be eliminated by embedding the impedance sheet (here represented by a dielectric slab) in a metal ground plane; if the symmetry is of the type shown in Fig. 10c, the impedance sheet can be embedded in a 'magnetic' ground plane (with zero surface admittance). The latter case differs from those just treated only in that the requirement $A(z) = 0$ outside $0 \leq z \leq \ell$ forces the choice of tangential $\underline{H}$-(rather than $\underline{E}$-) component as $A(z)$.

c) Surface-Wave Radiation: The Young and Fresnel Points of View

On the basis of Secs. 4(a) and (b), we can now describe the radiation mechanism of surface wave antennas. Typical one-dimensional structures are the waveguide-fed dielectric rod shown in Fig. 31a, the Yagi (linear array of dipoles parasitically fed by a dipole-reflector combination at one end), the 'cigar' antenna (like the Yagi, but with metal discs rather than dipoles as individual elements); two-dimensional examples are the finite-length corrugated surface, and the center-fed dielectric disc (annular cylindrical geometry), both usually embedded in

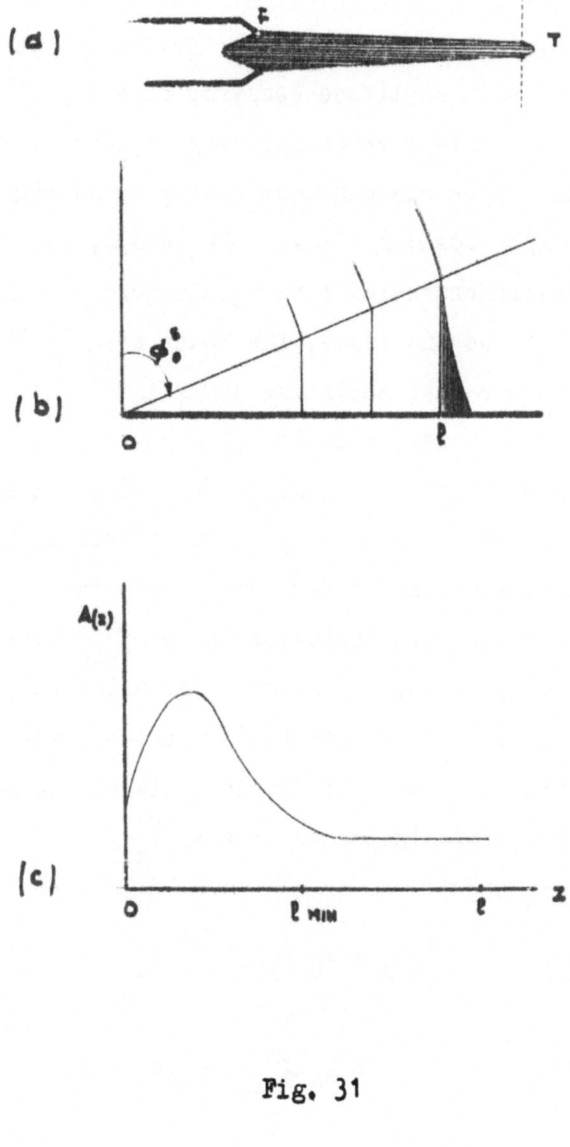

Fig. 31

a metal ground plane.[23]

Figs. 31b and c summarize what we have learned concerning the quasi-near field of the interface: in the region $0 \leq \phi \leq \phi_0^s$ in b, the space wave (56) propagates with circular wavefronts and with amplitude decaying radially as $1/\sqrt{k_0 r}$. As a function of ϕ, this wave decays very rapidly as ϕ approaches $\pi/2$ [which can be checked most easily by noting that $\Gamma(\phi) = (Z_s - \omega\mu/k_0 \cos\phi) / (Z_s + \omega\mu/k_0 \cos\phi)$, so that at $\phi = \pi/2$ the entire bracketed term equals zero]. In the immediate vicinity of the interface, the space wave (57) with circular wavefronts and radial amplitude decay $1/\sqrt{(k_0 z)^3}$ is superimposed on the surface wave with its plane wave fronts (only the latter are shown in b), and finally swamped by it after a distance given approximately by (76c). The aperture distribution A(z) therefore looks as in Fig. 31c, where the 'peel out' distance of the surface wave is indicated by ℓ_{min} (for a reason that will shortly be made apparent). To allow comparison with the Hansen-Woodyard condition (82) we express the peel-out distance in terms of the phase difference between the surface wave and the velocity-of-light wave at $z = \ell_{min}$:

$$\beta_z^s \ell_{min} - k_0 \ell_{min} = k_0 \ell_{min}(\beta_z^s/k_0 - 1)$$

$$\cong k_0 \ell_{min}(\alpha_x^s/k_0)^2/2,$$

where we assume that the surface wave is loosely bound; using (76c),

$$\beta_z^s \ell_{min} - k_o \ell_{min} \cong 1.25, \qquad (83)$$

i.e., ℓ_{min} is the point at which the phase of the surface wave leads that of the space wave by approximately 75 degrees.

When the space wave arrives at $z = \ell$, the impedance discontinuity will partly reflect and partly radiate it. Kay[24] has calculated both effects by solving a Wiener-Hopf integral equation of the type described in the companion lecture by C. Angulo, with the result that the reflection coefficient is negligibly small if the wave is very loosely bound, and that the radiated field, under the same condition, is endfire with a pattern $T(\theta)$ given approximately by

$$T(\theta) \cong 1/\left[(\alpha_x^s/k_o)^2 + \theta^2\right] \qquad (84)$$

in the region of the main lobe ($\theta \ll \pi$); θ is the complement of ϕ, already introduced in Fig. 4b. We can derive (84) by making the crude assumption that the field distribution in the terminal x-y plane through $z = \ell$ is the unperturbed incident surface wave and its mirror image, $A(x) = \exp(-\alpha_x^s |x|)$, and by integrating over dx in analogy with (77b):

$$T(\theta) = k_o \int_{-\infty}^{\infty} A(x) e^{jk_o x \sin\theta} dx \; .$$

The result is $2\alpha_x^s/k_o \left[(\alpha_x^s/k_o)^2 + \sin^2\theta\right]$, and under the assumptions made this agrees with (84). Within this degree of approximation, the pattern in the half space $-\pi/2 \leq \theta \leq \pi/2$ has no nulls, and is more sharply peaked the looser the binding of the surface wave; the latter characteristic we would expect on the basis that the transverse extent of the surface wave (say the value of $|x|$ at which the amplitude is one-tenth of its value at the interface) increases the smaller α_x^s, i.e., the closer β_z^s is to k_0.

The radiation mechanism of a surface wave antenna can be viewed in two ways. The most natural one, on the basis of the foregoing, is to picture the total radiation pattern as the superposition of radiation from the feed F in Fig. 31a and the radiation (84) from the terminal plane T. The far field pattern of the feed is given by (56) for the case of a line source. In practice, the feed has to be unidirectional so as to excite the surface wave in one direction only, for example a waveguide horn, or a dipole-reflector combination (whose two-dimensional equivalent was shown in Fig. 21a; note in Fig. 21b that $F(-\beta_z^s)$ is almost zero if $|\beta_z^s|$ is very close to $|k_0|$). Since the surface wave radiation from T is much more directive than the direct feed radiation, the antenna must be made at least long enough so that the surface wave is as fully established as possible in the terminal aperture (granted that it can never extend to $x = \pm\infty$, as Fig. 31b makes clear). In the cross section through ℓ_{min}, the surface wave was found to dominate the space wave at the interface, and on checking the ϕ-dependence near $\phi = \pi/2$ of (56) and (57), one finds that this holds throughout the portion of the cross section lying between the interface and ϕ_0^s. Thus the peel-out distance ℓ_{min} in (83) is the minimum length of a surface wave antenna. If the feed excites the surface wave with 100 percent efficiency, i.e., if all of the energy goes into the surface wave and none into the direct feed radiation, ℓ_{min} is also the optimum length, since a further increase in ℓ produces no further increase in directivity. If the excitation efficiency is less than 100 percent (we have already mentioned that in most antenna applications it is about 70 percent), the feed and terminal radiation interfere, and their combined pattern will not necessarily produce maximum directivity at $\ell = \ell_{min}$. Experiments by Ehrenspeck and Poehler[23] with Yagi antennas show that, for a prescribed β_z^s, maximum directivity is obtained when

$$\beta_z^s \ell - k_o \ell \cong 2\pi/3 , \tag{85}$$

i.e., when the phase difference is 120 degrees [rather than the 75 degrees of (83)] . This result is basic for the design of most practical surface wave antennas; it can of course also be obtained by differentiating with respect to ℓ the directivity of the combined feed and terminal-plane pattern. This combined pattern, the total radiation pattern of the surface wave antenna, resembles the highly directive terminal-plane pattern (84) in the region of the peak ($\theta \ll \alpha_x^s/k_o$); as θ increases and (84) drops to the level of the much broader feed pattern, the two produce a typical interference pattern with relative minima (but no sharp nulls as in Fig. 30) and maxima (the antenna 'side-lobes').[23]

The second, entirely different, way of calculating the surface wave antenna pattern is to integrate the antenna aperture distribution from 0 to ℓ as in (77b), with A(z) given by Fig. 31c or analytically, by the superposition of the interface space wave (57) and the surface wave (59). Within the approximations made, this method yields the same antenna pattern as the first method. It resembles the sin ψ/ψ pattern to the left of ψ_e in Fig. 30, except that the main lobe is slightly narrower and the nulls are partly filled in. The deviation from sin ψ/ψ is of course due to the fact that A(z) now has a large initial hump instead of being constant throughout. The hump also explains why the maximum-directivity relation between β_z^s and ℓ is no longer the Hansen-Woodyard condition (82) but (85), usually referred to as the 'modified' Hansen-Woodyard condition.

In the early Nineteenth Century, two views of diffraction phenomena vied with each other: that of Young, who pictured diffraction in terms of waves emanating from the edge of an aperture, and that of Fresnel, who superimposed Huygens wavelets emanating from the aperture itself. Our two ways of calculating the surface wave antenna radiation correspond respectively to Young's and Fresnel's approaches to diffraction in general. That they lead to identical results (within the approximations made) is not surprising in view of the work of Rubinowicz[25] and others who showed that both Young's and Fresnel's approach, when appropriately reformulated in modern terms, can be rigorously derived from Maxwell's equations. In the early days of surface wave antenna theory, this analogy was not recognized and the two viewpoints concerning surface wave radiation were felt to be incompatible. The adherents of aperture integration, for example, could easily explain, as we did in Sec. 4(b), why the beamwidth of the antenna decreases (and the directivity increases) as the antenna length increases. How, they asked, could this be understood in terms of radiation from F and T? Doesn't a source excite the residue wave eveywhere, and couldn't one therefore place the terminal plane arbitrarily close to the feed and thus, by choosing β_z^s sufficiently close to k_o, produce unlimited directivity with an infinitesimally short antenna? We know from Sec 4(a) that the premise is correct, and yet the conlusions do not follow; but the surface wave peel-out process had not then been studied. Conversely, adherents of the two-point radiation picture asked how an aperture on which a surface wave had somehow been excited with 100 percent efficiency, as shown in Fig. 32a, could possibly radiate when it is well known that a surface wave is trapped along the interface. It is worth examining the answer to this objection. A surface wave propagating from minus to plus infinity obviously does not radiate, but the wave in Fig. 32a extends only from 0 to ℓ and therefore produces the $\sin\psi/\psi$ pattern in (80b). The sine can be broken up into two exponentials, and omitting absolute phase and amplitude terms the pattern is therefore

$$R(\phi) = \frac{1}{\psi} - \frac{1}{\psi}e^{-j2\psi} \quad . \tag{86}$$

Remembering that $\psi = \ell(k_o \sin\phi - \beta_z^s)/2$, the phase of the second term is seen to be precisely the difference between the phase of a wave coming from z = 0 and one from z = ℓ; the two terms therfore correspond to radiation from F and from T, with individual radiation pattern $1/\psi$ and $-1/\psi$, respectively. For loose

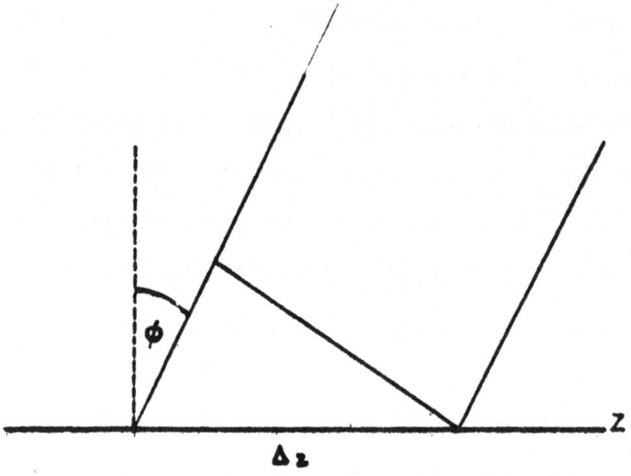

Fig. 33

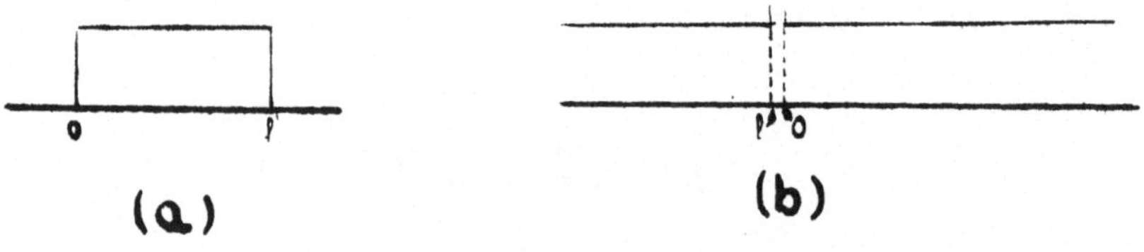

Fig. 32

binding and small θ , the pattern

$$\frac{1}{\psi} = \frac{1}{\beta_z^s/k_0 - \sin\theta} \cong \frac{1}{\frac{1}{2}\left(\frac{\Delta x^s}{k_0}\right)^2 + \frac{\theta^2}{2}}$$

evidently equals $T(\theta)$ in (84), as in fact it should for the radiation from $z = \ell$. That the radiation from $z = 0$ is equal in magnitude and opposite in sign to $T(\theta)$ is simply explained in terms of Fig. 32b, where the termination and excitation point of Fig. 32a are superimposed: since this wave extends from minus to plus infinity, its radiation is zero, and therefore the patterns from ℓ and 0 must cancel. This circumstance also throws light on the Hansen-Woodyard condition: because the feed and terminal radiation are 180 degrees out of phase, ℓ and β_z^s must be so related as to introduce an addition 180 degree phase shift between F and T, thus producing constructive interference in the forward direction. The error in the original objection lies in the assumption that the surface wave in Fig. 32a is excited with 100 percent efficiency, whereas (86) shows that the efficiency is only 50 percent. It is physically possible to construct a feed that radiates a $1/\psi$ pattern (it turns out to extend over fully one-third of antenna length),[23] but there is no point in doing so since a comparison between (82) and (85) shows that the aperture distribution in Fig. 32a produces less directivity than an equal-length aperture distribution of the type Fig. 31c. (Antennas whose directivity exceeds that of an equal-length constant-amplitude linear-phase aperture are said to be 'superdirective'; surface wave antennas are slightly so.)

Surface wave antennas of the type just described necessarily produce endfire beams only, and even with this restriction precise shaping of the pattern cannot be achieved since the only parameters at our disposal are the surface impedance and the feed configuration. In an attempt to introduce additional parameters, one tries to vary the surface impedance as a function of z, and this has led to modest success in lowering sidelobes, decreasing the beamwidth, or increasing the pattern banwidth of the antenna. A crude approach to variable-impedance surfaces is to regard A(z) as the superposition of the feed radiation and a variable-phase and -amplitude surface wave, the latter being computed on the assumption that the impedance variations are so slow that surface wave energy is being conserved in successive cross sections as the wave travels along z. For a TM wave, for example, we write, in analogy with (15),

$$H_y(x,z) = H_{yo} e^{-\alpha_x x} e^{-j\int_0^z \beta_z dz'} , \qquad (87)$$

for the magnetic field very near the interface, where H_{yo}, α_x, and β_z are functions of z. [The form of (87) implies that the wave equation is considered separable, which it will certainly not be far from the interface.] Energy conservation imples

$$\int_0^\infty S_z dx = \text{const.},$$

with S_z the z-component of the Poynting vector, which is given by the first term of $\underline{S}_{TM}$ in (27) [where $\hat{\beta} = \hat{z}$, $\beta = \beta_z$, $|H| = H_{yo}\exp(-\alpha_x x)$ from (87)]. Therefore

$$H_{yo}^2 \int_0^\infty \beta_z e^{-2\alpha_x x} dx = \text{const.},$$

or $H_{yo} \sim \sqrt{\alpha_x / \beta_z}$. An impedance variation often employed is a simple taper along the antenna length, which can be approximated by $X_s = a/z$. Then

$$\alpha_x / \omega \varepsilon_0 = X_s \quad [\text{from } (33\underline{b} \text{ and } \underline{d})]$$

$$\alpha_x = \omega \varepsilon_0 a/z.$$

Substituting in (87), and assuming very loose binding (so $\beta_z \cong k_0 \left[1 + \frac{1}{2}(\alpha_x/k_0)^2\right]$), we obtain

$$H_y(x,z) \sim \frac{1}{\sqrt{k_0 z}} e^{-a\varepsilon_0 \omega x/z} e^{-jk_0 z}.$$

Near the interface, where $z = \rho$ and $x/z = \theta$ in Fig. 13, this agrees exactly with the asymptotic result (40) quoted in Sec. 2(e). With other impedance variations, this technique does not fare as well; a better approximation can be found by using Keller's geometric theory of diffraction.[26] Because α_x and β_z are interrelated through (33a), independent control over the amplitude and phase of A(z) cannot be obtained and the pattern control achievable is therefore still limited. A more promising approach consists in 'modulating' the surface, i.e., varying the impedance periodically, with results that were described in Sec. 3(f). With either method, beams tilted at an angle away from the interface can in principle be obtained.

d) Leaky Wave Radiation: Pattern Control

The calculation of leaky wave parameters in terms of the leaky waveguide's dimensions was mentioned at the end of Sec. 2(e). Several leaky structures are known[23] on which β_z and α_z are independently controllable through distinct sets of geometric parameters, and as will now be shown, the phase and amplitude of A(z) can then also be separately controlled. With this freedom of choice among aperture distributions, a beam can be pointed in any wanted direction, and prescribed pattern shapes produced with great precision. Two methods of pattern synthesis through controlled leaky wave radiation will be described.

To separate the phase and amplitude of the aperture distribution, we write $A(z) = A_0(z)\exp\left[-j\int_0^z \beta_z(z')dz'\right]$. The phase as a function of z is directly controlled by $\beta_z(z)$. Its choice is limited by the range of β_z values that can be produced on a given structure; in particular, the sign of β_z cannot be reversed by any simple method, and the phase must therefore decrease monotonically with increasing z. The amplitude $A_0(z)$ can be related to $\alpha_z(z)$ as follows. Consider a leaky structure with input power P_0, which decreases steadily as a function of z until at $z = \ell$ only P_ℓ is left, the power $P_0 - P_\ell$ having been radiated along the way. We have

$$P(z) = P_o e^{-2\int_0^z \alpha_z(z')dz'} \tag{88a}$$

(which looks more familiar if α_z = const.). The rate of power loss due to radiation is proportional to the square of the aperture illumination:

$$-\frac{dP(z)}{dz} = aA_o^2(z) \; . \tag{88b}$$

Differentiating (88a),

$$-\frac{dP(z)}{dz} = 2\alpha_z(z)P(z) \; ; \tag{88c}$$

integrated in turn from $z = 0$ to $z = z'$, and from $z=0$ to $z = \ell$, (88c) becomes

$$-P(z) + P_o = a\int_0^z A_o^2(z')dz'$$

$$-P_\ell + P_o = a\int_0^\ell A_o^2(z')dz' \; . \tag{88d}$$

From (88b) and (88c),

$$2\,\alpha_z(z)P(z) = aA_o^2(z) \; ,$$

and with the help of (88d), which eliminates the constant a, we arrive at the equation

$$\alpha_z(z) = \frac{A_0^2(z)/2}{\frac{1}{1-P_\ell/P_0}\int_0^\ell A_0^2(z')dz' - \int_0^z A_0^2(z')dz'},$$

which is basic in the design of all leaky wave antennas. To give an example: if one desires a constant amplitude $A_0(z) = 1$, and allows, say, 10 percent of the power to be wasted in the terminal load ($P_\ell/P_0 = 0.1$), one finds

$$\alpha_z(z) = \frac{1/2}{1.1\,\ell - z},$$

i.e., α_z must increase monotonically with z; in a slotted waveguide, this is achieved by flaring the slot at an appropiate rate from feed to termination, as noted on intuitive grounds in Sec. 4(a). The choice of $A_0(z)$ is again restricted only by the range of α_z values that can be produced through parameter variations on the given structure.

With the aperture phase and amplitude reasonably under our control, we can now synthesize a very large variety of patterns. The discussion will be limited to two methods: in the first, β_z is kept constant (linearly progressive phase) and the amplitude is varied; in the second, the amplitude is prescribed (for example as a constant) and β_z is varied. The methods are used to illustrate different aspects of pattern synthesis: the first, to show the effect on the pattern of symmetric and asymmetric amplitude tapers in the aperture; the second, to show that points along the leaky-wave aperture can under certain circumstances be uniquely related to specific directions in the far field.

Let β_z be chosen in accordance with (80d) so as to point the beam in the desired peak direction. It is convenient to rewrite (77b) in terms of $\xi = 2z/\ell - 1$, the normalized distance measured from the aperture center: omitting a constant phase and amplitude term,

$$R(\phi) = k_o \int_{-1}^{1} A_o(\xi) e^{j\psi\xi} d\xi \,, \tag{89}$$

with ψ defined by (80c). The effect of an amplitude taper that is symmetric with respect to the aperture center $\xi = 0$ is best seen by considering an example, say

$$A_o(\xi) = 1 - |\xi| \,, \tag{90a}$$

a linearly gabled distribution that drops to zero at the aperture ends. Substituting in (89) and integrating by parts, one finds (omitting constants)

$$R(\phi) = \frac{\sin^2(\psi/2)}{\psi^2} .$$

The envelope $1/\psi^2$ of this function decays more rapidly than the $1/\psi$ envelope of the $\sin\psi/\psi$ pattern in Fig. 30, and the distribution therefore leads to lower sidelobes than the constant-amplitude distribution, at the expense of a broader main beam. That the effect of increasing the amplitude toward the ends is just the reverse can be seen from the extreme case of pure end excitation (as in interferometry), which produces a pattern of many narrow lobes all of equal amplitude.

Asymmetric tapering is illustrated by

$$A_o(\xi) = 1 - \xi , \qquad (90\underline{b})$$

which is maximum at $z = 0$ and decays linearly to zero at $z = \ell$. Substitution in (89) results in

$$R(\phi) = \frac{\sin\psi}{\psi} + j\left(\frac{\cos\psi}{\psi} - \frac{\sin\psi}{\psi^2}\right) .$$

The magnitude of this function has no sharply defined maxima or minima, since $\sin\psi$ is zero when $\cos\psi$ is 1 and vice versa; such a pattern is said to be 'shaped'. [Note that in the absence of a well-defined main beam, the effect of the individual element pattern $I(\phi)$ cannot be omitted and (77a) rather than (b) must be used in duscussing the far-field characteristics.]

By expanding the exponential[22] or $A_o(\xi)$[27] in (89) in a power series, one obtains general expressions relating the pattern to the amplitude distribution, with β_z held constant. These expressions verify for all symmetric and asymmetric tapers the observations we made for (90a) and (b).

In the second method, (77a) is written in the form

$$F(\phi) = k_o I(\phi) \int_0^{\ell} A_o(z) e^{j \int_0^z \psi(z')dz'} dz. \qquad (91a)$$

If $A_o(z)$ and $\psi(z)$ vary sufficiently smoothly, a saddle point exists at $\psi(z) = 0$, or

$$\beta_z(z) = k_o \sin \phi(z). \qquad (91b)$$

Expression (91b) relates every point z along the aperture to a specific direction $\phi(z)$ in the far field; in the interval Δz in Fig. 33, for example, β_z radiates at the angle ϕ, i.e., Δz is a quasi constant-amplitude linear-phase aperture as in (80a), and (91b) over that region is identical with (80d). A one-to-one correspondence between points on two separate wave-fronts allows the drawing of rays connecting them, and the saddle point evaluation of (91a) is therefore a geometric-optics approximation.[22] For $k_o z \gg 1$, this approximation is good, except that the aperture distribution at $z = 0$ and $z = \ell$ cannot be smooth so that diffraction lobes inviariably arise there. [Note the transition from the Fresnel point of view in (91a) to that of Young implicit in the last statement.]

The power density in the aperture distribution over Δz is proportional to $A_0^2(z)\Delta z$, or to $A_0^2(z)\Delta z/\cos\phi$ in the projected aperture $\Delta z \cos\phi$ in Fig.33. Since geometric optics holds, the power in Δz must flow directly into the incremental far-field angle $\Delta\phi$:

$$I^2(\phi)\,|F(\phi)|^2\Delta\phi = aA_0^2(z)\Delta z/\cos\phi, \qquad (92\underline{a})$$

or

$$\frac{d\phi}{dz} = \frac{aA_0^2(z)}{I^2(\phi)\,|F(\phi)|^2\cos\phi} \qquad (92\underline{b})$$

where a is a proportionally constant which must be chosen positive if ϕ increases with z, or negative if ϕ decreases as z increases. [Saddle point integration of (91a) would have produced the same result.][27]. The differential equation (92b) governs the rate at which aperture power is 'sprayed' into the far field; in conjunction with (91b), it fixes $\beta_z(z)$ when the pattern $|F(\phi)|$ and the aperture distribution $A_0(z)$ are prescribed.

As an example, let a leaky-wave antenna of length ℓ be embedded in an infinite ground plane. The desired amplitude of the far-field pattern is constant $\left[|F(\phi)| = 1\right]$ in the range $0 \leq \phi \leq \pi/2$ and zero for $-\pi/2 \leq \phi < 0$, i.e., the pattern has the shape of a circle in the right quadrant of the upper half space and is zero in the left. For the sake of simplicity, choose $I(\phi) = A_0(z) = 1$. Conservation of energy requires that the total power in aperture and far field be equal: from (92a),

$$\int_0^{\pi/2} |F(\phi)|^2 \cos\phi \, d\phi = \int_0^{\ell} a A_0^2 dz ,$$

and therefore $|a| = 1/\ell$. (92b) now reads

$$\frac{d\phi}{dz} = \pm \frac{1}{\ell \cos\phi} .$$

Integrating, and using (91b),

$$z = \pm \ell \beta_z(z)/k_0 + b.$$

The sign and the constant b are fixed by the choice of β_z at the feed point; if we start with $\beta_z(0) = k_0$, i.e., if the initial interval Δz radiates endfire, then ϕ must decrease with increasing z, and $b = \ell$, so that

$$\beta_z(z) = k_0(1-z/\ell), \tag{93}$$

which is the desired relation. As z increases from 0, $\beta_z(z)$

decreases and the beam generated by successive sections of the leaky guide moves off endfire until at $z = \ell$, $\beta_z = 0$ and the radiation is broadside. (Special techniques are needed for the physical realization of the latter.)[23] To discover how close the pattern of (93) is to the desired quarter circle, (93) must be substituted back in the far-field integral (91a); the fit will be found to improve the longer the antenna, but diffraction lobes caused by the antenna ends can never be quite eliminated from the quadrant in which $F(\phi) = 0$.

Acknowledgements

The author is indebted to Prof.A.A.Oliner and to Dr.A.F.Kay for numerous discussions, over a period of several years, on surface and leaky waves. Prof.L.Ronchi Abbozzo corrected many mistakes in the typescript of these lectures.

F.J.Zucker

BIBLIOGRAPHY

Because of the extensive bibliography appended to the companion paper by M.A. Miller and V.I.Talanov, we restrict ours to references needed directly in the rext. (A full bibliography specializing in surface- and leaky-wave antennas is given by Zucker.[23])

The reader's attention is especially invited to the new monograph by Barlow and Brown,[12] and to the treatise by Brekhovskikh,[8] which cover in detail some of the topics we only touched on.

1. C.T. Tai, "The Effect of a Grounded Slab on the Radiation from a Line Source", J. Appl. Physics, vol. 22, p. 405 (1951).

2. R.S. Elliott,"On the Theory of Corrugated Plane Surfaces," IRE Trans. on Antennas and Propagation, AP-2, p. 71 (1954).

3. R.S. Elliott,"Azimuthal Surface Waves on Circular Cylinders", J. Appl. Phys., vol. 26, p. 368 (1955).

4. L.B. Felsen, "Electromagnetic Properties of Wedges and Cone Surfaces With a Linearly Varying Surface Impedance," IRE Trans. on Antennas and Propagation, AP-7 (Special Supplement), p. S231 (1959).

5. V.I. Talanov, "On Surface Electromagnetic Waves in Systems With Nonuniform Impedance", Izvest. VUZ MVO (Radio fizika), vol. 2, p. 32 (1959).

6. N. Marcuvitz, Waveguide Handbook, McGraw-Hill Book Company, New York, 1951.

7. L.O. Goldstone and A.A. Oliner, "Leaky Wave Antennas I: Rectangular Waveguides", IRE Trans. on Antennas and Propagation, AP-7, p. 307 (1959).

8. L. Brekhovskikh, Waves in Layered Media, Academic Press, New York, 1960.

9. J.R. Wait, "An Approach to the Classification of Electromagnetic Surface Waves", Proceedings of the 1960 URSI General Assembly in London, England (to be published by Elsevier Publishing Company, Amsterdam, 1963).

10. N. Marcuvitz and J. Schwinger, "On the Representation of the Electric and Magnetic Fields Produced by Currents and Discontinuities in Wave Guides. I", J. Appl. Phys., vol. 22, p. 806 (1951).

11. A.L. Cullen, "The Excitation of Plane Surface Waves", Monograph No. 93, Proc. I.E.E., vol. 101, Part IV (1954).

12. H.E.M. Barlow and J. Brown, Radio Surface Waves, Oxford University Press, London, 1962.

13. G. Goubau, "On the Excitation of Surface Waves", Proc. IRE, vol. 40, p. 865 (1952).

14. R.E. Collin, Field Theory of Guided Waves, McGraw-Hill Book Company, New York, 1960.

15. E. Snitzer and H. Osterberg, "Observed Dielectric Waveguide Modes in the Visible Spectrum," J. of the Opt. Soc. of America, vol. 51, p. 499 (1961).

16. J.A. Stratton, Electromagnetic Waves, McGraw-Hill Book Company, New York, 1941.

17. F.J. Zucker, "Theory and Application of Surface Waves", Nuovo Cimento, vol. 9, Suppl. No. 3, p. 450 (1952).

18. G.M. Roe, Ph.D. Thesis, University of Minnesota, 1947.

19. A.A. Oliner and A. Hessel, "Guided Waves on Sinusoidally-Modulated Reactance Surfaces," IRE Trans. on Antennas and Propagation, vol. AP-7, Special Supplement, p. S201 (1959).

20. S. Barone, "Leaky Wave Contributions to the Field of a Line Source Above a Dielectric Slab", Report R- 532-56, PIB-462, Polytechinic Institute of Brooklyn (1956).

21. E.S. Cassey and M. Cohn, "On the Existence of Leaky Waves Due to a Line Source Above a Grounded Dielectric Slab", IRE Trans. on Microwave Theory and Techinques, vol. MTT-9, p. 243 (1961).

22. S. Silver, <u>Microwave Antenna Theory and Design</u>, McGraw-Hill Book Company, New York, 1949.

23. F.J. Zucker, "Surface- and Leaky-Wave Antennas", Ch. 16 in <u>Antenna Engineering Handbook</u>, H. Jasik, Ed., McGraw-Hill Book Company, New York, 1961.

24. A.F. Kay, "Scattering of a Surface Wave by a Discontinuity in Reactance", IRE Trans. on Antennas and Propagation, vol. AP-7, p. 22 (1959).

25. A. Rubinowicz, <u>Die Beugungswelle in der Kirchhoffschen Theorie der Beugung,</u> Panstwowe Wydawnictwo Naukowe, Warsaw, 1957.

26. J.B. Keller and F.C. Karal, "Excitation and Propagation of Surface Waves", J. Appl. Physics, vol. 31, p. 1039 (1960).

27. A.S. Dunbar, "On the Theory of Antenna Beam Shaping", J. Appl. Physics, vol. 23, p. 847 (1952).